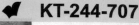

© Josh Timonen

Richard Dawkins first catapulted to fame with his iconic book *The Selfish Gene*, which he followed with a string of bestselling books: *The Extended Phenotype*, *The Blind Watchmaker*, *River Out of Eden*, *Climbing Mount Improbable*, *Unweaving the Rainbow*, *The Ancestor's Tale*, *The God Delusion* and a collection of his shorter writings, *A Devil's Chaplain*. *The Greatest Show on Earth* is his latest highly-acclaimed work which also became a Number One bestseller upon hardcover publication.

Dawkins is a Fellow of both the Royal Society and the Royal Society of Literature. He is the recipient of numerous honours and awards, including the Royal Society of Literature Award (1987), the Michael Faraday Award of the Royal Society (1990), the International Cosmos Prize for Achievement in Human Science (1997), the Kistler Prize (2001), the Shakespeare Prize (2005), the Lewis Thomas Prize for Writing about Science (2006), the Galaxy British Book Awards Author of the Year Award (2007) and the Nierenberg Prize for Science and the Public Interest (2009). He retired from his position as the Charles Simonyi Professor for the Public Understanding of Science at Oxford University in 2008 and remains a fellow of New College.

In 2008 Channel 4 broadcast a three-part television series, *The Genius of Charles Darwin*, written and presented by Richard Dawkins, examining the legacy of the great scientist and some of the issues covered in this timely book.

www.rbooks.co.uk

Also by Richard Dawkins

The Selfish Gene
The Extended Phenotype
The Blind Watchmaker
River Out of Eden
Climbing Mount Improbable
Unweaving the Rainbow
A Devil's Chaplain
The Ancestor's Tale
The God Delusion

THE GREATEST SHOW ON EARTH

THE EVIDENCE FOR EVOLUTION

RICHARD DAWKINS

BLACK SWAN

TRANSWORLD PUBLISHERS
61–63 Uxbridge Road, London W5 5SA
A Random House Group Company
www.rbooks.co.uk

THE GREATEST SHOW ON EARTH
A BLACK SWAN BOOK: 9780552775243

First published in Great Britain
in 2009 by Bantam Press
an imprint of Transworld Publishers
Black Swan edition published 2010

A CIP catalogue record for this book
is available from the British Library.

Addresses for Random House Group Ltd companies outside the UK
can be found at: www.randomhouse.co.uk
The Random House Group Ltd Reg. No. 954009

The Random House Group Limited supports The Forest Stewardship Council (FSC), the
leading international forest certification organisation. All our titles that are printed on
Greenpeace approved FSC certified paper carry the FSC logo. Our paper procurement policy
can be found at www.rbooks.co.uk/environment

Mixed Sources
Product group from well-managed
forests and other controlled sources
www.fsc.org Cert no. TT-COC-2139
© 1996 Forest Stewardship Council
FSC

Typeset in Minion by Falcon Oast Graphic Art Ltd.
Printed in the UK by CPI Cox & Wyman, Reading, RG1 8EX.

4 6 8 10 9 7 5 3

CONTENTS

For
Josh Timonen

PREFACE

THE evidence for evolution grows by the day, and has never been stronger. At the same time, paradoxically, ill-informed opposition is also stronger than I can remember. This book is my personal summary of the evidence that the 'theory' of evolution is actually a fact – as incontrovertible a fact as any in science.

It is not the first book I have written about evolution, and I need to explain what's different about it. It could be described as my missing link. *The Selfish Gene* and *The Extended Phenotype* offered an unfamiliar vision of the familiar theory of natural selection, but they didn't discuss the evidence for evolution itself. My next three books, in their different ways, sought to identify, and dissolve, the main barriers to understanding. These books, *The Blind Watchmaker*, *River Out of Eden* and (my favourite of the three) *Climbing Mount Improbable*, answered questions like, 'What is the use of half an eye?' 'What is the use of half a wing?' 'How can natural selection work, given that most mutations have negative effects?' Once again, however, these three books, although they cleared away stumbling blocks, did not present the actual evidence that evolution is a fact. My largest book, *The Ancestor's Tale*, laid out the full course of the history of life, as a sort of ancestor-seeking Chaucerian pilgrimage going backwards in time, but it again assumed that evolution is true.

Looking back on those books, I realized that the evidence for evolution itself was nowhere explicitly set out, and that this was a serious gap that I needed to close. The year 2009 seemed like a good time, it being the bicentennial year of Darwin's birth and the 150th anniversary of *On the Origin of Species*. Not surprisingly, the same thought occurred to others, and the year has seen some excellent volumes, most notably Jerry Coyne's *Why Evolution is True*. My highly favourable review of his book in the *Times Literary Supplement* is reproduced at http://richarddawkins.net/article,3594,Heat-the-Hornet,Richard-Dawkins.

The working title under which my literary agent, the visionary and indefatigable John Brockman, offered my book to publishers was *Only a Theory*. It later turned out that Kenneth Miller had already pre-empted that title for his book-length response to one of those remarkable courtroom trials by which scientific syllabuses are occasionally decided (a trial in which he played a heroic part). In any case, I had always doubted the title's suitability for my book, and I was ready to shelve it when I found that the perfect title had been lurking on another shelf all along. Some years ago, an anonymous well-wisher had sent me a T-shirt bearing the Barnumesque slogan: 'Evolution, the Greatest Show on Earth, the Only Game in Town'. From time to time I have worn it to give a lecture with that title, and I suddenly realized that it was ideal for this book even if, in its entirety, it was too long. I shortened it to *The Greatest Show on Earth*. 'Only a Theory', with a precautionary question mark to guard against creationist quote-mining, would do nicely as the heading to Chapter 1.

I have been helped in various ways by many people, including Michael Yudkin, Richard Lenski, George Oster, Caroline Pond, Henri D. Grissino-Mayer, Jonathan Hodgkin, Matt Ridley, Peter Holland, Walter Joyce, Aaron Galonsky, David Noakes, Elisabeth Cornwell, Yan Wong, Will Atkinson, Latha Menon, Christopher Graham, Paula Kirby, Lisa Bauer, Owen Selly, Victor Flynn, Michael Kettlewell, Karen Owens, John Endler, Iain Douglas-Hamilton, Sheila Lee, Phil Lord, Christine DeBlase and Rand Russell. Sally Gaminara and Hilary Redmon, and their teams in (respectively) Britain and America, have been wonderfully supportive and can-do-ish. On three occasions while the book was going through the final stages of production, exciting new discoveries were reported in the scientific literature. Each time, I diffidently asked if the orderly and complex procedures of publication might be violated to accommodate the new find. On all three occasions, far from grumbling at such disruptive last-minutemanship, as any normal publisher might, Sally and Hilary greeted the suggestion with cheerful enthusiasm and moved mountains to make it happen. Equally eager and helpful was Gillian Somerscales, who copy-edited and collated the book with literate intelligence and sensitivity.

My wife Lalla Ward has once again sustained me with unfailing encouragement, helpful stylistic criticisms and characteristically stylish suggestions. The book was conceived and begun during my last months in the professorship that bears the name of Charles Simonyi, and completed after I retired. In signing off as Simonyi Professor, fourteen years and seven books after our momentous first meeting, I would once again like to express my grateful appreciation to Charles. Lalla joins me in hoping that our friendship will long continue.

This book is dedicated to Josh Timonen, with thanks to him and to the small and dedicated band who originally worked with him to set up RichardDawkins.net. The web knows Josh as an inspired site designer, but that is just the tip of an amazing iceberg. Josh's creative talent runs deep, but the image of the iceberg captures neither the versatile breadth of his contributions to our joint endeavour, nor the warm good humour with which he makes them.

CHAPTER 1

ONLY
A THEORY?

IMAGINE that you are a teacher of Roman history and the Latin language, anxious to impart your enthusiasm for the ancient world – for the elegiacs of Ovid and the odes of Horace, the sinewy economy of Latin grammar as exhibited in the oratory of Cicero, the strategic niceties of the Punic Wars, the generalship of Julius Caesar and the voluptuous excesses of the later emperors. That's a big undertaking and it takes time, concentration, dedication. Yet you find your precious time continually preyed upon, and your class's attention distracted, by a baying pack of ignoramuses (as a Latin scholar you would know better than to say 'ignorami') who, with strong political and especially financial support, scurry about tirelessly attempting to persuade your unfortunate pupils that the Romans never existed. There never was a Roman Empire. The entire world came into existence only just beyond living memory. Spanish, Italian, French, Portuguese, Catalan, Occitan, Romansh: all these languages and their constituent dialects sprang spontaneously and separately into being, and owe nothing to any predecessor such as Latin. Instead of devoting your full attention to the noble vocation of classical scholar and teacher, you are forced to divert your time and energy to a rearguard defence of the proposition that the Romans existed at all: a defence against an exhibition of ignorant prejudice that would make you weep if you weren't too busy fighting it.

If my fantasy of the Latin teacher seems too wayward, here's a more realistic example. Imagine you are a teacher of more recent history, and your lessons on twentieth-century Europe are boycotted, heckled or otherwise disrupted by well-organized, well-financed and politically muscular groups of Holocaust-deniers. Unlike my hypothetical Rome-deniers, Holocaust-deniers really exist. They are vocal, superficially plausible, and adept at seeming learned. They are supported by the president of at least one currently powerful state, and they include at least one bishop of the Roman Catholic Church. Imagine that, as a teacher of European history, you are continually faced with belligerent demands to 'teach the controversy', and to give 'equal time' to the 'alternative theory' that the Holocaust never happened but was invented by a bunch of Zionist fabricators. Fashionably relativist intellectuals chime in to insist that there is no absolute truth: whether the Holocaust happened is a matter of personal belief; all points of view are equally valid and should be equally 'respected'.

The plight of many science teachers today is not less dire. When they attempt to expound the central and guiding principle of biology; when they honestly place the living world in its historical context – which means evolution; when they explore and explain the very nature of life itself, they are harried and stymied, hassled and bullied, even threatened with loss of their jobs. At the very least their time is wasted at every turn. They are likely to receive menacing letters from parents, and have to endure the sarcastic smirks and close-folded arms of brainwashed children. They are supplied with state-approved textbooks that have had the word 'evolution' systematically expunged, or bowdlerized into 'change over time'. Once, we were tempted to laugh this kind of thing off as a peculiarly American phenomenon. Teachers in Britain and Europe now face the same problems, partly because of American influence, but more significantly because of the growing Islamic presence in the classroom – abetted by the official commitment to 'multiculturalism' and the terror of being thought racist.

It is frequently, and rightly, said that senior clergy and theologians have no problem with evolution and, in many cases, actively support scientists in this respect. This is often true, as I know from the agreeable experience of collaborating with the then Bishop of

Oxford, now Lord Harries, on two separate occasions. In 2004 we wrote a joint article in the *Sunday Times* whose concluding words were: 'Nowadays there is nothing to debate. Evolution is a fact and, from a Christian perspective, one of the greatest of God's works.' The last sentence was written by Richard Harries, but we agreed about all the rest of our article. Two years previously, Bishop Harries and I had organized a joint letter to the then Prime Minister, Tony Blair, which read as follows:

Dear Prime Minister,

We write as a group of scientists and Bishops to express our concern about the teaching of science in the Emmanuel City Technology College in Gateshead.

Evolution is a scientific theory of great explanatory power, able to account for a wide range of phenomena in a number of disciplines. It can be refined, confirmed and even radically altered by attention to evidence. It is not, as spokesmen for the college maintain, a 'faith position' in the same category as the biblical account of creation which has a different function and purpose.

The issue goes wider than what is currently being taught in one college. There is a growing anxiety about what will be taught and how it will be taught in the new generation of proposed faith schools. We believe that the curricula in such schools, as well as that of Emmanuel City Technology College, need to be strictly monitored in order that the respective disciplines of science and religious studies are properly respected.

Yours sincerely

The Rt Revd Richard Harries, Bishop of Oxford; Sir David Attenborough FRS; The Rt Revd Christopher Herbert, Bishop of St Albans; Lord May of Oxford, President of the Royal Society;

Professor John Enderby FRS, Physical Secretary, Royal Society; The Rt Revd John Oliver, Bishop of Hereford; The Rt Revd Mark Santer, Bishop of Birmingham; Sir Neil Chalmers, Director, Natural History Museum; The Rt Revd Thomas Butler, Bishop of Southwark; Sir Martin Rees FRS, Astronomer Royal; The Rt Revd Kenneth Stevenson, Bishop of Portsmouth; Professor Patrick Bateson FRS, Biological Secretary, Royal Society; The Rt Revd Crispian Hollis, Roman Catholic Bishop of Portsmouth; Sir Richard Southwood FRS; Sir Francis Graham-Smith FRS, Past Physical Secretary, Royal Society; Professor Richard Dawkins FRS

Bishop Harries and I organized this letter in a hurry. As far as I remember, the signatories to the letter constituted 100 per cent of those we approached. There was no disagreement either from scientists or from bishops.

The Archbishop of Canterbury has no problem with evolution, nor does the Pope (give or take the odd wobble over the precise palaeontological juncture when the human soul was injected), nor do educated priests and professors of theology. This is a book about the positive evidence that evolution is a fact. It is not intended as an anti-religious book. I've done that, it's another T-shirt, this is not the place to wear it again. Bishops and theologians who have attended to the evidence for evolution have given up the struggle against it. Some may do so reluctantly, some, like Richard Harries, enthusiastically, but all except the woefully uninformed are forced to accept the fact of evolution. They may think God had a hand in starting the process off, and perhaps didn't stay his hand in guiding its future progress. They probably think God cranked the universe up in the first place, and solemnized its birth with a harmonious set of laws and physical constants calculated to fulfil some inscrutable purpose in which we were eventually to play a role. But, grudgingly in some cases, happily in others, thoughtful and rational churchmen and women accept the evidence for evolution.

What we must not do is complacently assume that, because bishops and educated clergy accept evolution, so do their congregations. Alas, as I have documented in the Appendix, there is ample evidence to the contrary from opinion polls. More than 40 per cent of Americans deny that humans evolved from other animals, and think that we – and by implication all of life – were created by God within the last 10,000 years. The figure is not quite so high in Britain, but it is still worryingly large. And it should be as worrying to the churches as it is to scientists. This book is necessary. I shall be using the name 'history-deniers' for those people who deny evolution: who believe the world's age is measured in thousands of years rather than thousands of millions of years, and who believe humans walked with dinosaurs. To repeat, they constitute more than 40 per cent of the American population. The equivalent figure is higher in some countries, lower in others, but 40 per cent is a good average and I shall from time to time refer to the history-deniers as the '40-percenters'.

To return to the enlightened bishops and theologians, it would be nice if they'd put a bit more effort into combating the anti-scientific nonsense that they deplore. All too many preachers, while agreeing that evolution is true and Adam and Eve never existed, will then blithely go into the pulpit and make some moral or theological point about Adam and Eve in their sermons without once mentioning that,

"I still say it's only a theory."

of course, Adam and Eve never actually existed! If challenged, they will protest that they intended a purely 'symbolic' meaning, perhaps something to do with 'original sin', or the virtues of innocence. They may add witheringly that, obviously, nobody would be so foolish as to take their words literally. But do their congregations know that? How is the person in the pew, or on the prayer-mat, supposed to know which bits of scripture to take literally, which symbolically? Is it really so easy for an uneducated churchgoer to guess? In all too many cases the answer is clearly no, and anybody could be forgiven for feeling confused. If you don't believe me, look at the Appendix.

Think about it, Bishop. Be careful, Vicar. You are playing with dynamite, fooling around with a misunderstanding that's waiting to happen – one might even say almost bound to happen if not forestalled. Shouldn't you take greater care, when speaking in public, to let your yea be yea and your nay be nay? Lest ye fall into condemnation, shouldn't you be going out of your way to counter that already extremely widespread popular misunderstanding and lend active and enthusiastic support to scientists and science teachers?

The history-deniers themselves are among those that I am trying to reach in this book. But, perhaps more importantly, I aspire to arm those who are not history-deniers but know some – perhaps members of their own family or church – and find themselves inadequately prepared to argue the case.

Evolution is a fact. Beyond reasonable doubt, beyond serious doubt, beyond sane, informed, intelligent doubt, beyond doubt evolution is a fact. The evidence for evolution is at least as strong as the evidence for the Holocaust, even allowing for eye witnesses to the Holocaust. It is the plain truth that we are cousins of chimpanzees, somewhat more distant cousins of monkeys, more distant cousins still of aardvarks and manatees, yet more distant cousins of bananas and turnips ... continue the list as long as desired. That didn't have to be true. It is not self-evidently, tautologically, obviously true, and there was a time when most people, even educated people, thought it wasn't. It didn't have to be true, but it is. We know this because a rising flood of evidence supports it. Evolution is a fact, and this book will demonstrate it. No reputable scientist disputes it, and no unbiased reader will close the book doubting it.

Why, then, do we speak of 'Darwin's *theory* of evolution', thereby, it seems, giving spurious comfort to those of a creationist persuasion – the history-deniers, the 40-percenters – who think the word 'theory' is a concession, handing them some kind of gift or victory?

WHAT IS A THEORY? WHAT IS A FACT?

Only a theory? Let's look at what 'theory' means. The *Oxford English Dictionary* gives two meanings (actually more, but these are the two that matter here).

> **Theory, Sense 1:** A scheme or system of ideas or statements held as an explanation or account of a group of facts or phenomena; a hypothesis that has been confirmed or established by observation or experiment, and is propounded or accepted as accounting for the known facts; a statement of what are held to be the general laws, principles, or causes of something known or observed.

> **Theory, Sense 2:** A hypothesis proposed as an explanation; hence, a mere hypothesis, speculation, conjecture; an idea or set of ideas about something; an individual view or notion.

Obviously the two meanings are quite different from one another. And the short answer to my question about the theory of evolution is that the scientists are using Sense 1, while the creationists are – perhaps mischievously, perhaps sincerely – opting for Sense 2. A good example of Sense 1 is the Heliocentric Theory of the Solar System, the theory that Earth and the other planets orbit the sun. Evolution fits Sense 1 perfectly. Darwin's theory of evolution is indeed a 'scheme or system of ideas or statements'. It does account for a massive 'group of facts or phenomena'. It is 'a hypothesis that has been confirmed or established by observation or experiment' and, by generally informed consent, it is 'a statement of what are held to be the general laws, principles, or causes of something known or observed'. It is certainly very far from 'a mere hypothesis, speculation, conjecture'. Scientists and creationists are understanding the word 'theory' in two very different senses. Evolution is a theory in the same sense as the heliocentric theory. In neither case should the word 'only' be used, as in 'only a theory'.

As for the claim that evolution has never been 'proved', proof is a notion that scientists have been intimidated into mistrusting. Influential philosophers tell us we can't prove anything in science. Mathematicians can prove things – according to one strict view, they are the only people who can – but the best that scientists can do is fail to disprove things while pointing to how hard they tried. Even the undisputed theory that the moon is smaller than the sun cannot, to the satisfaction of a certain kind of philosopher, be proved in the way that, for example, the Pythagorean Theorem can be proved. But massive accretions of evidence support it so strongly that to deny it the status of 'fact' seems ridiculous to all but pedants. The same is true of evolution. Evolution is a fact in the same sense as it is a fact that Paris is in the Northern Hemisphere. Though logic-choppers rule the town,* some theories are beyond sensible doubt, and we call them facts. The more energetically and thoroughly you try to disprove a theory, if it survives the assault, the more closely it approaches what common sense happily calls a fact.

I could carry on using 'Theory Sense 1' and 'Theory Sense 2' but numbers are unmemorable. I need substitute words. We already have a good word for 'Theory Sense 2'. It is 'hypothesis'. Everybody understands that a hypothesis is a tentative idea awaiting

---

* Not my favourite Yeats line, but apt in this case.

confirmation (or falsification), and it is precisely this tentativeness that evolution has now shed, although it was still burdened with it in Darwin's time. 'Theory Sense 1' is harder. It would be nice simply to go on using 'theory', as though 'Sense 2' didn't exist. Indeed, a good case could be made that Sense 2 *shouldn't* exist, because it is confusing and unnecessary, given that we have 'hypothesis'. Unfortunately Sense 2 of 'theory' is in common use and we can't by fiat ban it. I am therefore going to take the considerable, but just forgivable, liberty of borrowing from mathematics the word 'theorem' for Sense 1. It is actually a mis-borrowing, as we shall see, but I think the risk of confusion is outweighed by the benefits. As a gesture of appeasement towards affronted mathematicians, I am going to change my spelling to 'theorum'.* First, let me explain the strict mathematical usage of theorem, while at the same time clarifying my earlier statement that, strictly speaking, only mathematicians are licensed to *prove* anything (lawyers aren't, despite well-remunerated pretensions).

To a mathematician, a proof is a logical demonstration that a conclusion necessarily follows from axioms that are assumed. Pythagoras' Theorem is necessarily true, provided only that we assume Euclidean axioms, such as the axiom that parallel straight lines never meet. You are wasting your time measuring thousands of right-angled triangles, trying to find one that falsifies Pythagoras' Theorem. The Pythagoreans proved it, anybody can work through the proof, it's just true and that's that. Mathematicians use the idea of proof to make a distinction between a 'conjecture' and a 'theorem', which bears a superficial resemblance to the *OED*'s distinction between the two senses of 'theory'. A conjecture is a proposition that looks true but has never been proved. It will become a theorem when it has been proved. A famous example is the Goldbach Conjecture, which states that any even integer can be expressed as the sum of

---

* For the sake of decorum / Pronounce it theorum.

two primes. Mathematicians have failed to disprove it for all even numbers up to 300 thousand million million million, and common sense would happily call it Goldbach's Fact. Nevertheless it has never been proved, despite lucrative prizes being offered for the achievement, and mathematicians rightly refuse to place it on the pedestal reserved for theorems. If anybody ever finds a proof, it will be promoted from Goldbach's Conjecture to Goldbach's Theorem, or maybe X's Theorem where X is the clever mathematician who finds the proof.

Carl Sagan made sarcastic use of the Goldbach Conjecture in his riposte to people who claim to have been abducted by aliens.

> Occasionally, I get a letter from someone who is in 'contact' with extraterrestrials. I am invited to 'ask them anything'. And so over the years I've prepared a little list of questions. The extraterrestrials are very advanced, remember. So I ask things like, 'Please provide a short proof of Fermat's Last Theorem'. Or the Goldbach Conjecture . . . I never get an answer. On the other hand, if I ask something like 'Should we be good?' I almost always get an answer. Anything vague, especially involving conventional moral judgements, these aliens are extremely happy to respond to. But on anything specific, where there is a chance to find out if they actually know anything beyond what most humans know, there is only silence.

Fermat's Last Theorem, like the Goldbach Conjecture, is a proposition about numbers to which nobody has found an exception. Proving it has been a kind of holy grail for mathematicians ever since 1637, when Pierre de Fermat wrote in the margin of an old mathematics book, 'I have a truly marvellous proof . . . which this margin is too narrow to contain.' It was finally proved by the English mathematician Andrew Wiles in 1995. Before that, some mathematicians think it should have been called a conjecture. Given the length and complication of Wiles's successful proof, and his reliance on advanced twentieth-century methods and knowledge, most mathematicians think Fermat was (honestly) mistaken in his claim to have proved it. I tell the story only to illustrate the difference between a conjecture and a theorem.

As I said, I am going to borrow the mathematicians' term 'theorem', but I'm spelling it 'theorum' to differentiate it from a mathematical theorem. A scientific theorum such as evolution or heliocentrism is a theory that conforms to the Oxford dictionary's 'Sense 1'.

> [It] has been confirmed or established by observation or experiment, and is propounded or accepted as accounting for the known facts; [it is] a statement of what are held to be the general laws, principles, or causes of something known or observed.

A scientific theorum has not been – cannot be – proved in the way a mathematical theorem is proved. But common sense treats it as a fact in the same sense as the 'theory' that the Earth is round and not flat is a fact, and the theory that green plants obtain energy from the sun is a fact. All are scientific theorums: supported by massive quantities of evidence, accepted by all informed observers, undisputed facts in the ordinary sense of the word. As with all facts, if we are going to be pedantic, it is undeniably possible that our measuring instruments, and the sense organs with which we read them, are the victims of a massive confidence trick. As Bertrand Russell said, 'We may all have come into existence five minutes ago, provided with ready-made memories, with holes in our socks and hair that needed cutting.' Given the evidence now available, for evolution to be anything other than a fact would require a similar confidence trick by the creator, something that few theists would wish to credit.

It is time now to examine the dictionary definition of a 'fact'. Here is what the *OED* has to say (again there are several definitions, but this is the relevant one):

Fact: Something that has really occurred or is actually the case; something certainly known to be of this character; hence, a particular truth known by actual observation or authentic testimony, as opposed to what is merely inferred, or to a conjecture or fiction; a datum of experience, as distinguished from the conclusions that may be based upon it.

Notice that, like a theorum, a fact in this sense doesn't have the same rigorous status as a proved mathematical theorem, which follows inescapably from a set of assumed axioms. Moreover, 'actual observation or authentic testimony' can be horribly fallible, and is over-rated in courts of law. Psychological experiments have given us some stunning demonstrations, which should worry any jurist inclined to give superior weight to 'eye-witness' evidence. A famous example was prepared by Professor Daniel J. Simons at the University of Illinois. Half a dozen young people standing in a circle were filmed for 25 seconds tossing a pair of basketballs to each other, and we, the experimental subjects, watch the film. The players weave in and out of the circle and change places as they pass and bounce the balls, so the scene is quite actively complicated. Before being shown the film, we are told that we have a task to perform, to test our powers of observation. We have to count the total number of times balls are passed from person to person. At the end of the test, the counts are duly written down, but – little does the audience know – this is not the real test!

After showing the film and collecting the counts, the experimenter drops his bombshell. 'And how many of you saw the gorilla?' The majority of the audience looks baffled: blank. The experimenter then replays the film, but this time tells the audience to watch in a relaxed fashion without trying to count anything. Amazingly, nine seconds into the film, a man in a gorilla suit strolls nonchalantly to the centre of the circle of players, pauses to face the camera, thumps his chest as if in belligerent contempt for eye-witness evidence, and then strolls off with the same insouciance as before (see colour page 8). He is there in full view for nine whole seconds – more than one-third of the film – and yet the majority of the witnesses never see him. They would swear an oath in a court of law that no man in a gorilla suit was present, and they would swear that they had been watching with more than usually acute concentration for the whole 25 seconds, precisely because they were counting ball-passes. Many experiments along these lines have been performed, with similar results, and with similar reactions of stupefied disbelief when the audience is finally shown the truth. Eye-witness testimony, 'actual observation', 'a datum of experience' – all are, or at least can be, hopelessly unreliable. It is, of course, exactly this unreliability among observers that stage conjurors exploit with their techniques of deliberate distraction.

The dictionary definition of a fact mentions 'actual observation or authentic testimony, as opposed to what is merely inferred' (emphasis added). The implied pejorative of that 'merely' is a bit of a cheek. Careful inference can be more reliable than 'actual observation', however strongly our intuition protests at admitting it. I myself was flabbergasted when I failed to see the Simons gorilla, and frankly incredulous that it had really been there. Sadder and wiser after my second viewing of the film, I shall never again be tempted to give eye-witness testimony an automatic preference over indirect scientific inference. The gorilla film, or something like it, should perhaps be shown to all juries before they retire to consider their verdicts. All judges too.

Admittedly, inference has to be based ultimately on observation by our sense organs. For example, we use our eyes to observe the printout from a DNA sequencing machine, or from the Large Hadron Collider. But – all intuition to the contrary – direct observation of an alleged event (such as a murder) as it actually happens is not necessarily more reliable than indirect observation of its consequences (such as DNA in a bloodstain) fed into a well-constructed inference engine. Mistaken identity is more likely to arise from direct eye-witness testimony than from indirect inference derived from DNA evidence. And, by the way, there is a distressingly long list of people who have been wrongly convicted on eye-witness testimony and subsequently freed – sometimes after many years – because of new evidence from DNA. In Texas alone, thirty-five condemned people have been exonerated since DNA evidence became admissible in court. And that's just the ones who are still alive. Given the gusto with which the State of Texas enforces the death penalty (during his six years as Governor, George W. Bush signed a death warrant once a fortnight on average), we have to assume that a substantial number of executed people would have been exonerated if DNA evidence had been available in time for them.

This book will take inference seriously – not *mere* inference but proper scientific inference – and I shall show the irrefragable power of the inference that evolution is a fact. Obviously, the vast majority of evolutionary change is invisible to direct eye-witness observation. Most of it happened before we were born, and in any case it is usually too slow to be seen during an individual's lifetime. The same is true of the relentless pulling apart of Africa and South America, which occurs, as we shall see in Chapter 9, too slowly for us to notice. With evolution, as with continental drift, inference after the event is all that is available to us, for the obvious reason that we don't exist until after the event. But do not for one nanosecond underestimate the power of such inference. The slow drifting apart of South America and Africa is now an established fact in the ordinary language sense of 'fact', and so is our common ancestry with porcupines and pomegranates.

We are like detectives who come on the scene after a crime has been committed. The murderer's actions have vanished into the past. The detective has no hope of witnessing the actual crime with his own eyes. In any case, the gorilla-suit experiment and others of its kind have taught us to mistrust our own eyes. What the detective *does* have is traces that remain, and there is a great deal to trust there. There are footprints, fingerprints (and nowadays DNA fingerprints too), bloodstains, letters, diaries. The world is the way the world should be if this and this history, but not that and that history, led up to the present.

The distinction between the two dictionary meanings of 'theory' is not an unbridgeable chasm, as many historical examples show. In the history of science, theorums often start off as 'mere' hypotheses. Like the theory of continental drift, an idea may even begin its career mired in ridicule, before progressing by painful steps to the status of a theorum or undisputed fact. This is not a philosophically difficult point. The fact that some widely held past beliefs have been conclusively proved erroneous doesn't mean we have to fear that future evidence will always show our present beliefs to be wrong. How vulnerable our present beliefs are depends, among other things, on how strong the evidence for them is. People used to think the sun was smaller than the Earth, because they had inadequate evidence. Now we have evidence, which was not previously available, that shows conclusively that it is much larger, and we can be totally confident that this evidence will never, ever be superseded. This is not a temporary hypothesis that has so far survived disproof. Our present beliefs about many things may be disproved, but we can with complete confidence make a list of certain facts that will never be disproved. Evolution and the heliocentric theory weren't always among them, but they are now.

Biologists often make a distinction between the *fact* of evolution (all living things are cousins), and the *theory* of what drives it (they usually mean natural selection, and they may contrast it with rival theories such as Lamarck's theory of 'use and disuse' and the 'inheritance of acquired characteristics'). But Darwin himself thought of both as theories in the tentative, hypothetical, conjectural sense. This was because, in those days, the available evidence was less compelling and it was still possible for reputable scientists to dispute both evolution and natural selection. Nowadays it is no longer possible to dispute the fact of evolution itself – it has graduated to become a theorum or obviously supported fact – but it could still (just) be doubted that natural selection is its major driving force.

Darwin explained in his autobiography how in 1838 he was reading Malthus's *On Population* 'for amusement' (under the influence, Matt Ridley suspects, of his brother Erasmus's formidably intelligent friend, Harriet Martineau) and received the inspiration for natural selection: 'Here, then I had at last got a theory by which to work.' For Darwin, natural selection was a hypothesis, which might have been right or might have been wrong. He thought the same of evolution itself. What we now call the fact of evolution was, in 1838, a hypothesis for which evidence needed to be collected. By the time Darwin came to publish *On the Origin of Species* in 1859, he had amassed enough evidence to propel evolution itself, though still not natural selection, a long way towards the status of fact. Indeed, it was this elevation from hypothesis towards fact that occupied Darwin for most of his great book. The elevation has continued until, today, there is no longer a doubt in any serious mind, and scientists speak, at least informally, of the *fact* of evolution. All reputable biologists go on to agree that natural selection is one of its most important driving forces, although – as some biologists insist more than others – not the only one. Even if it is not the only one, I have yet to meet a serious biologist who can point to an alternative to natural selection as a driving force of *adaptive* evolution – evolution towards positive improvement.

In the rest of this book, I shall demonstrate that evolution is an inescapable fact, and celebrate its astonishing power, simplicity and beauty. Evolution is within us, around us, between us, and its workings are embedded in the rocks of aeons past. Given that, in most cases, we don't live long enough to watch evolution happening before our eyes, we shall revisit the metaphor of the detective coming upon the scene of a crime after the event and making inferences. The aids to inference that lead scientists to the fact of evolution are far more numerous, more convincing, more incontrovertible, than any eye-witness reports that have ever been used, in any court of law, in any century, to establish guilt in any crime. Proof beyond reasonable doubt? *Reasonable* doubt? That is the understatement of all time.

CHAPTER 2

DOGS, COWS AND CABBAGES

WHY did it take so long for a Darwin to arrive on the scene? What delayed humanity's tumbling to that luminously simple idea which seems, on the face of it, so much easier to grasp than the mathematical ideas given us by Newton two centuries earlier – or, indeed, by Archimedes two millennia earlier? Many answers have been suggested. Perhaps minds were cowed by the sheer time it must take for great change to occur – by the mismatch between what we now call geological deep time and the lifespan and comprehension of the person trying to understand it. Perhaps it was religious indoctrination that held us back. Or perhaps it was the daunting complexity of a living organ such as an eye, freighted as it is with the beguiling illusion of design by a master engineer. Probably all those played a role. But Ernst Mayr, grand old man of the neo-Darwinian synthesis, who died in 2005 at the age of 100, repeatedly voiced a different suspicion. For Mayr, the culprit was the ancient philosophical doctrine of – to give it its modern name – *essentialism*. The discovery of evolution was held back by the dead hand of Plato.*

THE DEAD HAND OF PLATO

For Plato, the 'reality' that we think we see is just shadows cast on the wall of our cave by the flickering light of the camp fire. Like other

---

* This isn't Mayr's phrase, though it expresses his idea.

classical Greek thinkers, Plato was at heart a geometer. Every triangle drawn in the sand is but an imperfect shadow of the true *essence* of triangle. The lines of the essential triangle are pure Euclidean lines with length but no breadth, lines defined as infinitely narrow and as never meeting when parallel. The angles of the essential triangle really do add up to exactly two right angles, not a picosecond of arc more or less. This is not true of a triangle drawn in the sand: but the triangle in the sand, for Plato, is but an unstable shadow of the ideal, essential triangle.

Biology, according to Mayr, is plagued by its own version of essentialism. Biological essentialism treats tapirs and rabbits, pangolins and dromedaries, as though they were triangles, rhombuses, parabolas or dodecahedrons. The rabbits that we see are wan shadows of the perfect 'idea' of rabbit, the ideal, essential, Platonic rabbit, hanging somewhere out in conceptual space along with all the perfect forms of geometry. Flesh-and-blood rabbits may vary, but their variations are always to be seen as flawed deviations from the ideal essence of rabbit.

How desperately unevolutionary that picture is! The Platonist regards any change in rabbits as a messy departure from the essential rabbit, and there will always be resistance to change – as if all real rabbits were tethered by an invisible elastic cord to the Essential Rabbit in the Sky. The evolutionary view of life is radically opposite. Descendants can depart indefinitely from the ancestral form, and each departure becomes a potential ancestor to future variants. Indeed, Alfred Russel Wallace, independent co-discoverer with Darwin of evolution by natural selection, actually called his paper 'On the tendency of varieties to depart indefinitely from the original type'.

If there is a 'standard rabbit', the accolade denotes no more than the centre of a bell-shaped distribution of real, scurrying, leaping, variable bunnies. And the distribution shifts with time. As generations go by, there may gradually come a point, not clearly defined, when the norm of what we call rabbits will have departed so far as to deserve a different name. There is no permanent rabbitiness, no essence of rabbit hanging in the sky, just populations of furry, long-eared, coprophagous, whisker-twitching individuals, showing a statistical distribution of variation in size, shape, colour and proclivities. What used to be the longer-eared end of the old distribution may find itself the centre of a new distribution later in geological time. Given a sufficiently large number of generations, there may be no overlap between ancestral and descendant distributions: the longest ears among the ancestors may be shorter than the shortest ears among the descendants. All is fluid, as another Greek philosopher, Heraclitus, said; nothing fixed. After a hundred million years it may be hard to believe that the descendant animals ever had rabbits for ancestors. Yet in no generation during the evolutionary process was the predominant type in the population far from the modal type in the previous generation or the following generation. This way of thinking is what Mayr called *population thinking*. Population thinking, for him, was the antithesis of essentialism. According to Mayr, the reason Darwin was such an unconscionable time arriving on the scene was that we all – whether because of Greek influence or for some other reason – have essentialism burned into our mental DNA.

For the mind encased in Platonic blinkers, a rabbit is a rabbit is a rabbit. To suggest that rabbitkind constitutes a kind of shifting cloud of statistical averages, or that today's typical rabbit might be different from the typical rabbit of a million years ago or the typical rabbit of a million years hence, seems to violate an internal taboo. Indeed, psychologists studying the development of language tell us that children are natural essentialists. Maybe they have to be if they are to remain sane while their developing minds divide things into discrete categories each entitled to a unique noun. It is no wonder that Adam's first task, in the Genesis myth, was to give all the animals names.

And it is no wonder, in Mayr's view, that we humans had to wait for our Darwin until well into the nineteenth century. To dramatize how very anti-essentialist evolution is, consider the following. On the 'population-thinking' evolutionary view, every animal is linked to every other animal, say rabbit to leopard, by a chain of intermediates, each so similar to the next that every link could in principle mate with its neighbours in the chain and produce fertile offspring. You can't violate the essentialist taboo more comprehensively than that. And it is not some vague thought-experiment confined to the imagination. On the evolutionary view, there really is a series of intermediate animals connecting a rabbit to a leopard, every one of whom lived and breathed, every one of whom would have been placed in exactly the same species as its immediate neighbours on either side in the long, sliding continuum. Indeed, every one of the series was the child of its neighbour on one side and the parent of its neighbour on the other. Yet the whole series constitutes a continuous bridge from rabbit to leopard – although, as we shall see later, there never was a 'rabbipard'. There are similar bridges from rabbit to wombat, from leopard to lobster, from every animal or plant to every other. Maybe you have reasoned for yourself why this startling result follows necessarily from the evolutionary world-view, but let me spell it out anyway. I'll call it the hairpin thought experiment.

Take a rabbit, any female rabbit (arbitrarily stick to females, for convenience: it makes no difference to the argument). Place her mother next to her. Now place the grandmother next to the mother and so on back in time, back, back, back through the megayears, a seemingly endless line of female rabbits, each one sandwiched between her daughter and her mother. We walk along the line of rabbits, backwards in time, examining them carefully like an inspecting general. As we pace the line, we'll eventually notice that the ancient rabbits we are passing are just a little bit different from the modern rabbits we are used to. But the rate of change will be so slow that we shan't notice the trend from generation to generation, just as we can't see the motion of the hour hand on our watches – and just as we can't see a child growing, we can only see later that she has become a teenager, and later still an adult. An additional reason why we don't notice the change in rabbits from one generation to another is that, in any one century, the variation within the current population will normally be greater than the variation between mothers and daughters. So if we try to discern the movement of the 'hour hand' by comparing mothers with daughters, or indeed grandmothers with granddaughters, such slight differences as we may see will be swamped by the differences among the rabbits' friends and relations gambolling in the meadows round about.

Nevertheless, steadily and imperceptibly, as we retreat through time, we shall reach ancestors that look less and less like a rabbit and more and more like a shrew (and not very like either). One of these creatures I'll call the hairpin bend, for reasons that will become apparent. This animal is the most recent common ancestor (in the female line, but that is not important) that rabbits share with leopards. We don't know exactly what it looked like, but it follows from the evolutionary view that it definitely had to exist. Like all animals, it was a member of the same species as its daughters and its mother. We now continue our walk, except that we have turned the bend in the hairpin and are walking forwards in time, aiming towards the leopards (among the hairpin's many and diverse descendants, for we shall continually meet forks in the line, where we consistently choose the fork that will eventually lead to leopards). Each shrew-like animal along our forward walk is now followed by her daughter. Slowly, by imperceptible degrees, the shrew-like animals will change, through intermediates that might not resemble any modern animal much but strongly resemble each other, perhaps passing through vaguely stoat-like intermediates, until eventually, without ever noticing an abrupt change of any kind, we arrive at a leopard.

Various things must be said about this thought experiment. First, we happen to have chosen to walk from rabbit to leopard, but I repeat that we could have chosen porcupine to dolphin, wallaby to giraffe or human to haddock. The point is that for any two animals there has to be a hairpin path linking them, for the simple reason that every species shares an ancestor with every other species: all we have to do is walk backwards from one species to the shared ancestor, then turn through a hairpin bend and walk forwards to the other species.

Second, notice that we are talking only about locating a chain of animals that links a modern animal to another modern animal. We are most emphatically not *evolving* a rabbit into a leopard. I suppose you could say we are *de*-evolving back to the hairpin, then evolving forwards to the leopard from there. As we'll see in a later chapter, it is unfortunately necessary to explain, again and again, that modern species don't evolve into other modern species, they just share ancestors: they are cousins. This, as we shall see, is also the answer to that disquietingly common plaint: 'If humans have evolved from chimpanzees, how come there are still chimpanzees around?'

Third, on our forward march from the hairpin animal, we arbitrarily choose the path leading to the leopard. This is a real path of evolutionary history, but, to repeat this important point, we choose to ignore numerous branch points where we could have followed evolution to countless other end points – for the hairpin animal is the grand ancestor not only of rabbits and leopards but of a large fraction of modern mammals.

The fourth point, which I have already emphasized, is that, however radical and extensive the differences between the ends of the hairpin – rabbit and leopard, say – each step along the chain that links them is very, very small. Every individual along the chain is as similar to its neighbours in the chain as mothers and daughters are expected to be. And *more* similar to its neighbours in the chain, as I have also mentioned, than to typical members of the surrounding population.

You can see how this thought experiment drives a coach and horses through the elegant Greek temple of Platonic ideal forms. And you can see how, if Mayr is right that humans are deeply imbued with essentialist preconceptions, he might well also be right about why we historically found evolution so hard to stomach.

The word 'essentialism' itself wasn't invented till 1945 and so was not available to Darwin. But he was only too familiar with the biological version of it in the form of the 'immutability of species', and much of his effort was directed towards combating it under that name. Indeed, in several of Darwin's books – more so in others than *On the Origin of Species* itself – you'll understand fully what he's on about only if you shed modern presuppositions about evolution, and remember that a large part of his audience would have been essentialists who never doubted the immutability of species. One of Darwin's most telling weapons in arguing against this supposed immutability was the evidence from domestication, and it is domestication that will occupy the rest of this chapter.

SCULPTING THE GENE POOL

Darwin knew plenty about animal and plant breeding. He communed with pigeon fanciers and horticulturalists, and he loved dogs.* Not only is the first chapter of *On the Origin of Species* all about domestic varieties of animals and plants; Darwin also wrote a whole book on the subject. *The Variation of Animals and Plants under Domestication* has chapters on dogs and cats, horses and asses, pigs, cattle, sheep and goats, rabbits, pigeons (two chapters; pigeons were a particular love of Darwin), chickens and various other birds, and plants, including the amazing cabbages. Cabbages are a vegetable affront to essentialism and the immutability of species. The wild cabbage, *Brassica oleracea*, is an undistinguished plant, vaguely like a weedy version of a domestic cabbage. In just a few centuries, wielding the fine and coarse chisels furnished by the toolbox of selective breeding techniques, horticulturalists have sculpted this rather nondescript plant into vegetables as strikingly different from each other and from the wild ancestor as broccoli, cauliflower, kohlrabi, kale, Brussels sprouts, spring greens, romanescu and, of course, the various kinds of vegetables that are still commonly called cabbage.

Another familiar example is the sculpting of the wolf, *Canis lupus*, into the two hundred or so breeds of dog, *Canis familiaris*, that are recognized as separate by the UK Kennel Club, and the

---

* Who could not love dogs, they are such good sports?

larger number of breeds that are genetically isolated from one another by the apartheid-like rules of pedigree breeding.

Incidentally, the wild ancestor of all domestic dogs really does seem to be the wolf and only the wolf (although its domestication may have happened independently in different places around the world). Evolutionists haven't always thought so. Darwin, along with many of his contemporaries, suspected that several species of wild canid, including wolves and jackals, had contributed ancestry to our domestic dogs. The Nobel Prize-winning Austrian ethologist Konrad Lorenz was of the same view. His *Man Meets Dog*, published in 1949, pushes the notion that domestic dog breeds fall into two main groups: those derived from jackals (the majority) and those derived from wolves (Lorenz's own favourites, including Chows). Lorenz seems to have had no evidence at all for his dichotomy, other than the differences that he thought he saw in the personalities and characters of the breeds. The matter remained open until molecular genetic evidence came along to clinch it. There is now no doubt. Domestic dogs have no jackal ancestry at all. All breeds of dogs are modified wolves: not jackals, not coyotes and not foxes.

The main point I want to draw out of domestication is its astonishing power to change the shape and behaviour of wild animals, and the speed with which it does so. Breeders are almost like modellers with endlessly malleable clay, or like sculptors wielding chisels, carving dogs or horses, or cows or cabbages, to their whim. I shall return to this image shortly. The relevance to natural evolution is that, although the selecting agent is man and not nature, the process is otherwise exactly the same. This is why Darwin gave so much prominence to domestication at the beginning of *On the Origin of Species*. Anybody can understand the principle of evolution by artificial selection. Natural selection is the same, with one minor detail changed.

Strictly speaking, it is not the body of the dog or the cabbage that is carved by the breeder/sculptor but the gene pool of the breed or species. The idea of a gene pool is central to the body of knowledge and theory that goes under the name of the 'Neo-Darwinian Synthesis'. Darwin himself knew nothing of it. It was not a part of his intellectual world, nor indeed were genes. He was aware, of course, that characteristics run in families; aware that offspring tend to resemble their parents and siblings; aware that particular characteristics of dogs and pigeons breed true. Heredity was a central plank of his theory of natural selection. But a gene pool is something else. The concept of a gene pool has meaning only in the light of Mendel's law of the independent assortment of hereditary particles. Darwin never knew Mendel's laws, for although Gregor Mendel, the Austrian monk who was the father of genetics, was Darwin's contemporary, he published his findings in a German journal which Darwin never saw.

A Mendelian gene is an all-or-nothing entity. When you were conceived, what you received from your father was not a substance, to be mixed with what you received from your mother as if mixing blue paint and red paint to make purple. If this were really how heredity worked (as people vaguely thought in Darwin's time) we'd all be a middling average, halfway between our two parents. In that case, all variation would rapidly disappear from the population (no matter how assiduously you mix purple paint with purple paint, you'll never reconstitute the original red and blue). In fact, of course, anybody can plainly see that there is no such intrinsic tendency for variation to decrease in a population. Mendel showed that this is because when paternal genes and maternal genes are combined in a child (he didn't use the word 'gene', which wasn't coined until 1909), it is not like blending paints, it is more like shuffling and reshuffling cards in a pack. Nowadays, we know that genes are lengths of DNA code, not physically separate like cards, but the principle remains valid. Genes don't blend; they shuffle. You could say they are shuffled badly, with groups of cards sticking together for several generations of shuffling before chance happens to split them.

Any one of your eggs (or sperms if you are male) contains either your father's version of a particular gene or your mother's version, not a blend of the two. And that particular gene came from one and only one of your four grandparents; and from one and only one of your eight great-grandparents.[*]

Hindsight says this should have been obvious all along. When you cross a male with a female, you expect to get a son or a daughter, not a hermaphrodite.[†] Hindsight says anybody in an armchair could have generalized the same all-or-none principle to the inheritance of each and every characteristic. Fascinatingly, Darwin himself was glimmeringly close to this, but he stopped just short of making the full connection. In 1866 he wrote, in a letter to Alfred Wallace:

My dear Wallace

I do not think you understand what I mean by the non-blending of certain varieties. It does not refer to fertility. An instance will explain. I crossed the Painted Lady and Purple sweet peas, which are very differently coloured varieties, and got, even out of the same pod, both varieties perfect but none intermediate. Something of this kind, I should think, must occur at first with your butterflies . . . Though these cases are in appearance so wonderful, I do not know that they are really more so than every female in the world producing distinct male and female offspring.

---

[*] This would be strictly true on the model of genetics that Mendel offered us, and the model of genetics that all biologists followed until the Watson–Crick revolution of the 1950s. It is *nearly* but not quite true, given what we now know about genes as long stretches of DNA. For all practical purposes we can take it as true.

[†] On the farm where I spent my childhood, we had one especially obstreperous and aggressive cow called Arusha. Arusha was 'a character' and a problem. One day the herdsman, Mr Evans, ruefully remarked: 'Seems to me, Arusha is more like a cross between a bull and a cow.'

Darwin came *that close* to discovering Mendel's law of the non-blending of (what we would now call) genes.* The case is analogous to the claim, by various aggrieved apologists, that other Victorian scientists, for example Patrick Matthew and Edward Blyth, had discovered natural selection before Darwin did. In a sense that is true, as Darwin acknowledged, but I think the evidence shows that they didn't understand how *important* it is. Unlike Darwin and Wallace, they didn't see it as a *general* phenomenon with universal significance – with the power to drive the evolution of all living things in the direction of positive improvement. In the same way, this letter to Wallace shows that Darwin got tantalizingly close to grasping the point about the non-blending nature of heredity. But he didn't see its generality, and in particular he failed to see it as the answer to the riddle of why variation didn't automatically disappear from populations. That was left to twentieth-century scientists, building on Mendel's before-his-time discovery.†

So now the concept of the gene pool starts to make sense. A sexually reproducing population, such as, say, all the rats on Ascension Island, remotely isolated in the South Atlantic, is continually shuffling all the genes on the island. There is no intrinsic tendency for each generation to become less variable than the previous generation, no tendency towards ever more boringly grey, middling intermediates. The genes

---

* There is a persistent, but false, rumour that Darwin possessed a bound copy of the German journal in which Mendel published his results but that the relevant pages were found uncut on Darwin's death. The meme probably originates from the fact that he possessed a book called *Die Pflanzen-mischlinge* by W. O. Focke. Focke did briefly refer to Mendel, and the page where he did so was indeed uncut in Darwin's copy. But Focke laid no special emphasis on Mendel's work and showed no evidence of understanding its profound significance, so it is not obvious that Darwin would have picked it out even if he had cut the relevant page. In any case, Darwin's German was not great. If he had read Mendel's paper, the history of biology would have been very different. It is arguable that even Mendel himself did not understand the full importance of his findings. If he had, he might have written to Darwin. In the library of Mendel's monastery in Brno, I have held in my hand Mendel's own copy (in German) of *On the Origin of Species* and seen his marginalia, which indicate that he read it.

† Beginning in 1908 with the endearingly eccentric, cricket-loving mathematician G. H. Hardy and, independently, the German doctor Wilhelm Weinberg, the theory culminated in the work of the great geneticist and statistician Ronald Fisher, and, again largely independently, his co-founders of population genetics, J. B. S. Haldane and Sewall Wright.

remain intact, shuffled about from individual body to individual body as the generations go by, but not *blending* with one another, never contaminating each other. At any one time, the genes are all sitting in the bodies of individual rats, or they are moving into new rat bodies via sperms. But if we take a long view across many generations, we see all the rat genes on the island being mixed up as though they were cards in a single well-shuffled pack: one single pool of genes.

I'm guessing that the rat gene pool on a small and isolated island such as Ascension is a self-contained and rather well-stirred pool, in the sense that the recent ancestors of any one rat could have lived anywhere on the island, but probably not anywhere other than on the island, give or take the occasional stowaway on a ship. But the gene pool of the rats on a large land mass such as Eurasia would be much more complicated. A rat living in Madrid would derive most of its genes from ancestors living in the western end of the Eurasian continent rather than, say, Mongolia or Siberia, not because of specific barriers to gene flow (though those exist too) but because of the sheer distances involved. It takes time for sexual shuffling to work a gene from one side of a continent to the other. Even if there are no physical barriers such as rivers or mountain ranges, gene flow across such a large land mass will still be slow enough for the gene pool to deserve the name 'viscous'. A rat living in Vladivostok would trace most of its genes back to ancestors in the east. The Eurasian gene pool would be shuffled, as on Ascension Island, but not homogeneously shuffled because of the distances involved. Moreover, partial barriers such as mountain ranges, large rivers or deserts would further get in the way of homogeneous shuffling, thereby structuring and complicating the gene pool. These complications don't devalue the idea of the gene pool. The perfectly stirred gene pool is a useful abstraction, like a mathematician's abstraction of a perfect straight line. Real gene pools, even on small islands like Ascension, are imperfect approximations, only partially shuffled. The smaller and less broken-up the island, the better the approximation to the abstract ideal of the perfectly stirred gene pool.

Just to round off the thought about gene pools, each individual animal that we see in a population is a *sampling* of the gene pool of its time (or rather its parents' time). There is no intrinsic tendency in gene pools for particular genes to increase or decrease in frequency. But when there *is* a systematic increase or decrease in the frequency with which we see a particular gene in a gene pool, that is precisely and exactly what is meant by evolution. The question, therefore, becomes: *why* should there be a systematic increase or decrease in a gene's frequency? That, of course, is where things start to get interesting, and we shall come to it in due course.

Something funny happens to the gene pools of domestic dogs. Breeders of pedigree Pekineses or Dalmatians go to elaborate lengths to stop genes crossing from one gene pool to another. Stud books are kept, going back many generations, and miscegenation is the worst thing that can happen in the book of a pedigree breeder. It is as though each breed of dog were incarcerated on its own little Ascension Island, kept apart from every other breed. But the barrier to interbreeding is not blue water but human rules. Geographically the breeds all overlap, but they might as well be on separate islands because of the way their owners police their mating opportunities. Of course, from time to time the rules are broken. Like a rat stowing away on a ship to Ascension Island, a whippet bitch, say, escapes the leash and mates with a spaniel. But the mongrel puppies that result, however loved they may be as individuals, are cast off the island labelled Pedigree Whippet. The island itself remains a pure whippet island. Other pure-bred whippets ensure that the gene pool of the virtual island labelled Whippet continues uncontaminated. There are hundreds of man-made 'islands', one for each breed of pedigree dog. Each one is a virtual island, in the sense that it is not geographically localized. Pedigree whippets or Pomeranians are to be found in many different places around the world, and cars, ships and planes are used to ferry the genes from one geographical place to another. The virtual genetic island that is the Pekinese gene pool overlaps geographically, but not genetically (except when a bitch breaks cover), with the virtual genetic island that is the boxer gene pool and the virtual island that is the St Bernard gene pool.

Now let's return to the remark that opened my discussion of gene pools. I said that if human breeders are to be seen as sculptors, what they are carving with their chisels is not dog flesh but gene pools. It appears to be dog flesh because the breeder might announce an intention to, say, shorten the snouts of future generations of boxers. And the end product of such an intention would indeed be a shorter snout, as though a chisel had been taken to the ancestor's face. But, as we have seen, a typical boxer in any one generation is a sampling of the contemporary gene pool. It is the gene pool that has been carved and whittled over the years. Genes for long snouts have been chiselled out of the gene pool and replaced by genes for short snouts. Every breed of dog, from dachshund to Dalmatian, from boxer to borzoi, from poodle to Pekinese, from Great Dane to chihuahua, has been carved, chiselled, kneaded, moulded, not literally as flesh and bone but in its gene pool.

It isn't all done by carving. Many of our familiar breeds of dog were originally derived as hybrids of other breeds, often quite recently, for example in the nineteenth century. Hybridization, of course, represents a deliberate violation of the isolation of the gene pools on virtual islands. Some hybridization schemes are designed with such care that the breeders would resent their products being described as mongrels or mutts (as President Obama delightfully described himself). The 'Labradoodle' is a hybrid between a standard poodle and a Labrador retriever, the result of a carefully crafted quest for the best virtues of both breeds. Labradoodle owners have established societies and associations just like those of pure-bred pedigree dogs. There are two schools of thought in the Labradoodle Fancy, and those of other such designer hybrids. There are those who are happy to go on making Labradoodles by mating poodles and Labradors together. And there are those who are trying to initiate a new Labradoodle gene pool that will breed true, when Labradoodles are mated together. At present, second-generation Labradoodle genes recombine to produce more variety than pure-bred pedigree dogs are supposed to show. This is how many 'pure' breeds got their start: they went through an intermediate stage of high variation, subsequently trimmed down through generations of careful breeding.

Sometimes, new breeds of dog get their start with the adoption of a single major mutation. Mutations are the random changes in genes that constitute the raw material for evolution by non-random selection. In nature, large mutations seldom survive, but geneticists like them in the laboratory because they are easy to study. Breeds of dog with very short legs, like basset hounds and dachshunds, acquired them in a single step with the genetic mutation called achondroplasia, a classic example of a large mutation that would be unlikely to survive in nature. A similar mutation is responsible for the commonest kind of human dwarfism: the trunk is of nearly normal size, but the legs and arms are short. Other genetic routes produce miniature breeds that retain the proportions of the original. Dog breeders can achieve changes in size and shape by selecting combinations of a few major mutations such as achondroplasia and lots of minor genes. Nor do they need to understand the genetics in order to achieve change effectively. Without any understanding at all, just by choosing who mates with whom, you can breed for all kinds of desired characteristics. This is what dog breeders, and animal and plant breeders generally, achieved for centuries before anybody understood anything about genetics. And there's a lesson in that about natural selection, for nature, of course, has no understanding or awareness of anything at all.

The American zoologist Raymond Coppinger makes the point that puppies of different breeds are much more similar to each other than adult dogs are. Puppies can't afford to be different, because the main thing they have to do is suck,* and sucking presents pretty much the same challenges for all breeds. In particular, in order to

* Not suckle: mothers suckle, babies suck.

be good at sucking, a puppy can't have a long snout like a borzoi or a retriever. That's why all puppies look like pugs. You could say that an adult pug is a puppy whose face didn't properly grow up. Most dogs, after they are weaned, develop a relatively longer snout. Pugs, bulldogs and Pekineses don't; they grow in other departments, while the snout retains its infantile proportions. The technical term for this is *neoteny*, and we'll meet it again when we come on to human evolution in Chapter 7.

If an animal grows at the same rate in all its parts, so that the adult is just a uniformly inflated replica of the infant, it is said to grow isometrically. Isometric growth is quite rare. In allometric growth, by contrast, different parts grow at different rates. Often, the rates of growth of different parts of an animal bear some simple mathematical relation to each other, a phenomenon that was investigated especially by Sir Julian Huxley in the 1930s. Different breeds of dog achieve their different shapes by means of genes that change the allometric growth relationships between the parts of the body. For example, bulldogs get their Churchillian scowl from a genetic tendency towards slower growth of the nasal bones. This has knock-on effects on the relative growth of the surrounding bones, and indeed all the surrounding tissues. One of these knock-on effects is that the palate is pulled up into an awkward position, so the bulldog's teeth stick out and it has a tendency to dribble. Bulldogs also have breathing difficulties, which are shared by Pekineses. Bulldogs even have difficulty being born because the head is disproportionately big. Most if not all the bulldogs you see today were born by caesarian section.

Borzois are the opposite. They have extra long snouts. Indeed, they are unusual in that the elongation of the snout begins before they are born, which probably makes borzoi puppies less proficient suckers than other breeds. Coppinger speculates that the human desire to breed borzois for long snouts has reached a limit imposed by the survival capacity of puppies trying to suck.

What lessons do we learn from the domestication of the dog? First, the great variety among breeds of dogs, from Great Danes to

Yorkies, from Scotties to Airedales, from ridgebacks to dachshunds, from whippets to St Bernards, demonstrates how easy it is for the non-random selection of genes – the 'carving and whittling' of gene pools – to produce truly dramatic changes in anatomy and behaviour, and so fast. Surprisingly few genes may be involved. Yet the changes are so large – the differences between breeds so dramatic – that you might expect their evolution to take millions of years instead of just a matter of centuries. If so much evolutionary change can be achieved in just a few centuries or even decades, just think what might be achieved in ten or a hundred million years.

Viewing the process over centuries, it is no empty fancy that human dog breeders have seized dog flesh like modelling clay and pushed it, pulled it, kneaded it into shape, more or less at will. Of course, as I pointed out earlier, we have really been kneading not dog flesh but dog gene pools. And 'carved' is a better metaphor than 'kneaded'. Some sculptors work by taking a lump of clay and kneading it into shape. Others take a lump of stone or wood, and carve it by *subtracting* bits with a chisel. Obviously dog fanciers don't carve dogs into shape by subtracting bits of dog flesh. But they do something close to carving dog gene pools by subtraction. It is more complicated than pure subtraction, however. Michelangelo took a single chunk of marble, and then subtracted marble from it to reveal David lurking inside. Nothing was added. Gene pools, on the other hand, are continually added to, for example by mutation, while at the same time non-random death subtracts. The analogy to sculpture breaks down here, and should not be pushed too tenaciously, as we'll see again in Chapter 8.

The idea of sculpture calls to mind the over-muscled physiques of human body-builders, and non-human equivalents such as the Belgian Blue breed of cattle. This walking beef factory has been contrived via a particular genetic alteration called 'double muscling'. There is a substance called myostatin, which limits muscle growth. If the gene that makes myostatin is disabled, muscles grow larger than usual. It is quite often the case that a given gene can mutate in more than one way to produce the same outcome, and indeed there are various ways in which the myostatin-producing gene can be disabled, with the same effect. Another example is the breed of pig called the Black Exotic, and there are individual dogs of various breeds that show the same exaggerated musculature for the same reason. Human body-builders achieve a similar physique by an extreme regime of exercise, and often by the use of anabolic steroids: both environmental manipulations that mimic the genes of the Belgian Blue and the Black Exotic. The end result is the same, and that is a lesson in itself. Genetic and environmental changes can produce identical outcomes. If you wanted to rear a human child to win a body-building contest and you had a few centuries to spare, you could start by genetic manipulation, engineering exactly the same freak gene as characterizes Belgian Blue cattle and Black Exotic pigs. Indeed, there are some humans known to have deletions of the myostatin gene, and they tend to be abnormally well muscled. If you started with a mutant child and made it pump iron as well (presumably the cattle and pigs could not be cajoled into this), you could probably end up with something more grotesque than Mr Universe.

Political opposition to eugenic breeding of humans sometimes spills over into the almost certainly false assertion that it is impossible. Not only is it immoral, you may hear it said, it wouldn't work. Unfortunately, to say that something is morally wrong, or politically undesirable, is not to say that it wouldn't work. I have no doubt that, if you set your mind to it and had enough time and enough political power, you could breed a race of superior body-builders, or high-jumpers, or shot-putters; pearl fishers, sumo wrestlers, or sprinters; or (I suspect, although now with less confidence because there are no animal precedents) superior musicians, poets, mathematicians or wine-tasters. The reason I am confident about selective breeding for athletic prowess is that the qualities needed are so similar to those that demonstrably work in the breeding of racehorses and carthorses, of greyhounds and sledge dogs. The reason I am still pretty confident about the practical feasibility (though not the moral or political desirability)

'And God blessed them, and God said unto them, Be fruitful, and multiply, and replenish the earth, and subdue it: and have dominion over the fish of the sea, and over the fowl of the air, and over every living thing that moveth upon the earth.'

Opinion polls show that many people are creationists who believe that all living things came into existence in a single week six thousand years ago.

What artificial selection
can do in a very short time:
wild cabbage (**a**) and its useful
(**b**) and monstrous (**c**) descendants.
Sunflowers were (**d**) artificially selected long ago by Native
Americans and (**e**) enhanced by modern horticulturalists.

The Belgian Blue mound of beef (**f**) is artificially
mutated. The beefcake woman (**g**) is artificially
nurtured and exercised. Environmentally induced
change can closely mimic genetic change.

(**h**) Chihuahua and Great Dane: both are wolves under the skin, but who would guess it from their appearance, after a few centuries of artificial selection?

3

h

4

(a) The long nectary of this Madagascar orchid led both Darwin and Wallace to predict the eventual discovery of a long tongue to match. Years later, it was found: *Xanthopan morgani praedicta*, Darwin's hawk moth. (b) Bucket orchid: one of the most elaborate exponents of 'silver bullet' pollination. (c) Euglossine bee, struggling to leave bucket orchid and picking up pollen as it does so.

a

b

c

(d) Moth that thinks it's a hummingbird? The hummingbird hawk moth, a wonderful example of convergent evolution. (e) Hummingbird in exquisite action. Bright red flowers are usually bird-pollinated, for birds, unlike insects, see well at the red end of the spectrum. (f) Sunbird sucking nectar from a red flower in Africa. (g) Bucking bronco ride of a thynnid wasp on a hammer orchid. (h) Honey trap? This orchid is a deceiver, relying on its resemblance to a female bee to lure a male into attempted copulation. (i) Evening primrose as we see it. (j) Evening primrose as an insect sees it? Not quite, but with false colours to show the patterns that an insect, with its ultraviolet vision, might see. (k) Spider orchid. Is the resemblance to a spider shaped by natural selection?

a

The bright colours of the cock pheasant (**a**) have been selected by generations of hens. (**b**) Underwater cock pheasants? Male guppies in predator-free waters are free to evolve bright colours that attract predators. As with roses and tulips, human breeders have latched on and taken the trend further. These guppies appeal to aquarists as well as females.

(**c**) Danger lurks in beauty. Purple praying mantis lies in wait for insects lured by the flower that it mimics. (**d**) Other mantises mimic leaves; this is the nymph (a juvenile stage) of one of them. Some animals, such as this gecko (**e**) from Madagascar, mimic dead leaves. (**f**) This is not the front end of a snake but the back end of a caterpillar. Its ancestors survived because a significant number of would-be predators were frightened by the resemblance.

b

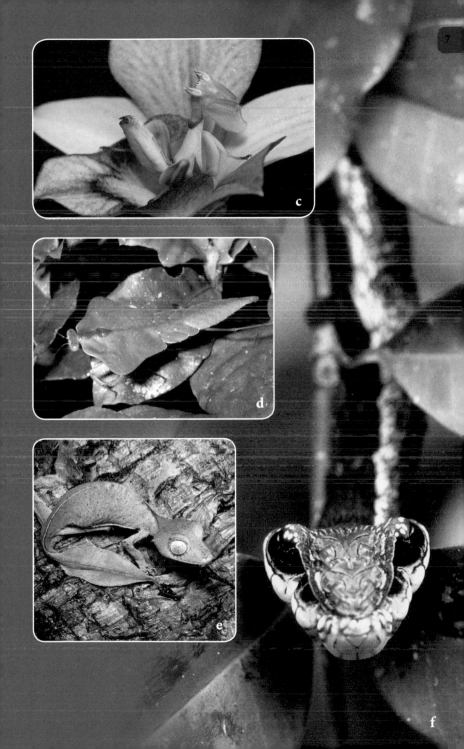

8

(**a**) Gorilla in our midst. Stunning evidence of the unreliability of eyewitness testimony (see text p. 14)

© 2005, Daniel J. Simons

Figure provided by Daniel Simons. The video depicted in this figure is available as part of a DVD from Viscog Productions (http://www.viscog.com).

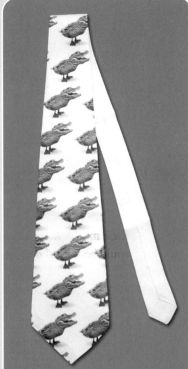

(**b**) If evolution is true, why isn't the world full of crocoducks and fronkeys, of doggypotamuses and kangarabbits? To celebrate this knockdown argument (see text pp. 152–3) Josh Timonen kindly made me a crocoduck tie, to wear in honour of creationists everywhere.

(**c**) Tempting bait to hook an eager creationist (see text p. 154)

of selective breeding for mental or otherwise un[...]
that there are so few examples where an attempt at se[...]
in animals has ever failed, even for traits that might have bee[...]
surprising. Who would have thought, for example, that dogs cou[...]
bred for sheep-herding skills, or 'pointing', or bull-baiting?

You want high milk yield in cows, orders of magnitude more
gallons than could ever be needed by a mother to rear her babies?
Selective breeding can give it to you. Cows can be modified to grow vast
and ungainly udders, and these continue to yield copious quantities
of milk indefinitely, long after the normal weaning period of a calf.
As it happens, dairy horses have not been bred in this way, but will
anyone contest my bet that we could do it if we tried? And of course,
the same would be true of dairy humans, if anyone wanted to try. All
too many women, bamboozled by the myth that breasts like melons
are attractive, pay surgeons large sums of money to implant silicone,
with (for my money) unappealing results. Does anyone doubt that,
given enough generations, the same deformity could be achieved by
selective breeding, after the manner of Friesian cows?

About twenty-five years ago I developed a computer simulation
to illustrate the power of artificial selection: a kind of computer game
equivalent to breeding prize roses or dogs or cattle. The player is faced
with an array of nine shapes on the screen 'computer biomorphs'
– the middle one of which is the 'parent' of the surrounding eight. All
the shapes are constructed under the influence of a dozen or so 'genes',
which are simply numbers handed down from 'parent' to 'offspring',
with the possibility of small 'mutations' intervening on the way. A
mutation is just a slight increment or decrement in the numerical value
of the parent's gene. Each shape is constructed under the influence of
a particular set of numbers, which are its own particular values of
the dozen genes. The player looks over the array of nine shapes and
sees no genes but chooses the preferred 'body' shape she wants to
breed from. The other eight biomorphs disappear from the screen,
the chosen one glides to the centre, and 'spawns' eight new mutant
'children'. The process repeats for as many 'generations' as the player

Biomorphs from the 'Blind Watchmaker' program

has time for, and the average shape of the 'organisms' on the screen gradually 'evolves' as the generations go by. Only genes are passed from generation to generation, so, by directly choosing biomorphs by eye, the player is inadvertently choosing genes. That is just what happens when breeders choose dogs or roses to breed from.

So much for the genetics. The game starts to get interesting when we consider the 'embryology'. The embryology of a biomorph on the screen is the process by which its 'genes' – those numerical values – influence its shape. Many very different embryologies can be imagined, and I have tried out quite a few of them. My first program, called 'Blind Watchmaker', uses a tree-growing embryology. A main 'trunk' sprouts two 'branches', then each branch sprouts two branches of its own, and so on. The number of branches, and their angles and lengths, are all under genetic control, determined by the numerical values of the genes. An important feature of the branching tree embryology is that it is *recursive*. I won't expound that idea here, but it means that a single mutation typically has an effect all over the tree, rather than just in one corner of it.

Although the Blind Watchmaker program starts off with a simple branching tree, it rapidly wanders off into a wonderland of evolved forms, many with a strange beauty, and some – depending on the intentions of the human player – coming to resemble familiar creatures such as insects, spiders or starfish. On the left is a 'safari park' of creatures that just one player of the game (me) found in the byways and backwaters of this strange computer wonderland. In a later version of the program, I expanded the embryology to allow for genes controlling the colour and shape of the 'branches' of the tree.

A more elaborate program, called 'Arthromorphs', which I wrote jointly with Ted Kaehler, then working for the Apple Computer Company, embodies an 'embryology' with some interesting biological features specifically geared to breeding 'insects', 'spiders', 'centipedes' and other creatures resembling arthropods. I have explained the arthromorphs in detail, along with the biomorphs, 'conchomorphs' (computer molluscs) and other programs in this vein, in *Climbing Mount Improbable*.

As it happens, the mathematics of shell embryology are well understood, so artificial selection using my 'conchomorph' program is capable of generating extremely lifelike forms (see over). I shall refer back to these programs, to make a completely different point, in the final chapter. Here I have introduced them for the purpose of illustrating the power of artificial selection, even in an extremely over-simplified computer environment. In the real world of agriculture and horticulture, the world of the pigeon fancier or dog breeder, artificial selection can achieve so much more. Biomorphs, arthromorphs

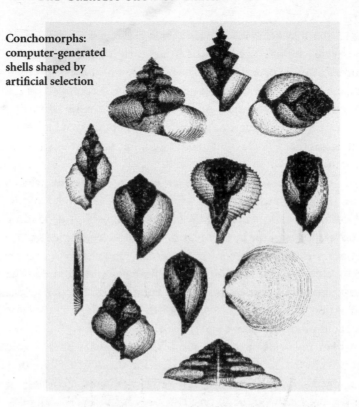

Conchomorphs: computer-generated shells shaped by artificial selection

and conchomorphs just illustrate the principle, in something like the same way that artificial selection itself is going to illustrate the principle behind natural selection – in the next chapter.

Darwin had first-hand experience of the power of artificial selection and he gave it pride of place in Chapter 1 of *On the Origin of Species*. He was softening his readers up to take delivery of his own great insight, the power of natural selection. If human breeders can transform a wolf into a Pekinese, or a wild cabbage into a cauliflower, in just a few centuries or millennia, why shouldn't the non-random survival of wild animals and plants do the same thing over millions of years? That will be the conclusion of my next chapter; but my strategy first will be to continue the softening-up process, to ease the passage towards understanding of natural selection.

CHAPTER 3

THE PRIMROSE PATH TO MACRO-EVOLUTION

CHAPTER 2 showed how the human eye, working by selective breeding over many generations, sculpted and kneaded dog flesh to assume a bewildering variety of forms, colours, sizes and behaviour patterns. But we are humans, accustomed to making choices that are deliberate and planned. Are there other animals that do the same thing as human breeders, perhaps without deliberation or intention but with similar results? Yes, and they carry this book's softening-up program steadily forward. This chapter embarks on a step-by-step seduction of the mind as we pass from the familiar territory of dog breeding and artificial selection to Darwin's giant discovery of natural selection, via some colourful intermediate stages. The first of these intermediate steps along the path of seduction (is it over the top to call it a primrose path?) takes us into the honeyed world of flowers.

Wild roses are agreeable little flowers, pretty enough, but nothing to write home about in the terms one might lavish on, say, 'Peace' or 'Lovely Lady' or 'Ophelia'. Wild roses have a delicate aroma, unmistakable, but not to-swoon-for like 'Memorial Day' or 'Elizabeth Harkness' or 'Fragrant Cloud'. The human eye and the human nose went to work on wild roses, enlarging them, shaping them, doubling up the petals, tinting them, refining the bloom, boosting natural fragrances to heady extremes, adjusting habits of growth, eventually entering them in sophisticated hybridization programs until, today, after decades of skilful selective breeding, there are hundreds of prized varieties, each with its own evocative or commemorative name. Who would not like to have a rose named after her?

INSECTS WERE THE FIRST DOMESTICATORS

Roses tell the same story as dogs, but with one difference, which is relevant to our softening-up strategy. The flower of the rose, even before human eyes and noses embarked on their work of genetic chiselling, owed its very existence to millions of years of very similar sculpting by insect eyes and noses (well, antennae, which is what insects smell with). And the same is true of all the flowers that beautify our gardens.

The sunflower, *Helianthus annuus*, is a North American plant whose wild form looks like an aster or large daisy. Cultivated sunflowers today have been domesticated to the point where their flowers are the size of a dinner plate.* 'Mammoth' sunflowers, originally bred in Russia, are 12 to 17 feet high, the head diameter is close to one foot, which is more than ten times the size of a wild sunflower's disc, and there is normally only one head per plant, instead of the many, much smaller, flowers of the wild plant. The Russians started breeding this American flower, by the way, for religious reasons. During Lent and Advent, the use of oil in cooking was banned by the Orthodox Church. Conveniently, and for a reason that I – untutored in the profundities of theology – shall not presume to fathom, sunflower seed oil was deemed to be exempt from this prohibition.† This provided one of the economic pressures that drove the recent selective breeding of the sunflower. Long before the modern era, however, native Americans had been cultivating these nutritious and spectacular flowers for food,

---

* As in all members of the daisy family, each 'flower' is actually many little flowers (florets), bundled together in the dark disk in the middle. The yellow petals that surround the sunflower are in fact the petals of just the florets around the edge. The florets in the rest of the disk have petals, but too small to be noticed.

† Perhaps because – being a New World plant – the sunflower is not mentioned explicitly in the Bible. The theological mind takes a delight in the niceties of dietary laws and the ingenuity required to dodge them. In South America, capybaras (sort of giant guinea pigs) were deemed to be honorary fish for the purposes of Catholic dietary laws on Fridays, presumably because they live in water. According to the food writer Doris Reynolds, French Catholic gourmets discovered a loophole that enabled them to eat meat on Fridays. Lower a leg of lamb into a well and then 'fish' it out. They must think God is awfully easily fooled.

for dyes and for decoration, and they achieved results intermediate between the wild sunflower and the extravagant extremes of modern cultivars. But before that again, sunflowers, like all brightly coloured flowers, owed their very existence to selective breeding by insects.

The same is true of most of the flowers we are aware of – probably all the flowers that are coloured anything other than green and whose smell is anything more than just vaguely plant-like. Not all the work was done by insects – for some flowers the pollinators that did the initial selective breeding were hummingbirds, bats, even frogs – but the principle is the same. Garden flowers have been further enhanced by us, but the wild flowers with which we started only caught our attention in the first place because insects and other selective agents had been there before us. Generations of ancestral flowers were chosen by generations of ancestral insects or hummingbirds or other natural pollinators. It is a perfectly good example of selective breeding, with the minor difference that the breeders were insects and hummingbirds, not humans. At least, I think the difference is minor. You may not, in which case I still have some softening up to do.

What might tempt us to think it a major difference? For one thing, humans *consciously* set out to breed, say, the darkest, most blackish purple rose they can, and they do it to satisfy an aesthetic whim, or because they think other people will pay money for it. Insects do it not for aesthetic reasons but for reasons of . . . well, here we need to back up and look at the whole matter of flowers and their relationship with their pollinators. Here's the background. For reasons I won't go into now, it is of the essence of sexual reproduction that you shouldn't fertilize yourself. If you did that, after all, there'd be little point in bothering with sexual reproduction in the first place. Pollen must somehow be transported from one plant to another. Hermaphroditic plants that have male and female parts within one flower often go to elaborate lengths to stop the male half from fertilizing the female half. Darwin himself studied the ingenious way this is achieved in primroses.

Taking the need for cross-fertilization as a given, how do flowers achieve the feat of moving pollen across the physical gap that separates them from other flowers of the same species? The obvious way is by the wind, and plenty of plants use it. Pollen is a fine, light powder. If you release enough of it on a breezy day, one or two grains may have the luck to land on the right spot in a flower of the right species. But wind pollination is wasteful. A huge surplus of pollen needs to be manufactured, as hay fever sufferers know. The vast majority of pollen grains land somewhere other than where they should, and all that energy and costly *matériel* is wasted. There is a more directed way for pollen to be targeted.

Why don't plants choose the animal option, and walk around looking for another plant of the same species, then copulate with it? That's a harder question to deal with than you might think. It's circular simply to assert that plants don't walk, but I'm afraid that will have to do for now.* The fact is, plants don't walk. But animals walk. And animals fly, and they have nervous systems capable of directing them towards particular targets, with sought-for shapes and colours. So if only there were some way to persuade an animal to dust itself with pollen and then walk or preferably fly to another plant of the right species . . .

Well, the answer's no secret: that's exactly what happens. The story is in some cases highly complex and in all cases fascinating. Many flowers use a bribe of food, usually nectar. Maybe bribe is too loaded a word. Would you prefer 'payment for services rendered'? I'm happy with both, so long as we don't misunderstand them in a human way. Nectar is sugary syrup, and it is manufactured by plants specifically and only for paying, and fuelling, bees, butterflies, hummingbirds, bats and other hired transport. It is costly to make, funnelling off a proportion of the sunshine energy trapped by the leaves, the solar panels of the plant. From the point of view of the bees and hummingbirds, it is high-energy aviation fuel. The energy

---

* Oliver Morton discusses this and related issues in his provokingly lyrical book *Eating the Sun*.

locked up in the sugars of nectar could have been used elsewhere in the economy of the plant, perhaps to make roots, or to fill the underground storage magazines that we call tubers, bulbs and corms, or even to make huge quantities of pollen for broadcasting to the four winds. Evidently, for a large number of plant species, the trade-off works out in favour of paying insects and birds for their wings, and fuelling their flight muscles with sugar. It's not a totally overwhelming advantage, however, because some plants do use wind pollination, presumably because details of their economic circumstances tip their balance that way. Plants have an energy economy and, as with any economy, trade-offs may favour different options under different circumstances. That's an important lesson in evolution, by the way. Different species do things in different ways, and we often won't understand the differences until we have examined the whole economy of the species.

If wind pollination is at one end of a continuum of cross-fertilization techniques – shall we call it the profligate end? – what is at the other end, the 'magic bullet' end? Very few insects can be relied upon to fly like a magic bullet straight from the flower where they have picked up pollen to another flower of exactly the right species. Some just go to any old flower, or possibly any flower of the right colour, and it is still a matter of luck whether it happens to be the same species as the flower that has just paid it in nectar. Nevertheless, there are some lovely examples of flowers that lie far out towards the magic bullet end of the continuum. High on the list are orchids, and it's no wonder that Darwin devoted a whole book to them.

Both Darwin and his co-discoverer of natural selection, Wallace, called attention to an amazing orchid from Madagascar, *Angraecum sesquipedale* (see colour page 4), and both men made the same remarkable prediction, which was later triumphantly vindicated. This orchid has tubular nectaries that reach down more than 11 inches by Darwin's own ruler. That's nearly 30 centimetres. A related species, *Angraecum longicalcar*, has nectar-bearing spurs that are even longer, up to 40 centimetres (more than 15 inches). Darwin, purely on the strength of *A. sesquipedale*'s existence in Madagascar, predicted in his orchid book of 1862 that there must be 'moths capable of extension to a length of between ten and eleven inches'. Wallace, five years later (it isn't clear whether he had read Darwin's book) mentioned several moths whose probosces were nearly long enough to meet the case.

> I have carefully measured the proboscis of a specimen of *Macrosila cluentius* from South America in the collection of the British Museum, and find it to be nine inches and a quarter long! One from tropical Africa (*Macrosila morganii*) is seven inches and a half. A species having a proboscis two or three inches longer could reach the nectar in the largest flowers of *Angræcum sesquipedale*, whose nectaries vary in length from ten to fourteen inches. That such a moth exists in Madagascar may be safely predicted; and naturalists who visit that island should search for it with as much confidence as astronomers searched for the planet Neptune, – and they will be equally successful!

In 1903, after Darwin's death but well within Wallace's long lifetime, a hitherto unknown moth was discovered which turned out to fulfil the Darwin/Wallace prediction, and was duly honoured with the sub-specific name *praedicta*. But even *Xanthopan morgani praedicta*, 'Darwin's hawk moth', is not sufficiently well endowed to pollinate *A. longicalcar*, and the existence of this flower encourages us to suspect the existence of an even longer-tongued moth, with the same confidence as Wallace invoked the predicted discovery of the planet Neptune. By the way, this little example gives the lie, yet again, to the allegation that evolutionary science cannot be predictive because it concerns past history. The Darwin/Wallace prediction was still a perfectly valid one, even though the *praedicta* moth must already have existed before they made it. They were predicting that, at some time in the future, somebody would discover a moth with a tongue long enough to reach the nectar in *A. sesquipedale*.

Insects have good colour vision, but their whole spectrum is shifted towards the ultraviolet and away from the red. Like us, they see yellow, green, blue and violet. Unlike us, however, they also see well into the ultraviolet range; and they don't see red, at 'our' end of the spectrum. If you have a red tubular flower in your garden it is a good bet, though not a certain prediction, that in the wild it is pollinated not by insects but by birds, who see well at the red end of the spectrum – perhaps hummingbirds if it is a New World plant, or sunbirds if an Old World plant. Flowers that look plain to us may actually be lavishly decorated with spots or stripes for the benefit of insects, ornamentation that we can't see because we are blind to ultraviolet. Many flowers guide bees in to land by little runway markings, painted on the flower in ultraviolet pigments, which the human eye can't see.

The evening primrose (*Oenothera*) looks yellow to us. But a photograph taken through an ultraviolet filter shows a pattern for the benefit of bees, which we can't see with normal vision (see colour page 5). In the photograph it appears as red, but that is a 'false colour': an arbitrary choice by the photographic process. It doesn't mean that bees would see it as red. Nobody knows how ultraviolet (or yellow or any other colour) looks to a bee (I don't even know how red looks to you – an old philosophical chestnut).

A meadow full of flowers is nature's Times Square, nature's Piccadilly Circus. A slow-motion neon sign, it changes from week to week as different flowers come into season, carefully prompted by cues from, for example, the changing length of days to synchronize with others of their own species. This floral extravaganza, splashed across the green canvas of a meadow, has been shaped and coloured, magnified and titivated by the past choices made by animal eyes: bee eyes, butterfly eyes, hoverfly eyes. In New World forests we'd have to add hummingbird or in African forests sunbird eyes to the list.

Hummingbirds and sunbirds are not particularly closely related, by the way. They look and behave like each other because they have converged upon the same way of life, largely revolving around flowers and nectar (although they eat insects as well as nectar). They have long beaks for probing nectaries, extended by even longer tongues. Sunbirds are less accomplished hoverers than hummingbirds, who can even go backwards like a helicopter. Also convergent, although from a far distant vantage point in the animal kingdom, are the hummingbird hawk moths, again consummate hoverers with spectacularly long tongues (all three types of nectar junkie are illustrated on colour page 5).

We shall return to convergent evolution later in the book, after properly understanding natural selection. Here, in this chapter, flowers are seducing us, drawing us in, step by step, lining our path to that understanding. Hummingbird eyes, hawk-moth eyes, butterfly eyes, hoverfly eyes, bee eyes are critically cast over wild flowers, generation after generation, shaping them, colouring them, swelling them, patterning and stippling them, in almost exactly the same way as human eyes later did with our garden varieties; and with dogs, cows, cabbages and corn.

For the flower, insect pollination represents a huge advance in economy over the wasteful scattergun of wind pollination. Even if a bee visits flowers indiscriminately, lurching promiscuously from buttercup to cornflower, from poppy to celandine, a pollen grain clinging to its hairy abdomen has a much greater chance of hitting the right target – a second flower of the same species – than it would have if scattered on the wind. Slightly better would be a bee with a preference for a particular colour, say blue. Or a bee that, while not having any long-term colour preference, tends to form colour habits, so that it chooses colours in runs. Better still would be an insect that visits flowers of only one species. And there are flowers, like the Madagascar orchid that inspired the Darwin/Wallace prediction, whose nectar is available only to certain insects that specialize in that kind of flower and benefit from their monopoly over it. Those Madagascar moths are the ultimate magic bullets.

From a moth's point of view, flowers that reliably provide nectar are like docile, productive milch cows. From the flowers' point of view, moths that reliably transport their pollen to other flowers of the same species are like a well-paid Federal Express service, or like well-trained homing pigeons. Each side could be said to have domesticated the other, selectively breeding them to do a better job than they previously did. Human breeders of prize roses have had almost exactly the same kinds of effects on flowers as insects have – just exaggerated them a bit. Insects bred flowers to be bright and showy. Gardeners made them brighter and showier still. Insects made roses pleasantly fragrant. We came along and made them even more so. Incidentally, it is a fortunate coincidence that the fragrances that bees and butterflies prefer happen to appeal to us too. Flowers such as 'stinking Benjamin' (*Trillium erectum*) or the 'corpse flower' (*Amorphophallus titanum*), which use flesh flies or carrion beetles as pollinators, often nauseate us, because they mimic the smell of decaying meat. Such flowers have not, I presume, had their scents enhanced by human domesticators.

Of course, the relationship between insects and flowers is a two-way street, and we mustn't neglect to look in both directions. Insects may 'breed' flowers to be more beautiful, but not because they enjoy the beauty.* Rather, the flowers benefit from being perceived as attractive by insects. The insects, by choosing the most attractive flowers to visit, inadvertently 'breed for' floral beauty. At the same time, the flowers are breeding the insects for pollination ability. Then again, I have implied that insects breed flowers for high nectar yield, like dairymen breeding massively uddered Friesians. But it is in the flowers' interests to ration their nectar. Satiate an insect and it has no incentive to go on and look for a second flower – bad news for the first flower, for which the second visit, the pollinating visit, is the whole point of the exercise. From the flowers' point of view, a delicate balance must be struck between providing too much nectar (no visit to a second flower) and too little (no incentive to visit the first flower).

---

* At least there is no reason to think that they do, or indeed that they enjoy anything in the sense we understand. I shall return to this perennial temptation in Chapter 12.

Insects have milked flowers for their nectar, and bred them for increased yield – probably encountering resistance from the flowers, as we have just seen. Have beekeepers (or horticulturalists with the interests of beekeepers in mind) bred flowers to be even more productive of nectar, just as dairy farmers bred Friesian and Jersey cows? I'd be intrigued to know the answer. Meanwhile, there's no doubt of the close parallel between horticulturalists as breeders of pretty and fragrant flowers, and bees and butterflies, hummingbirds and sunbirds doing the same thing.

YOU ARE MY NATURAL SELECTION

Are there other examples of selective breeding by non-human eyes? Oh yes. Think of the dull, camouflaged plumage of a hen pheasant, compared with the splendiferous male of the same species. There seems little doubt that, if his individual survival were the only thing that mattered, the cock golden pheasant would 'prefer' to look like the female, or like a grown-up version of how he was as a chick. The female and the chicks are obviously well camouflaged, and that's the way the male would be too if individual survival were his priority. The same is true of other pheasants such as Lady Amherst's and the familiar ring-necked pheasant. The cocks look flamboyant and dangerously attractive to predators, but each species in a very different way. The hens are camouflaged and dull-coloured, each species in pretty much the same way. What is going on here?

One way to put it is Darwin's way: 'sexual selection'. But another way – and the one that better suits my primrose path – is 'selective breeding by females of males'. Bright colours may indeed attract predators, but they attract female pheasants too. Generations of hens chose to mate with bright, glowing males, rather than the dull brown creatures that the males would surely have remained but for selective breeding by females. The same thing happened with peahens selectively breeding peacocks, female birds of paradise breeding males, and numerous other examples of birds, mammals, fish, amphibians, reptiles and insects where females (it's usually females rather than males, for reasons we needn't go into) choose from among competing males. As with garden flowers, human pheasant-breeders have improved upon the selective handiwork of the hen pheasants that preceded them, producing spectacular variants of the golden pheasant, for example, although more by picking one or two major mutations rather than by gradually shaping the bird through generations of breeding. Humans have also selectively bred some amazing varieties of pigeons (as Darwin knew at first hand) and chickens, descended from a Far Eastern bird, the red jungle fowl *Gallus gallus.*

This chapter is mostly about selection by eyes, but other senses can do the same thing. Fanciers have bred canaries for their songs, as well as for their appearance. The wild canary is a yellowish brown finch, not spectacular to look at. Human selective breeders have taken the palette of colours thrown up by random genetic variation and manufactured a colour distinctive enough to be named after the bird: canary yellow. By the way, the bird itself is named after the islands,* not the other way around as with the Galapagos Islands, whose name comes from a Spanish word for tortoise.

Varieties of chicken: three illustrations from Darwin's *The Variation of Animals and Plants under Domestication*

---

* Which were in turn named after the 'multitude of dogs of a huge size' mentioned in Pliny's *Natural History.*

But canaries are best known for their song, and this too has been tuned up and enriched by human breeders. Various songsters have been manufactured, including Rollers, which have been bred to sing with the beak closed, Waterslagers, which sound like bubbling water, and Timbrados, which produce metallic, bell-like notes, together with a castanet-like chatter that befits their Spanish origins. Domestically bred songs are longer, louder and more frequent than the wild ancestral type. But all these highly prized songs are made up of elements that occur in wild canaries, just as the habits and tricks of various breeds of dogs come from elements to be found in the behavioural repertoire of wolves.*

Once again, human breeders have only been building on the earlier selective breeding efforts of female birds. Over generations, wild female canaries inadvertently bred males for their singing prowess by choosing to mate with males whose songs were especially appealing. In the particular case of canaries it happens that we know a little more. Canaries (and Barbary doves) have been favourite subjects for research on hormones and reproductive behaviour. It is known that in both species the sound of male vocalization (even from a tape recording) causes the females' ovaries to swell and secrete hormones that bring them into reproductive condition and make them more ready to mate. One could say that male canaries are manipulating females by singing to them. It is almost as though they were giving them hormone injections. One could also say that females are selectively breeding males to become better and better at singing. The two ways of looking at the matter are two sides of the same coin. As with other bird species, by the way, there is a complication: song is not only appealing to females, it is also a deterrent to rival males – but I'll leave that on one side.

Now, to move the argument on, look at the pictures opposite. The first is a woodcut of a Japanese kabuki mask, representing a

---

* For example, herding in sheepdogs is derived from stalking in wolves, with the killing removed from the end of the sequence.

Kabuki mask of samurai warrior

samurai warrior. The second is a crab of the species *Heikea japonica*, which is found in Japanese waters. The generic name, *Heikea*, comes from a Japanese clan called the Heike, who were defeated at sea in the battle of Danno-Ura (1185) by a rival clan called the Genji. Legend tells that the ghosts of drowned Heike warriors now inhabit the bottom of the sea, in the bodies of crabs – *Heikea japonica*. The myth is encouraged by the pattern on the back of this crab, which resembles the fiercely grimacing face of a samurai warrior. The famous zoologist Sir Julian Huxley was impressed enough by the resemblance to write, 'The resemblance of *Dorippe* to an angry Japanese warrior is far too specific and far too detailed to be accidental . . . It came about because those crabs with a more perfect resemblance to a warrior's face were less frequently eaten than the

Heikea japonica crab

others.' (*Dorippe* was what the crab was called in 1952 when Huxley wrote. It reverted to *Heikea* in 1990 when somebody rediscovered that it had been so named as early as 1824 – such are the strict priority rules of zoological nomenclature.)

This theory, that generations of superstitious fishermen threw back into the sea crabs that resembled human faces, received new legs in 1980 when Carl Sagan discussed it in his wonderful *Cosmos*. In his words,

> Suppose that, by chance, among the distant ancestors of this crab, one arose that resembled, even slightly, a human face. Even before the battle of Danno-ura, fishermen may have been reluctant to eat such a crab. In throwing it back, they set in motion an evolutionary process . . . As the generations passed, of crabs and fishermen alike, the crabs with patterns that most resembled a samurai face survived preferentially until eventually there was produced not just a human face, not just a Japanese face, but the visage of a fierce and scowling samurai.

It's a lovely theory, too good to die easily, and the meme has indeed replicated itself through the canon. I even found a website where you can vote on whether the theory is true (31 per cent of 1,331 voters), whether the photographs are fakes (15 per cent), whether Japanese craftsmen carve the shells to look that way (6 per cent), whether the resemblance is just a coincidence (38 per cent), or even whether the crabs really are manifestations of drowned samurai warriors (an amazing 10 per cent). Scientific truths are not, of course, decided by plebiscite, and I voted only because I was otherwise not allowed to see the voting figures. I'm afraid I voted with the killjoys. I think, on balance, that the resemblance is probably a coincidence. Not because, as one authoritative sceptic has pointed out, the ridges and grooves on the crab's back actually signify underlying muscle attachments. Even on the Huxley/Sagan theory, the superstitious fishermen would have to have begun by noticing some kind of original resemblance, however slight, and a symmetrical pattern of muscle attachments is exactly the kind of thing that would have provided that initial resemblance. I am more impressed by the same sceptic's observation that these crabs are too small to be worth eating anyway. According to him, all crabs of that size would have been thrown back, whether or not their backs resembled human faces, although I have to say that this more telling source of scepticism had a large bite taken out of it when I was taken out to dinner in Tokyo and my host ordered, for all the company, a dish of crabs. They were much larger than *Heikea*, and they were thickly encrusted in stout, calcified carapaces, but that didn't stop this superman picking up whole crabs, one by one, and biting into them like an apple, with a crunching sound that seemed to presage hideously bleeding gums. A crab as small as *Heikea* would be a doddle to such a gastronomic champion. He would surely swallow it whole without batting an eyelid.

My main reason for scepticism about the Huxley/Sagan theory is that the human brain is demonstrably eager to see faces in random patterns, as we know from scientific evidence, on top of the numerous legends about faces of Jesus, or the Virgin Mary, or Mother Teresa, being seen on slices of toast, or pizzas, or patches of damp on a wall. This eagerness is enhanced if the pattern departs from randomness in the specific direction of being symmetrical. All crabs (except hermit crabs) are symmetrical anyway. I reluctantly suspect that the resemblance of *Heikea* to a samurai warrior is no more than an accident, much as I would like to believe it has been enhanced by natural selection.

Never mind. There are plenty of other examples not involving humans, where animal 'fishermen', as it were, 'throw back' (or don't see in the first place) would-be food because of a resemblance to something sinister, and where the resemblance is certainly not due to chance. If you were a bird, out hunting caterpillars in the forest, what would you do if you were suddenly confronted with a snake? Leap back startled, would be my guess, and then give it a wide berth. Well, there is a caterpillar – to be precise, the rear end of a caterpillar

– that bears an unmistakable resemblance to a snake. It is truly alarming if you are frightened of snakes – as I shamefacedly confess I am. I even think I might be reluctant to pick this animal up, despite knowing perfectly well that it is in fact a harmless caterpillar. (A picture of this extraordinary creature appears on colour page 7.) I have the same problem with picking up wasp-mimicking or bee-mimicking hoverflies, even though I can see, from their possession of only one pair of wings, that they are stingless flies. These are among a vast list of animals that gain protection because they look like something else: something inedible like a pebble, a twig or a frond of seaweed, or something positively nasty like a snake or a wasp or the glaring eyes of a possible predator.

Have bird eyes, then, been breeding insects for their resemblance to unpalatable or venomous models? There's one sense in which we surely have to answer yes. What, after all, is the difference between this and peahens breeding peacocks for beauty, or humans breeding dogs or roses? Mainly, peahens are breeding *positively* for something attractive, by approaching it, while the caterpillar-hunting birds are breeding *negatively* for something repellent, by avoiding it. Right then, here's another example, and in this case the 'breeding' is positive, even though the selector doesn't benefit from its choice. Far from it.

Deep-sea angler fish sit on the bottom of the sea, waiting patiently for prey.* Like many deep-sea fish, anglers are spectacularly ugly by our standards. Maybe by fish standards too, although it probably doesn't matter because, down where they live, it is too dark to see much anyway. Like other denizens of the deep sea, female angler fish often make their own light – or rather, they have special receptacles in which they house bacteria which make light for them. Such 'bioluminescence' isn't bright enough to reflect any detail, but it is bright enough to attract other fish. A spine which, in a normal fish, would be just one of the rays in a fin, becomes elongated and stiffened to make a fishing rod. In some species the 'rod' is so long and flexible

---

* It doesn't affect the point I am making, but this story applies only to female angler fish. The males are usually tiny dwarfs, who attach themselves parasitically to a female's body, like a little extra fin.

that you'd call it a line rather than a rod. And on the end of the fishing rod or line is – what else? – a bait, or lure. The baits vary from species to species, but they always resemble small food items: perhaps a worm, or a small fish, or just a nondescript but temptingly jiggling morsel. Often the bait is actually luminous: another natural neon sign, and in this case the message being flashed is 'come and eat me'. Small fish are indeed tempted. They approach close to the bait. And it is the last thing they do for, at that moment, the angler opens her huge maw and the prey is engulfed with the inrush of water.

Now, would we say that the small prey fish are 'breeding for' more and more appealing lures, just as peahens breed for more appealing peacocks, and horticulturalists breed for more appealing roses? It's hard to see why not. In the case of the roses, the most attractive blooms are the ones deliberately chosen for breeding by the gardener. Much the same is true of peacocks chosen by peahens. It is possible that the peahens are not aware that they are choosing, whereas the rose-growers are. But that doesn't seem a very important distinction under the circumstances. Slightly more compelling is a distinction between the angler fish example and the other two. The prey fish are indeed choosing the most 'attractive' angler fish for breeding, via the indirect route of choosing them for survival by feeding them! Anglers with unattractive lures are more likely to starve to death and therefore less likely to breed. And the small prey fish are indeed doing the 'choosing'. But they are choosing with their lives! What we are homing in on here is true natural selection, and we are reaching the end of the progressive seduction that is this chapter.

Here's the progression laid out.

1 Humans deliberately choose attractive roses, sunflowers etc. for breeding, thereby preserving the genes that produce the attractive features. This is called artificial selection, it's something humans have known about since long before Darwin, and everybody understands that it is powerful enough to turn wolves into chihuahuas and to stretch maize cobs from inches to feet.

2   Peahens (we don't know whether consciously and deliberately, but let's guess not) choose attractive peacocks for breeding, again thereby preserving attractive genes. This is called sexual selection, and Darwin discovered it, or at least clearly recognized it and named it.

3   Small prey fish (definitely not deliberately) choose attractive angler fish for survival, by feeding the most attractive ones with their own bodies, thereby inadvertently choosing them for breeding and passing on, and therefore preserving, the genes that produce the attractive features. This is called – yes, we've finally got there – natural selection, and it was Darwin's greatest discovery.

Darwin's special genius realized that nature could play the role of selecting agent. Everybody knew about artificial selection,* or at least everybody with any experience of farms or gardens, dog shows or dovecotes. But it was Darwin who first spotted that you don't have to have a choosing *agent*. The choice can be made automatically by

---

* The popular *canard* about Hitler being inspired by Darwin comes partly from the fact that both Hitler and Darwin were impressed by something that everybody has known for centuries: you can breed animals for desired qualities. Hitler aspired to turn this common knowledge to the human species. Darwin didn't. His inspiration took him in a much more interesting and original direction. Darwin's great insight was that you don't need a breeding agent at all: nature – raw survival or differential reproductive success – can play the role of breeder. As for Hitler's 'Social Darwinism' – his belief in a struggle between races – that is actually very *un*-Darwinian. For Darwin, the struggle for existence was a struggle between individuals within a species, *not* between species, races or other groups. Don't be misled by the ill-chosen and unfortunate subtitle of Darwin's great book: *The preservation of favoured races in the struggle for life*. It is abundantly clear from the text itself that Darwin didn't mean races in the sense of 'A group of people, animals, or plants, connected by common descent or origin' (*Oxford English Dictionary*, definition 6.I). Rather, he intended something more like the *OED*'s definition 6.II: 'A group or class of people, animals, or things, having some common feature or features'. An example of sense 6.II would be 'All those individuals (regardless of their geographical race) who have blue eyes'. In the technical jargon of modern genetics, which was not available to Darwin, we would express the sense of 'race' in his subtitle as 'All those individuals who possess a certain allele.' The misunderstanding of the Darwinian struggle for existence as a struggle between *groups* of individuals – the so-called 'group selection' fallacy – is unfortunately not confined to Hitlerian racism. It constantly resurfaces in amateur misinterpretations of Darwinism, and even among some professional biologists who should know better.

survival – or failure to survive. Survival counts, Darwin realized, because only survivors reproduce and pass on the genes (Darwin didn't use the word) that helped them to survive.

I chose the angler fish as my example, because this can still be represented as an agent using its eyes to choose that which survives. But we have reached the point in our argument – Darwin's point – where we no longer need to talk about a choosing agent at all. Move now from angler fish to, say, tuna or tarpon, fish that actively pursue their prey. By no sensible stretch of language or imagination could we claim that the prey 'choose' which tarpon survive by being eaten. What we can say, however, is that the tarpon that are better equipped to catch prey, for whatever reason – fast swimming muscles, keen eyes, etc. – will be the ones that survive, and therefore the ones that reproduce and pass on the genes that made them successful. They are 'chosen' by the very act of staying alive, whereas another tarpon that was, *for whatever reason*, less well equipped would not survive. So, we can add a fourth step to our list.

4 Without any kind of choosing agent, those individuals that are 'chosen' by the fact that they happen to possess superior equipment to survive are the most likely to reproduce, and therefore to pass on the genes for possessing superior equipment. Therefore every gene pool, in every species, tends to become filled with genes for making superior equipment for survival and reproduction.

Notice how all-encompassing natural selection is. The other examples I have mentioned, steps 1, 2 and 3 and lots of others, can all be wrapped up in natural selection, as special cases of the more general phenomenon. Darwin worked out the most general case of a phenomenon that people already knew about in restricted form. Hitherto, they had known about it only in the special case of artificial selection. The general case is the non-random survival of randomly varying hereditary equipment. It doesn't matter how the non-random survival comes about. It can be deliberate, explicitly intentional choice by an agent (as with humans choosing pedigree greyhounds for breeding); it can be inadvertent choice by an agent without explicit intention (as with peahens choosing peacocks for breeding); it can be inadvertent choice which the chooser – with a hindsight that is granted to us but not the chooser itself – would prefer not to have made (as with prey fish choosing to approach an angler fish's lure); or it can be something that we wouldn't recognize as choice at all, as when a tarpon survives by virtue of, say, an obscure biochemical advantage buried deep within its muscles, which gives it an extra burst of speed when pursuing prey. Darwin himself said it beautifully, in a favourite passage from *On the Origin of Species*:

> It may be said that natural selection is daily and hourly scrutinising, throughout the world, every variation, even the slightest; rejecting that which is bad, preserving and adding up all that is good; silently and insensibly working, whenever and wherever opportunity offers, at the improvement of each organic being in relation to its organic and inorganic conditions of life. We see nothing of these slow changes in progress, until the hand of time has marked the long lapse of ages, and then so imperfect is our view into long past geological ages, that we see only that the forms of life are now different from what they formerly were.

I have here quoted, as is my usual practice, the *first* edition of Darwin's masterpiece. An interesting interpolation found its way into later editions: 'It may *metaphorically* be said that natural selection is daily and hourly . . .' (emphasis added). You might think that 'It may be said . . .' was cautious enough. But in 1866 Darwin received a letter from Wallace, co-discoverer of natural selection, suggesting that an even higher hedge against misunderstanding was regrettably necessary.

> My dear Darwin, – I have been so repeatedly struck by the utter inability of numbers of intelligent persons to see clearly, or at all, the self-acting and necessary effects of Natural Selection, that I am led to conclude that the term itself, and your mode of illustrating it, however clear and beautiful to many of us, are yet not the best adapted to impress it on the general naturalist public.

Wallace went on to quote a French author called Janet, who was evidently, unlike Wallace and Darwin, a deeply muddled individual:

> I see that he considers your weak point to be that you do not see that 'thought and direction are essential to the action of Natural Selection.' The same objection has been made a score of times by your chief opponents, and I have heard it as often stated myself in conversation. Now, I think this arises almost entirely from your choice of the term Natural Selection, and so constantly comparing it in its effects to man's selection, and also to your so frequently personifying nature as 'selecting', as 'preferring' . . . etc., etc. To the few this is as clear as daylight, and beautifully suggestive, but to many it is evidently a stumbling-block. I wish, therefore, to suggest to you the possibility of entirely avoiding this source of misconception in your great work, and also in future editions of the 'Origin,' and I think it may be done without difficulty and very effectually by adopting Spencer's term . . . 'Survival of the Fittest.' This term is the plain expression of the fact; 'Natural Selection' is a metaphorical expression of it . . .

Wallace had a point. Unfortunately, Spencer's term 'Survival of the Fittest' raises problems of its own, which Wallace couldn't have foreseen, and I won't go into them here. In spite of Wallace's warning, I prefer to follow Darwin's own strategy of introducing natural selection via domestication and artificial selection. I like to think that Monsieur Janet might have got the point this time around. But I did have another reason, too, for following Darwin's lead, and it is a good one. The ultimate test of a scientific hypothesis is experiment. Experiment specifically means that you don't just wait for nature to do something, and passively observe it and see what it correlates with. You go in there and *do* something. You *manipulate*. You *change* something, in a systematic way, and compare the result with a 'control' that lacks the change, or you compare it with a different change.

Experimental interference is of enormous importance, because without it you can never be sure that a correlation you observe has any causal significance. This can be illustrated by the so-called 'church clocks fallacy'. The clocks in the towers of two neighbouring churches chime the hours, but St A's a little before St B's. A Martian visitor, noting this, might infer that St A's chime *caused* St B's to chime. We, of course, know better, but the only real test of the hypothesis would be experimentally to ring the St A's chime at *random* times rather than once per hour. The Martian's prediction (which would of course be disproved in this case) is that St B's clock will still chime immediately after St A's. It is only experimental manipulation that can determine whether an observed correlation truly indicates causation.

If your hypothesis is that the non-random survival of random genetic variation has important evolutionary consequences, the *experimental* test of the hypothesis would have to be a deliberate human intervention. Go in and *manipulate* which variant survives and which doesn't. Go in there and *choose*, as a human breeder, which kinds of individuals get to reproduce. And that, of course, *is* artificial selection. Artificial selection is not just an *analogy* for natural selection. Artificial selection constitutes a true *experimental* – as opposed to observational – test of the hypothesis that selection causes evolutionary change.

Most of the known examples of artificial selection – for example, the manufacture of the various breeds of dog – are observed with the hindsight of history, rather than being deliberate tests of predictions under experimentally controlled conditions. But proper experiments

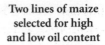

Two lines of maize
selected for high
and low oil content

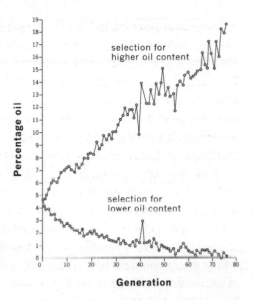

have been done, and the results have always been as expected from
the more anecdotal results on dogs, cabbages and sunflowers. Here
is a typical example, an especially good one because agronomists
at the Illinois Experimental Station began the experiment rather a
long time ago, in 1896 (Generation 1 in the graph). The diagram
above shows the oil content in maize seeds of two different artificially
selected lines, one selected for high oil yield, and the other for low
oil yield. This is a true experiment because we are comparing the
results of two deliberate manipulations or interventions. Evidently
the difference is dramatic, and it increases. It seems likely that both
the upward trend and the downward trend would eventually level off:
the low-yielding line because you can't drop below zero oil content,
and the high-yielding line for reasons that are nearly as plain.

Here's a further laboratory demonstration of the power of
artificial selection, which is instructive in another way. The diagram
overleaf shows some seventeen generations of rats, artificially
selected for resistance to tooth decay. The measure being plotted is
the time, in days, that the rats were free of dental caries. At the start

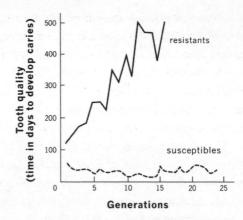

Two lines of rats selected for high and low resistance to tooth decay

Tooth quality (time in days to develop caries)

resistants

susceptibles

Generations

of the experiment, the typical period free of tooth decay was about 100 days. After only a dozen or so generations of systematic selection against caries, the decay-free period was about four times as long, or even more. Once again, a separate line was selected to evolve in the opposite direction: in this case the experiments systematically bred for susceptibility to tooth decay.

The example offers an opportunity to cut our teeth on natural selection thinking. Indeed, this discussion of rat teeth will be the first of three such excursions into natural selection proper, which we are now equipped to undertake. In the other two, as with the rats, we shall revisit creatures already met along the 'primrose path' from domestication, namely dogs and flowers.

RATS' TEETH

Why, if it is so easy to improve the teeth of rats by artificial selection, did natural selection apparently make such a poor job of it in the first place? Surely there is no benefit in tooth decay. Why, if artificial selection is capable of reducing it, didn't natural selection do the same job long ago? I can think of two answers, both instructive.

The first answer is that the original population that the human selectors used as their raw material consisted not of wild rats but of domesticated laboratory-bred white rats. It could be said that lab rats are feather-bedded, like modern humans, shielded from the cutting edge of natural selection. A genetic tendency to tooth decay would significantly reduce reproductive prospects in the wild, but might make no difference in a laboratory colony where the living is easy, and the decision on who breeds and who does not is taken by humans, with no eye to survival.

That's the first answer to the question. The second answer is more interesting, for it carries an important lesson about natural selection, as well as artificial selection. It is the lesson of trade-offs, and we have already adverted to it when talking about pollination strategies in plants. Nothing is free, everything comes with a price tag. It might seem obvious that tooth decay is to be avoided at all costs, and I do not doubt that dental caries significantly shortens life in rats. But let's think for a moment about what must happen in order to increase an animal's resistance to tooth decay. I don't know the details, but I am confident that it will be costly, and that is all I need to assume. Let us suppose it is achieved by a thickening of the wall of the tooth, and this requires extra calcium. It is not impossible to find extra calcium, but it has to come from somewhere, and it is not free. Calcium (or whatever the limiting resource might be) is not floating around in the air. It has to come into the body via food. And it is potentially useful for other things apart from teeth. The body has something we could call a calcium economy. Calcium is needed in bone, and it is needed in milk. (I'm assuming it is calcium we are talking about. Even if it is not calcium, there must be some costly limiting resource, and the argument will work just as well, whatever the limiting resource is. I'll continue to use calcium for the sake of argument.) An individual rat with extra strong teeth might well *tend* to live longer than a rat with rotten teeth, all other things being equal. But all other things are not equal, because the calcium needed to strengthen the teeth had to come from somewhere, say, bones. A rival individual whose genes did not predispose it to take calcium away from bones might consequently survive longer, because of its superior bones and in spite of its bad teeth. Or the rival individual might be better qualified to rear children because she makes more calcium-rich milk. As economists are fond of quoting from Robert Heinlein, there's no such thing as a free lunch. My rat example is hypothetical, but it is safe to say that, for economic reasons, there must be such a thing as a rat whose teeth are *too* perfect. Perfection in one department must be bought, in the form of a sacrifice in another department.

The lesson applies to all living creatures. We can expect bodies to be well equipped to survive, but this does not mean they should be perfect with respect to any one dimension. An antelope might run faster, and be more likely to escape a leopard, if its legs were a little longer. But a rival antelope with longer legs, although it might be better equipped to outsprint a predator, has to pay for its long legs in some other department of the body's economy. The materials needed to make the extra bone and muscle in the longer legs have to be taken from somewhere else, so the longer-legged individual is more likely to die for reasons other than predation. Or it may even be more likely to die from predation because its longer legs, although they can run faster when intact, are more likely to break, in which case it can't run at all. A body is a patchwork of compromises. I shall return to this point in the chapter on arms races.

What happens under domestication is that animals are artificially shielded from many of the risks that shorten the lives of wild animals. A pedigree dairy cow may yield prodigious quantities of milk, but its pendulously cumbersome udder would seriously impede it in any attempt to outrun a lion. Thoroughbred horses are superb runners and jumpers, but their legs are vulnerable to injury during races, especially over jumps, which suggests that artificial selection has pushed them into a zone that natural selection would not have tolerated. Moreover, Thoroughbreds thrive only on a rich diet supplied by humans. Whereas Britain's native ponies, for example, flourish on pasture, racehorses don't prosper unless they are fed a much richer diet of grains and supplements – which they would not find in the wild. Again, I'll return to such matters in the arms race chapter.

DOGS AGAIN

Having finally reached the topic of natural selection, we can turn back to the example of dogs for some other important lessons. I said that they are domesticated wolves, but I need to qualify this in the light of a fascinating theory of the evolution of the dog, which has again been most clearly articulated by Raymond Coppinger. The idea is that the evolution of the dog was not just a matter of artificial selection. It was at least as much a case of wolves adapting to the ways of man by natural selection. Much of the initial domestication of the dog was *self*-domestication, mediated by natural, not artificial, selection. Long before we got our hands on the chisels in the artificial selection toolbox, natural selection had already sculpted wolves into self-domesticated 'village dogs' without any human intervention. Only later did humans adopt these village dogs and transmogrify them, separately and comprehensively, into the rainbow spectrum of breeds that today grace (if grace is the word) Crufts and similar pageants of canine achievement and beauty (if beauty is the word).

Coppinger points out that when domestic animals break free and go feral for many generations, they usually revert to something close to their wild ancestor. We might expect feral dogs, therefore, to become rather wolf-like. But this doesn't happen. Instead, dogs left to go feral seem to become the ubiquitous 'village dogs' – 'pye-dogs' – that hang around human settlements all over the third world. This encourages Coppinger's belief that the dogs on which human breeders finally went to work were wolves no longer. They had already changed themselves into dogs: village dogs, pye-dogs, perhaps dingos.

Real wolves are pack hunters. Village dogs are scavengers that frequent middens and rubbish dumps. Wolves scavenge too, but they are not temperamentally suited to scavenging human rubbish because of their long 'flight distance'. If you see an animal feeding, you can measure its flight distance by seeing how close it will let you approach before fleeing. For any given species in any given situation, there will be an optimal flight distance, somewhere between too risky or foolhardy at the short end, and too flighty or risk-averse at the long end. Individuals that take off too late when danger threatens are more likely to be killed by that very danger. Less obviously, there is such a thing as taking off too soon. Individuals that are too flighty never get a square meal, because they run away at the first hint of danger on the horizon. It is easy for us to overlook the dangers of being too risk-averse. We are puzzled when we see zebras or antelopes calmly grazing in full view of lions, keeping no more than a wary eye on them. We are puzzled, because our own risk aversion (or that of our safari guide) keeps us firmly inside the Land Rover even though we have no reason to think there is a lion within miles. This is because we have nothing to set against our fear. We are going to get our square meals back at the safari lodge. Our wild ancestors would have had much more sympathy with the risk-taking zebras. Like the zebras, they had to balance the risk of being eaten against the risk of not eating. Sure, the lion might attack; but, depending on the size of your troop, the odds were that it would catch another member of it rather than you. And if you never ventured on to the feeding grounds, or down to the water hole, you'd die anyway, of hunger or thirst. It is the same lesson of economic trade-offs that we have already met, twice.*

The bottom line of that digression is that the wild wolf, like any other animal, will have an optimal flight distance, nicely poised – and potentially flexible – between too bold and too flighty. Natural selection will work on the flight distance, moving it one way or the other along the continuum if conditions change over evolutionary time. If a plenteous new food source in the form of village rubbish dumps enters the world of wolves, that is going to shift the optimum point towards the shorter end of the flight distance continuum, in the direction of reluctance to flee when enjoying this new bounty.

---

* Psychologists have analogous tests of risk-taking among humans, which show interesting differences. Entrepreneurs typically score highly on risk-taking measures, as do pilots, rock-climbers, motorcycle racers and other extreme sports enthusiasts. Women tend to be more risk-averse than men. Feminists will here point out that the causal arrow could go either way: women could become more risk-averse because of the occupations into which society thrusts them.

We can imagine wild wolves scavenging on a rubbish tip on the edge of a village. Most of them, fearful of men throwing stones and spears, have a very long flight distance. They sprint for the safety of the forest as soon as a human appears in the distance. But a few individuals, by genetic chance, happen to have a slightly shorter flight distance than the average. Their readiness to take slight risks – they are brave, shall we say, but not foolhardy – gains them more food than their more risk-averse rivals. As the generations go by, natural selection favours a shorter and shorter flight distance, until just before it reaches the point where the wolves really are endangered by stone-throwing humans. The optimum flight distance has shifted because of the newly available food source.

Something like this evolutionary shortening of the flight distance was, in Coppinger's view, the first step in the domestication of the dog, and it was achieved by natural selection, not artificial selection. Decreasing flight distance is a behavioural measure of what might be called increasing tameness. At this stage in the process, humans were not deliberately choosing the tamest individuals for breeding. At this early stage, the only interactions between humans and these incipient dogs were hostile. If wolves were becoming domesticated it was by self-domestication, not deliberate domestication by people. Deliberate domestication came later.

We can get an idea of how tameness, or anything else, can be sculpted – naturally or artificially – by looking at a fascinating experiment of modern times, on the domestication of Russian silver foxes for use in the fur trade. It is doubly interesting because of the lessons it teaches us, over and above what Darwin knew, about the domestication process, about the 'side-effects' of selective breeding, and about the resemblance, which Darwin well understood, between artificial and natural selection.

The silver fox is just a colour variant, valued for its beautiful fur, of the familiar red fox, *Vulpes vulpes*. The Russian geneticist Dimitri Belyaev was employed to run a fox fur farm in the 1950s. He was later sacked because his scientific genetics conflicted with the anti-scientific ideology of Lysenko, the charlatan biologist who managed to capture the ear of Stalin and so take over, and largely ruin, all of Soviet genetics and agriculture for some twenty years. Belyaev retained his love of foxes, and of true Lysenko-free genetics, and he was later able to resume his studies of both, as director of an Institute of Genetics in Siberia.

Wild foxes are tricky to handle, and Belyaev set out deliberately to breed for tameness. Like any other animal or plant breeder of his time, his method was to exploit natural variation (no genetic engineering in those days) and choose, for breeding, those males and females that came closest to the ideal he was seeking. In selecting for tameness, Belyaev could have chosen, for breeding, those dogs and bitches that most appealed to him, or looked at him with the cutest facial expressions. That might well have had the desired effect on the tameness of future generations. More systematically than that, however, he used a measure that was pretty close to the 'flight distance' I just mentioned in connection with wild wolves, but adapted for cubs. Belyaev and his colleagues (and successors, for the experimental program continued after his death) subjected fox cubs to standardized tests in which an experimenter would offer a cub food by hand, while trying to stroke or fondle it. The cubs were classified into three classes. Class III cubs were those that fled from or bit the person. Class II cubs would allow themselves to be handled, but showed no positive responsiveness to the experimenters. Class I cubs, the tamest of all, positively approached the handlers, wagging their tails and whining. When the cubs grew up, the experimenters systematically bred only from this tamest class.

After a mere six generations of this selective breeding for tameness, the foxes had changed so much that the experimenters felt obliged to name a new category, the 'domesticated elite' class, which were 'eager to establish human contact, whimpering to attract attention and sniffing and licking experimenters like dogs'. At the beginning of the experiment, none of the foxes were in the elite class. After ten generations of breeding for tameness, 18 per cent were 'elite'; after twenty generations, 35 per cent; and after thirty to thirty-five generations, 'domesticated elite' individuals constituted between 70 and 80 per cent of the experimental population.

Such results are perhaps not too surprising, except for the astonishing magnitude and speed of the effect. Thirty-five generations would pass unnoticed on the geological timescale. Even more interesting, however, were the unexpected side-effects of the selective breeding for tameness. These were truly fascinating and genuinely unforeseen. Darwin, the dog-lover, would have been entranced.

Belayev with his foxes, as they turn tame – and doglike

The tame foxes not only behaved like domestic dogs, they looked like them. They lost their foxy pelage and became piebald black and white, like Welsh collies. Their foxy prick ears were replaced by doggy floppy ears. Their tails turned up at the end like a dog's, rather than down like a fox's brush. The females came on heat every six months like a bitch, instead of every year like a vixen. According to Belyaev, they even sounded like dogs.

These dog-like features were side-effects. Belyaev and his team did not deliberately breed for them, only for tameness. Those other dog-like characteristics seemingly rode on the evolutionary coat-tails of the genes for tameness. To geneticists, this is not surprising. They recognize a widespread phenomenon called 'pleiotropy', whereby genes have more than one effect, seemingly unconnected. The stress is on the word 'seemingly'. Embryonic development is a complicated business. As we learn more about the details, that 'seemingly unconnected' turns into 'connected by a route that we now understand, but didn't before'. Presumably genes for floppy ears and piebald coats are pleiotropically linked to genes for tameness, in foxes as well as in dogs. This illustrates a generally important point about evolution. When you notice a characteristic of an animal and ask what its Darwinian survival value is, you may be asking the wrong question. It could be that the characteristic you have picked out is not the one that matters. It may have 'come along for the ride', dragged along in evolution by some other characteristic to which it is pleiotropically linked.

The evolution of the dog, then, if Coppinger is right, was not just a matter of artificial selection, but a complicated mixture of natural selection (which predominated in the early stages of domestication) and artificial selection (which came to the fore more recently). The transition would have been seamless, which again goes to emphasize the similarity – as Darwin recognized – between artificial and natural selection.

FLOWERS AGAIN

Let's now, in the third of our warm-up forays into natural selection, move on to flowers and pollinators and see something of the power of natural selection to drive evolution. Pollination biology furnishes us with some pretty amazing facts, and the high point of wondrousness is reached in the orchids. No wonder Darwin was so keen on them; no wonder he wrote the book I have already mentioned, *The Various Contrivances by which Orchids are Fertilised by Insects.* Some orchids, such as the 'magic bullet' Madagascar ones we met earlier, give nectar, but others have found a way to bypass the costs of feeding pollinators, by tricking them instead. There are orchids that resemble female bees (or wasps or flies) well enough to fool males into attempting to copulate with them. To the extent that such mimics resemble females of one particular insect species, to that extent will males of those species serve as magic bullets, going from flower to flower of just the one orchid species. Even if the orchid resembles 'any old bee' rather than one species of bee, the bees that it fools will still be 'fairly magic' bullets. If you or I were to look closely at a fly orchid or a bee orchid (see colour page 5), we would be able to tell that it was not a real insect; but we would be fooled at a casual glance out of the corner of our eye. And even looking at it head-on, I would say the bee orchid in the picture (h) is pretty clearly more of a bumble-bee orchid than a honey-bee orchid. Insects have compound eyes, which are not so acute as our camera eyes, and the shapes and colours of insect-mimicking orchids, reinforced by seductive scents that mimic those of female insects, are more than capable of tricking males. By the way, it is quite probable that the mimicry is enhanced when seen in the ultraviolet range, from which we are cut off.

The so-called spider orchid, *Brassia* (colour page 5 (k)), achieves pollination by a different kind of deception. The females of various species of solitary wasp ('solitary' because they don't live socially in large nests like the familiar autumn pests, called yellowjackets by

Americans) capture spiders, sting them to paralyse them, and lay their eggs on them as a living food supply for their larvae. Spider orchids resemble spiders sufficiently to fool female wasps into attempting to sting them. In the process they pick up pollinia – masses of pollen grains produced by the orchids. When they move on to try to sting another spider orchid, the pollinia are transferred. By the way, I can't resist adding the exactly backwards case of the spider *Epicadus heterogaster*, which mimics an orchid. Insects come to the 'flower' in search of nectar, and are promptly eaten by it.

Some of the most astonishing orchids that practise this seduction trick are to be found in Western Australia. Various species in the genus *Drakaea* are known as hammer orchids. Each species has a special relationship with a particular species of wasp of the type called thynnids. Part of the flower bears a crude resemblance to an insect, duping the male thynnid wasp into attempting to mate with it. So far in my description, *Drakaea* is not dramatically different from other insect-mimicking orchids. But *Drakaea* has a remarkable extra trick up its sleeve: the fake 'wasp' is borne on the end of a hinged 'arm', with a flexible 'elbow'. You can clearly see the hinge in the picture (colour page 5 (g)). The fluttering movement of the wasp gripping the dummy wasp causes the 'elbow' to bend, and the wasp is dashed repeatedly back and forth like a hammer against the other side of the flower – let's call it the anvil – where it keeps its sexual parts. The pollinia are dislodged and stick to the wasp, who eventually extricates himself and flies off, sadder but apparently no wiser: he goes on to repeat the performance on another hammer orchid, where he and the pollinia he bears are duly dashed against the anvil, so that his cargo finds its destined refuge on the female organs of the flower. I showed a film of this astounding performance in one of my Royal Institution Christmas Lectures for Children, and it can be seen in the recording of the lecture called 'The Ultraviolet Garden'.

In the same lecture I discussed the 'bucket orchids' of South America, which achieve pollination in an equally remarkable but rather different way. They too have specialized pollinators, not wasps but small bees, of the group called Euglossine. Again, these orchids provide no nectar. But the orchids don't fool the bees into mating with them either. Instead, they provide a vital piece of assistance for male bees, without which the bees would be unable to attract real females.

These little bees, which live only in South America, have a strange habit. They go to elaborate lengths to collect fragrant, or anyway smelly, substances, which they store in special containers attached to their enlarged hind legs. In different species these smelly substances can come from flowers, from dead wood, or even from faeces. It seems that they use the gathered perfumes to attract, or otherwise court, females. Many insects use particular scents to appeal to the opposite sex, and most of them manufacture the perfumes in special glands. Female silk moths, for example, attract males from an astonishingly long distance by releasing a unique scent, which they manufacture and which males detect – in minute traces from literally miles away – with their antennae. In the case of Euglossine bees, it is the males that use scent. And, unlike the female moths, they don't synthesize their own perfume but use the smelly ingredients that they have collected, not as pure substances but as carefully concocted blends which they put together like expert perfumiers. Each species mixes a characteristic cocktail of substances gathered from various sources. And there are some species of Euglossine bee that positively need, for manufacturing their characteristic species scent, substances that are supplied only by flowers of particular species of the orchid genus *Coryanthes* – bucket orchids. The common name of Euglossine bees is 'orchid bees'.

What an intricate picture of mutual dependence. The orchids need the Euglossine bees, for the usual 'magic bullet' reasons. And the bees need the orchids, for the rather weirder reason that they can't attract female bees without substances that are either impossible or at least too hard to find except through the good offices of bucket orchids. But the way in which pollination is achieved is even weirder still, and it superficially makes the bee look more like a victim than a cooperating partner.

A male Euglossine bee is attracted to the orchids by the smell of the substances that he needs in order to manufacture his sexual perfumes. He alights on the rim of the bucket and starts to scrape the waxy perfume into the special scent pockets in his legs. But the rim of the bucket is slippery underfoot – and there's a reason for this. The bee falls into the bucket, which is filled with liquid, in which he swims. He cannot climb up the slippery sides of the bucket. There is only one escape route, and this is a special bee-sized hole in the side of the bucket (not visible in the picture that appears on colour page 4). He is guided by 'stepping stones' to the hole and starts to crawl through it. It's a tight fit, and it becomes even tighter as the 'jaws' (these you can see in the picture: they look like the chuck of a lathe or electric drill) contract and trap him. While he is held in their grip, they glue two pollinia to his back. The glue takes a while to set, after which the jaws again relax and release the bee, who flies off, complete with pollinia on his back. Still in search of the precious ingredients for his perfumery, the bee lands on another bucket orchid and the process repeats itself. This time, however, as the bee struggles through the hole in the bucket, the pollinia are scraped off, and they fertilize the stigma of this second orchid.

The intimate relationship between flowers and their pollinators is a lovely example of what is called co-evolution – evolution together. Co-evolution often occurs between organisms that have something to gain from each other, partnerships in which each side contributes something to the other, and both gain from the cooperation. Another beautiful example is the set of relationships that have grown up around coral reefs, independently in many different parts of the world, between cleaner fish and larger fish. The cleaners belong to several different species, and some are not even fish at all but shrimps – a nice case of convergent evolution. Cleaning, among coral-reef fish, is a well-established way of life, like hunting or grazing or anteating among mammals. Cleaners make their living by picking parasites off the bodies of their larger 'clients'. That the clients benefit has been elegantly demonstrated by removing all the cleaners from an experimental area of reef, whereupon the health of lots of species of fish declines. I have discussed the cleaning habit elsewhere, so will say no more here.

Co-evolution also occurs between species that don't benefit from each other's presence, like predators and prey, or parasites and hosts. These kinds of co-evolution are sometimes called 'arms races' and I postpone discussing them to Chapter 12.

NATURE AS THE SELECTING AGENT

Let me draw this chapter, and the previous one, to a conclusion. Selection – in the form of artificial selection by human breeders – can turn a pye-dog into a Pekinese, or a wild cabbage into a cauliflower, in a few centuries. The difference between any two breeds of dog gives us a rough idea of the quantity of evolutionary change that can be achieved in less than a millennium. The next question we should ask is, how many millennia do we have available to us in accounting for the whole history of life? If we imagine the sheer quantity of difference that separates a pye-dog from a peke, which took only a few centuries of evolution, how much longer is the time that separates us from the beginning of evolution or, say, from the beginning of the mammals? Or from the time when fish emerged on to the land? The answer is that life began not just centuries ago but tens of millions of centuries ago. The measured age of our planet is about 4.6 billion years, or about 46 million centuries. The time that has elapsed since the common ancestor of all today's mammals walked the Earth is about two million centuries. A century seems a pretty long time to us. Can you imagine two million centuries, laid end to end? The time that has elapsed since our fish ancestors crawled out of the water on to the land is about three and a half million centuries: that is to say, about twenty thousand times as long as it took to make all the different – really very different – breeds of dogs from the common ancestor that they all share.

Hold in your head an approximate picture of the quantity of difference between a peke and a pye-dog. We aren't talking precise measurements here: it would do just as well to think about the difference between any one breed of dog and any other, for that is on average double the amount of change that has been wrought, by artificial selection, from the common ancestor. Bear in mind this order of evolutionary change, and then extrapolate backwards twenty thousand times as far into the past. It becomes rather easy to accept that evolution could accomplish the amount of change that it took to transform a fish into a human.

But all this presupposes that we know the age of the Earth, and of the various landmark points in the fossil record. This is a book about evidence, so I can't just assert dates but must justify them. How, actually, do we know the age of any particular rock? How do we know the age of a fossil? How do we know the age of the Earth? How, for that matter, do we know the age of the universe? We need clocks, and clocks are the subject of the next chapter.

CHAPTER 4

SILENCE AND SLOW TIME

IF the history-deniers who doubt the fact of evolution are ignorant of biology, those who think the world began less than ten thousand years ago are worse than ignorant, they are deluded to the point of perversity. They are denying not only the facts of biology but those of physics, geology, cosmology, archaeology, history and chemistry as well. This chapter is about how we know the ages of rocks and the fossils embedded in them. It presents the evidence that the timescale on which life has operated on this planet is measured not in thousands of years but in thousands of millions of years.

Remember, evolutionary scientists are in the position of detectives who come late to the scene of a crime. To pinpoint when things happened, we depend upon traces left by time-dependent processes – clocks, in a broad sense. One of the first things a detective does when investigating a murder is ask a doctor or pathologist to estimate the time of death. Much follows from this information, and in detective fiction an almost mystical reverence is accorded to the pathologist's estimate. The 'time of death' is a baseline fact, an inerrant pivot around which more or less far-fetched speculations by the detective revolve. But that estimate is, of course, subject to error, an error that can be measured and can be quite large. The pathologist uses various time-dependent processes to estimate the time of death: the body cools at a characteristic rate, rigor mortis sets in at a particular time, and so on. These are the rather crude 'clocks' available to the investigator of a murder. The clocks available to the evolutionary scientist are potentially much more accurate – in proportion to the timescale involved, of course, not more accurate to the nearest hour! The analogy to a precision clock is more persuasive for a Jurassic rock in the hands of a geologist than it is for a cooling corpse in the hands of a pathologist.

Man-made clocks work on timescales that are very short by evolutionary standards – hours, minutes, seconds – and the time-dependent processes they use are fast: the swinging of a pendulum, the swivelling of a hairspring, the oscillation of a crystal, the burning of a candle, the draining of a water vessel or an hourglass, the rotation of the earth (registered by a sundial). All clocks exploit some process that occurs at a steady and known rate. A pendulum swings at a very constant rate, which depends upon its length but not, at least in theory, on the amplitude of the swing or the mass of the bob on the end. Grandfather clocks work by linking a pendulum to an escapement which advances a toothed wheel, step by step; the rotation is then geared down to the speed of rotation of an hour hand, a minute hand and a second hand. Watches with hairspring wheels work in a similar way. Digital watches exploit an electronic equivalent of a pendulum, the oscillation of certain kinds of crystals when supplied with energy from a battery. Water clocks and candle clocks are much less accurate, but they were useful before the invention of event-counting clocks. They depend not on counting things, as a pendulum clock or a digital watch does, but on measuring some quantity. Sundials are inaccurate ways of telling the time.* But the rotation of the earth, which is the time-dependent process on which they rely, is accurate on the timescale of the slower clock that we call the calendar. This is because on that timescale it is no longer a measuring clock (a sundial measures the continuously varying angle of the sun) but a counting clock (counting day/night cycles).

Both counting clocks and measuring clocks are available to us

---

* I am a sundial, and I make a botch
  Of what is done far better by a watch
  Hilaire Belloc

on the immensely slow timescale of evolution. But to investigate evolution we don't need just a clock that tells the *present* time, as a sundial does, or a watch. We need something more like a stopwatch that can be *reset*. Our evolutionary clock needs to be *zeroed* at some point, so that we can calculate the elapsed time since a starting point, to give us, for example, the absolute age of some object such as a rock. Radioactive clocks for dating igneous (volcanic) rocks are conveniently zeroed at the moment the rock is formed by the solidification of molten lava.

Fortunately, a variety of zero-able natural clocks is available. This variety is a good thing, because we can use some clocks to check the accuracy of other clocks. Even more fortunately, they sensitively cover an astonishingly wide range of timescales, and we need this too because evolutionary timescales span seven or eight orders of magnitude. It's worth spelling out what this means. An order of magnitude means something precise. A change of one order of magnitude is one multiplication (or division) by ten. Since we use a decimal system,* the order of magnitude of a number is a count of the number of zeroes, before or after the decimal point. So a range of eight orders of magnitude constitutes a hundred millionfold. The second hand of a watch rotates 60 times as fast as the minute hand and 720 times as fast as the hour hand, so the three hands cover a range which is less than three orders of magnitude. This is tiny compared to the eight orders of magnitude spanned by our repertoire of geological clocks. Radioactive decay clocks are available for short timescales as well, even down to fractions of a second; but for evolutionary purposes, clocks that can measure centuries or perhaps decades are about the fastest we need. This fast end of the spectrum of natural clocks – tree rings and carbon dating – is useful for archaeological purposes,

---

* Which is presumably based on the evolutionary happenstance of our possessing ten fingers. Fred Hoyle has ingeniously speculated that, if we had been born with eight digits and therefore become accustomed to octal arithmetic instead of decimal, we might have invented binary arithmetic and hence electronic computers a century earlier than we did (since 8 is a power of 2).

and for dating specimens on the sort of timescale that covers the domestication of the dog or the cabbage. At the other end of the scale, we need natural clocks that can time hundreds of millions, even billions, of years. And, praise be, nature has provided us with just the wide range of clocks that we need. What's more, their ranges of sensitivity overlap with each other, so we can use them as checks on each other.

TREE RINGS

A tree-ring clock can be used to date a piece of wood, say a beam in a Tudor house, with astonishing accuracy, literally to the nearest year. Here's how it works. First, as most people know, you can age a newly felled tree by counting rings in its trunk, assuming that the outermost ring represents the present. Rings represent differential growth in different seasons of the year – winter or summer, dry season or wet season – and they are especially pronounced at high latitudes, where there is a strong difference between seasons. Fortunately, you don't actually have to cut the tree down in order to age it. You can peek at its rings without killing it, by boring into the middle of a tree and extracting a core sample. But just counting rings doesn't tell you in which century your house beam was alive, or your Viking longship's mast. If you want to pin down the date of old, long-dead wood you need to be more subtle. Don't just count rings, look at the pattern of thick and thin rings.

Just as the existence of rings signifies seasonal cycles of rich and poor growth, so some years are better than others, because the weather varies from year to year: there are droughts that retard growth, and bumper years that accelerate it; there are cold years and hot years, even years of freak El Niños or Krakatoa-type catastrophes. Good years, from the tree's point of view, produce wider rings than bad years. And the pattern of wide and narrow rings in any one region, caused by a particular trademark sequence of good years and bad years, is sufficiently characteristic – a fingerprint that labels the exact years in which the rings were laid down – to be recognizable from tree to tree.

Dendrochronologists measure rings on recent trees, where the exact date of every ring is known by counting backwards from the year in which the tree is known to have been felled. From these measurements, they construct a reference collection of ring patterns, to which you can compare the ring patterns of an archaeological sample of wood whose date you want to know. So you might get the report: 'This Tudor beam contains a signature sequence of rings that matches a sequence from the reference collection, which is known to have been laid down in the years 1541 to 1547. The house was therefore built after AD 1547.'

All very well, but not many of today's trees were alive in Tudor times, let alone in the stone age or beyond. There are some trees – bristlecone pines, some giant redwoods – that live for millennia, but most trees used for timber are felled when they are younger than a century or so. How, then, do we build up the reference collection of rings for more ancient times? For times so distant that not even the oldest surviving bristlecone pine goes back that far? I think you've already guessed the answer. Overlaps. A strong rope may be 100 yards long, yet no single fibre within it reaches more than a fraction of that total. To use the overlap principle in dendrochronology, you take the reference fingerprint patterns whose date is known from modern trees. Then you identify a fingerprint from the old rings of modern trees and seek the same fingerprint from the younger rings of long-dead trees. Then you look at the fingerprints from the older rings of those same long-dead trees, and look for the same pattern in the younger rings of even older trees. And so on. You can daisychain your way back, theoretically for millions of years using petrified forests, although in practice dendrochronology is only used on archaeological timescales over some thousands of years. And the amazing thing about dendrochronology is that, theoretically at least, you can be accurate to the nearest year, even in a petrified forest 100 million

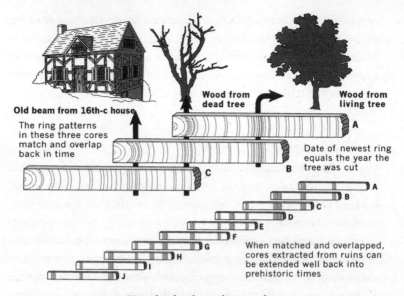

Old beam from 16th-c house

The ring patterns
in these three cores
match and overlap
back in time

Wood from
dead tree

Wood from
living tree

A

Date of newest ring
equals the year the
tree was cut

B

C

A
B
C
D
E
F
G
H
I
J

When matched and overlapped,
cores extracted from ruins can
be extended well back into
prehistoric times

How dendrochronology works

years old. You could literally say that *this* ring in a Jurassic fossil
tree was laid down exactly 257 years later than *this* other ring in
another Jurassic tree! If only there were enough petrified forests
to daisychain your way back continuously from the present, you
could say that this tree is not just of late Jurassic age: it was alive in
exactly 151,432,657 BC! Unfortunately, we don't have an unbroken
chain, and dendrochronology in practice takes us back only about
11,500 years. It is nevertheless a tantalizing thought that, if only we
could find enough petrified forests, we could date to the nearest
year over a timespan of hundreds of millions of years.

Tree rings are not quite the only system that promises total
accuracy to the nearest year. Varves are layers of sediment laid down
in glacial lakes. Like tree rings, they vary seasonally and from year to
year, so theoretically the same principle can be used, with the same
degree of accuracy. Coral reefs, too, have annual growth rings, just
like trees. Fascinatingly, these have been used to detect the dates
of ancient earthquakes. Tree rings too, by the way, tell us the dates

of earthquakes. Most of the other dating systems that are available to us, including all the radioactive clocks that we actually use over timescales of tens of millions, hundreds of millions or billions of years, are accurate only within an error range that is approximately proportional to the timescale concerned.

RADIOACTIVE CLOCKS

Let's now turn to radioactive clocks. There are quite a lot of them to choose from, and, as I said, they blessedly cover the gamut from centuries to thousands of millions of years. Each one has its own margin of error, which is usually about 1 per cent. So if you want to date a rock which is billions of years old, you must be satisfied with an error of plus or minus tens of millions of years. If you want to date a rock hundreds of millions of years old, you must be satisfied with an error of millions. To date a rock that is only tens of millions of years old, you must allow for an error of plus or minus hundreds of thousands of years.

To understand how radioactive clocks work, we first need to understand what is meant by a radioactive isotope. All matter is made up of elements, which are usually chemically combined with other elements. There are about 100 elements, slightly more if you count elements that are only ever detected in laboratories, slightly fewer if you count only those elements that are found in nature. Examples of elements are carbon, iron, nitrogen, aluminium, magnesium, fluorine, argon, chlorine, sodium, uranium, lead, oxygen, potassium and tin. The atomic theory, which I think everybody accepts, even creationists, tells us that each element has its own characteristic atom, which is the smallest particle into which you can divide an element without it ceasing to be that element. What does an atom look like, say an atom of lead, or copper, or carbon? Well, it certainly looks nothing like lead or copper or carbon. It doesn't *look* like anything, because it is too small to form any kind of image on your retina, even with an ultra-powerful microscope. We can use analogies or models to help us visualize an atom. The most famous model was proposed by the great Danish physicist Niels Bohr. The Bohr model, which is now rather out of date, is a miniature solar system. The role of the sun is played by the nucleus, and around it orbit the electrons, which play the role of planets. As with the solar system, almost all the mass of the atom is contained in the nucleus ('sun'), and almost all the volume is contained in the empty space that separates the electrons ('planets') from the nucleus. Each electron is tiny compared with the nucleus, and the space between them and the nucleus is huge compared with the size of either. A favourite analogy portrays the nucleus as a fly in the middle of a sports stadium. The nearest neighbouring nucleus is another fly, in the middle of an adjacent stadium. The electrons of each atom are buzzing about in orbit around their respective flies, smaller than the tiniest gnats, too small to be seen on the same scale as the flies. When we look at a solid lump of iron or rock, we are 'really' looking at what is almost entirely empty space. It looks and feels solid and opaque because our sensory systems and brains find it convenient to treat it as solid and opaque. It is convenient for the brain to represent a rock as solid because we can't walk through it. 'Solid' is our way of experiencing things that we can't walk through or fall through, because of the electromagnetic forces between atoms. 'Opaque' is the experience we have when light bounces off the surface of an object, and none of it goes through.

Three kinds of particle enter into the makeup of an atom, at least as envisaged in the Bohr model. Electrons we have already met. The other two, vastly larger than electrons but still tiny compared with anything we can imagine or experience with our senses, are called protons and neutrons, and they are found in the nucleus. They are almost the same size as each other. The number of protons is fixed for any given element and equal to the number of electrons. This number is called the atomic number. It is uniquely characteristic of an element, and there are no gaps in the list of atomic numbers – the famous periodic table.[*] Every

---

[*] Alas, the popular legend that it came to Dmitri Mendeleev in a dream may be false.

number in the sequence corresponds to exactly one, and only one, element. The element with 1 for its atomic number is hydrogen, 2 is helium, 3 lithium, 4 beryllium, 5 boron, 6 carbon, 7 nitrogen, 8 oxygen, and so on up to high numbers like 92, which is the atomic number of uranium.

Protons and electrons carry an electric charge, of opposite sign – we call one of them positive and the other negative by arbitrary convention. These charges are important when elements form chemical compounds with each other, mostly mediated by electrons. The neutrons in an atom are bound into the nucleus together with the protons. Unlike protons they carry no charge, and they play no role in chemical reactions. The protons, neutrons and electrons in any one element are exactly the same as those in every other element. There is no such thing as a gold-flavoured proton or a copper-flavoured electron or a potassium-flavoured neutron. A proton is a proton is a proton, and what makes a copper atom copper is that there are exactly 29 protons (and exactly 29 electrons). What we ordinarily think of as the nature of copper is a matter of chemistry. Chemistry is a dance of electrons. It is all about the interactions of atoms via their electrons. Chemical bonds are easily broken and remade, because only electrons are detached or exchanged in chemical reactions. The forces of attraction within atomic nuclei are much harder to break. That's why 'splitting the atom' has such a menacing ring to it – but it can happen, in 'nuclear' as opposed to chemical reactions, and radioactive clocks depend upon it.

Electrons have negligible mass, so the total mass of an atom, its 'mass number', is equal to the combined number of protons and neutrons. It is usually rather more than double the atomic number, because there are usually a few more neutrons than protons in a nucleus. Unlike the number of protons, the number of neutrons in an atom is not diagnostic of an element. Atoms of any given element can come in different versions called *isotopes*, which have differing numbers of neutrons, but always the same number of protons. Some elements, such as fluorine, have only one naturally occurring isotope. The atomic number of fluorine is 9 and its mass number is 19, from which you can deduce that it has 9 protons and 10 neutrons. Other elements have lots of isotopes. Lead has five commonly occurring isotopes. All have the same number of protons (and electrons), namely 82, which is the atomic number of lead, but the mass numbers range between 202 and 208. Carbon has three naturally occurring isotopes. Carbon-12 is the common one, with the same number of neutrons as protons: 6. There's also carbon-13, which is too rare to bother with, and carbon-14 which is rare but not too rare to be useful for dating relatively young organic samples, as we shall see.

Now for the next important background fact. Some isotopes are stable, others unstable. Lead-202 is an unstable isotope; lead-204, lead-206, lead-207 and lead-208 are stable isotopes. 'Unstable' means that the atoms spontaneously decay into something else, at a predictable rate, though not at predictable moments. The predictability of the rate of decay is the key to all radiometric clocks. Another word for 'unstable' is 'radioactive'. There are several kinds of radioactive decay, which offer possibilities for useful clocks. For our purposes it isn't important to understand them, but I explain them here to show the magnificent level of detail that physicists have achieved in working out such things. Such detail casts a sardonic light on the desperate attempts of creationists to explain away the evidence of radioactive dating, and keep the Earth young like Peter Pan.

All these kinds of instability involve neutrons. In one kind, a neutron turns into a proton. This means that the mass number stays the same (since protons and neutrons have the same mass) but the atomic number goes up by one, so the atom becomes a different element, one step higher in the periodic table. For example, sodium-24 turns itself into magnesium-24. In another kind of radioactive decay, exactly the reverse happens. A proton turns into a neutron. Again, the mass number stays the same, but this time the atomic number decreases by one, and the atom changes into the next element down in the periodic table. There are nuclear reactions that have the same result. For example, an energetic neutron hits a nucleus and knocks out one proton, taking its place. Again, there's no change in mass number; again, the atomic number goes down by one, and the atom turns into the next element down in the periodic table. There's also a more complicated kind of decay in which an atom ejects a so-called alpha particle. An alpha particle consists of two protons and two neutrons stuck together. This means that the mass number goes down by four and the atomic number goes down by two. The atom changes to whichever element is two below it in the periodic table. An example of alpha decay is the change of the very radioactive isotope uranium-238 (with 92 protons and 146 neutrons) to thorium-234 (with 90 protons and 144 neutrons).

Now we approach the nub of the whole matter. Every unstable or radioactive isotope decays at its own characteristic rate which is precisely known. Moreover, some of these rates are vastly slower than others. In all cases the decay is exponential. Exponential means that if you start with, say, 100 grams of a radioactive isotope, it is not the case that a fixed amount, say 10 grams, turns into another element in a given time. Rather, a fixed *proportion* of whatever is left turns into the second element. The favoured measure of decay rate is the 'half-life'. The half-life of a radioactive isotope is the time taken for half of its atoms to decay. The half-life is the same, no matter how many atoms have already decayed – that is what exponential decay means. You will appreciate that, with such successive halvings, we never really know when there is none left. However, we can say that after a sufficient time has elapsed – say ten half-lives – the number of atoms that remains is so small that, for practical purposes, it has all gone. For example, the half-life of carbon-14 is between 5,000 and 6,000 years. For specimens older than about 50,000–60,000 years, carbon dating is useless, and we need to turn to a slower clock.

The half-life of rubidium-87 is 49 billion years. The half-life of fermium-244 is 3.3 milliseconds. Such startling extremes serve to illustrate the stupendous *range* of clocks available. Although carbon-15's half-life of 2.4 seconds is too short for settling evolutionary questions, carbon-14's half-life of 5,730 years is just right for dating on the archaeological timescale, and we'll come to it presently. An isotope much used on the evolutionary timescale is potassium-40, with its half-life of 1.26 billion years, and I'm going to use it as my example, to explain the whole idea of a radioactive clock. It is often called the potassium argon clock, because argon-40 (one lower in the periodic table) is one of the elements to which potassium-40 decays (the other, resulting from a different kind of radioactive decay, is calcium-40, one higher in the periodic table). If you start with some quantity of potassium-40, after 1.26 billion years half of the potassium-40 will have decayed to argon-40. That's what half-life means. After another 1.26 billion years, half of what remains (a quarter of the original) will have decayed, and so on. After a shorter time than 1.26 billion years, a proportionately smaller quantity of the original potassium will have decayed. So, imagine that you start with some quantity of potassium-40 in an enclosed space with no argon-40. After a few hundreds of millions of years have elapsed, a scientist comes upon the same enclosed space and measures the relative proportions of potassium-40 and argon-40. From this proportion – regardless of the absolute quantities involved – knowing the half-life of potassium-40's decay and assuming there was no argon to begin with, one can estimate the time that has elapsed since the process started – since the clock was 'zeroed', in other words. Notice that we must know the ratio of parent (potassium-40) to daughter (argon-40) isotopes. Moreover, as we saw earlier in the chapter, it is necessary that our clock has the facility to be zeroed. But what does it mean to speak of a radioactive clock's being 'zeroed'? The process of crystallization gives it meaning.

Like all the radioactive clocks used by geologists, potassium/argon timing works only with so-called igneous rocks. Named after the Latin for fire, igneous rocks are solidified from molten rock – underground magma in the case of granite, lava from volcanoes in the case of basalt. When molten rock solidifies to form granite or basalt, it does so in the form of crystals. These are normally not big, transparent crystals like those of quartz, but crystals that are too small to look like crystals to the naked eye. The crystals are of various types, and several of these, such as some micas, contain potassium atoms. Among these are atoms of the radioactive isotope potassium-40. When a crystal is newly formed, at the moment when molten rock solidifies, there is potassium-40 but no argon. The clock is 'zeroed' in the sense that there are no argon atoms in the crystal. As the millions of years go by, the potassium-40 slowly decays and, one by one, atoms of argon-40 replace potassium-40 atoms in the crystal. The accumulating quantity of argon-40 is a measure of the time that has elapsed since the rock was formed. But, for the reason I have just explained, this quantity is meaningful only if expressed as the *ratio* of potassium-40 to argon-40. When the clock was zeroed, the ratio was 100 per cent in favour of potassium-40. After 1.26 billion years, the ratio will be 50–50. After another 1.26 billion years, half of the remaining potassium-40 will have been converted to argon-40, and so on. Intermediate proportions signify intermediate times since the crystal clock was zeroed. So geologists, by measuring the ratio between potassium-40 and argon-40 in a piece of igneous rock that they pick up today, can tell how long ago the rock first crystallized out of its molten state. Igneous rocks typically contain many different radioactive isotopes, not just potassium-40. A fortunate aspect of the way igneous rocks solidify is that they do so suddenly – so that *all* the clocks in a given piece of rock are zeroed simultaneously.

Only igneous rocks provide radioactive clocks, but fossils are almost never found in igneous rock. Fossils are formed in sedimentary rocks like limestone and sandstone, which are not solidified lava. They are layers of mud or silt or sand, gradually laid down on the floor of a sea or lake or estuary. The sand or mud becomes compacted over the ages and hardens as rock. Corpses that are trapped in the mud have a chance of fossilizing. Even though only a small proportion of corpses actually do fossilize, sedimentary rocks are the only rocks that contain any fossils worth speaking of.

Sedimentary rocks unfortunately cannot be dated by radioactivity.

Presumably the individual particles of silt or sand that go to make sedimentary rocks contain potassium-40 and other radioactive isotopes, and therefore could be said to contain radioactive clocks; but unfortunately these clocks are no use to us because they are not properly zeroed, or are zeroed at different times from each other. The particles of sand that are compacted to make sandstone may originally have been ground down from igneous rocks, but the igneous rocks from which they were ground all solidified at different times. Every grain of sand has a clock zeroed at its own time, and that time was probably long before the sedimentary rock formed and entombed the fossil we are trying to date. So, from a timekeeping point of view, sedimentary rock is a mess. It can't be used. The best we can do – and it is a pretty good best – is to use the dates of igneous rocks that are found near sedimentary rock, or embedded in it.

To date a fossil, you don't literally need to find it sandwiched between two slabs of igneous rock, although that is a neat way to illustrate the principle. The actual method used is more refined than that. Recognizably similar layers of sedimentary rock occur all over the world. Long before radioactive dating was discovered, these layers had been identified and given names: names like Cambrian, Ordovician, Devonian, Jurassic, Cretaceous, Eocene, Oligocene, Miocene. Devonian sediments are recognizably Devonian, not only in Devon (the county in south-west England that gave them their name) but in other parts of the world. They are recognizably similar to each other, and they contain similar lists of fossils. Geologists have long known the *order* in which these named sediments were laid down. It's just that, before the advent of radioactive clocks, we didn't know *when* they were laid down. We could arrange them in order because – obviously – older sediments tend to lie beneath younger sediments. Devonian sediments, for example, are older than Carboniferous (named after the coal which is frequently found in Carboniferous layers) and we know this because, in those parts of the world where the two layers coincide, the Devonian layer lies underneath the Carboniferous layer (the exceptions to this rule occur in places where we can tell, from other evidence, that the rocks have been tilted aslant, or even turned upside down). We aren't usually fortunate enough to find a complete run of layers, all the way from Cambrian at the bottom up to Recent at the top. But because the layers are so recognizable, you can work out their relative ages by daisychaining and jigsawing your way around the world.

So, long before we knew how old fossils were, we knew the *order* in which they were laid down, or at least the order in which the named sediments were laid down. We knew that Cambrian fossils, the world over, were older than Ordovician ones, which were older than Silurian; then came Devonian, then Carboniferous, Permian, Triassic, Jurassic, Cretaceous, and so on. And within these major named layers, geologists also distinguish sub-regions: upper Jurassic, middle Jurassic, lower Jurassic, and so on.

The named strata are usually identified by the fossils they contain. And we are going to use the ordering of the fossils as evidence for evolution! Is that in danger of turning into a circular argument? Certainly not. Think about it. Cambrian fossils are a characteristic assemblage, unmistakably recognizable as Cambrian. For the moment we are using a characteristic assemblage of fossils simply as *labels* for Cambrian rocks – indicator species – wherever we may find them. This, indeed, is why oil companies employ fossil experts to identify particular strata of rocks, usually by microfossils, tiny creatures called foraminifera, for example, or radiolaria.

A characteristic list of fossils is used to recognize Ordovician rocks, Devonian rocks, and so on. So far, all we are using these fossil assemblages for is to identify whether a slab of rock is, say, Permian or Silurian. Now we move on to use the order in which the named strata were laid down, helped by daisychaining around the world, as evidence of which strata are older or younger than which. Having established these two sets of information, we can then look at the fossils in successively younger strata, to see whether they constitute a sensible evolutionary sequence when compared with each other in sequence. Do they progress in a sensible direction? Do certain kinds of fossils, for example mammals, appear only *after* a given date, never before? The answer to all such questions is yes. Always yes. No exceptions. That is powerful evidence for evolution, for it was never a *necessary* fact, never something that had to follow from our method of identifying strata and our method of obtaining a temporal sequence.

It is a fact that literally nothing that you could remotely call a mammal has ever been found in Devonian rock or in any older stratum. They are not just statistically rarer in Devonian than in later rocks. They literally never occur in rocks older than a certain date. But this didn't have to be so. It could have been the case that, as we dug down lower and lower from the Devonian, through the Silurian and then even older, through the Ordovician, we suddenly found that the Cambrian era – older than any of them – teemed with mammals. That is in fact *not* what we find, but the possibility demonstrates that you can't accuse the argument of being circular: at any moment somebody might dig up a mammal in Cambrian rocks, and the theory of evolution would be instantly blown apart if they did. Evolution, in other words, is a falsifiable, and therefore scientific, theory. I shall return to this point in Chapter 6.

Creationist attempts to explain such findings often achieve high comedy. Noah's flood, we are told, is the key to understanding the order in which we find fossils of the major animal groups. Here's a direct quotation from a prizewinning creationist website.

Fossil sequence in geological strata shows:

(i) INVERTEBRATES (slow moving marine animals) would perish first followed by the more mobile fishes who would be overwhelmed by the flood silt

(ii) AMPHIBIA (close to the sea) would perish next as the waters rose.

(iii) REPTILES (slow moving land animals) next to die.

(iv) MAMMALS could flee from rising water, the larger, faster ones surviving the longest.

(v) MAN would exercise most ingenuity – clinging to logs, etc. to escape the flood.

This sequence is a perfectly satisfactory explanation of the order in which the various fossils are found in the strata. It is NOT the order in which they evolved but the order in which they were inundated at the time of Noah's flood.

Quite apart from all the other reasons to object to this remarkable explanation, there could only ever be a *statistical* tendency for mammals, for example, to be *on average* better at escaping the rising waters than reptiles. Instead, as we should expect on the evolution theory, there literally are *no* mammals in the lower strata of the geological record. The 'head for the hills' theory would be on more solid ground if there were a statistical tailing off of mammals as you move down through the rocks. There are literally *no* trilobites above Permian strata, literally *no* dinosaurs (except birds) above Cretaceous strata. Once again, the 'head for the hills' theory would predict a statistical tailing off.

Back to dating, and radioactive clocks. Because the relative ordering of the named sedimentary strata is well known, and the same order is found all over the world, we can use igneous rocks that overlie or underlie sedimentary strata, or are embedded in them, to date those named sedimentary strata, and hence the fossils within them. By a refinement of the method, we can date fossils that lie near the top of, say, the Carboniferous or the Cretaceous, as more recent than fossils that lie slightly lower in the same stratum. We don't need to find an igneous rock in the vicinity of any particular fossil we want to date. We can tell that our fossil is, say, late Devonian, from its position in a Devonian stratum. And we know, from the radioactive dating of igneous rocks found in association with Devonian strata all around the world, that the Devonian ended about 360 million years ago.

Unstable isotope	Decays to	Half-life (years)
Rubidium-87	Strontium	49,000,000,000
Rhenium-187	Osmium-187	41,600,000,000
Thorium-232	Lead-208	14,000,000,000
Uranium-238	Lead-206	4,500,000,000
Potassium-40	Argon-40	1,260,000,000
Uranium-235	Lead-207	704,000,000
Samarium-147	Neodymium-143	108,000,000
Iodine-129	Xenon-129	17,000,000
Aluminium-26	Magnesium-26	740,000
Carbon-14	Nitrogen-14	5,730

Radioactive clocks

The potassium argon clock is only one of many clocks that are available to geologists, all using the same principle on their different timescales. Above is a table of clocks, ranging from slow to fast. Notice, yet again, the astonishing range of half-lives, from 49 billion years at the slow end to less than 6,000 years at the fast end. The faster clocks, such as carbon-14, work in a somewhat different way. This is because the 'zeroing' of these higher-speed clocks is necessarily different. For isotopes with a short half-life, all the atoms that were present when the Earth was originally formed have long since disappeared. Before I turn to how carbon dating works, it is worth pausing to consider another piece of evidence in favour of an old Earth, a planet whose age is measured in billions of years.

Among all the elements that occur on Earth are 150 stable isotopes and 158 unstable ones, making 308 in all. Of the 158 unstable ones, 121 are either extinct or exist only because they are constantly renewed, like carbon-14 (as we shall see). Now, if we consider the 37 that have not gone extinct, we notice something significant. Every single one of them has a half-life greater than 700 million years. And if we look at the 121 that have gone extinct, every single one of them has a half-life less than 200 million years. Don't be misled, by the way. Remember we are talking *half-life* here, not life! Think of the fate of an isotope with a half-life of 100 million years. Isotopes whose half-life is less than a tenth or so of the age of the Earth are, for practical purposes, extinct, and don't exist except under special circumstances. With exceptions that are there for a special reason that we understand, the only isotopes that we find on Earth are those that have a half-life long enough to have survived on a very old planet. Carbon-14 is one of these exceptions, and it is exceptional for an interesting reason, namely that it is being continuously replenished. Carbon-14's role as a clock therefore needs to be understood in a different way from that of longer-lived isotopes. In particular, what does it mean to *zero* the clock?

CARBON

Of all the elements, carbon is the one that seems most indispensable to life – the one without which life on any planet is hardest to envisage. This is because of carbon's remarkable capacity for forming chains and rings and other complex molecular architectures. It enters the food web via photosynthesis, which is the process whereby green plants take in carbon dioxide molecules from the atmosphere and use energy from sunlight to combine the carbon atoms with water to make sugars. All the carbon in ourselves and in all other living creatures comes ultimately, via plants, from carbon dioxide in the atmosphere. And it is continually being recycled back to the atmosphere: when we breathe out, when we excrete, and when we die.

Most of the carbon in the atmosphere's carbon dioxide is carbon-12, which is not radioactive. However, about one atom in a trillion is carbon-14, which is radioactive. It decays rather rapidly, with a half-life of 5,730 years, as we have seen, to nitrogen-14. Plant biochemistry is blind to the difference between these two carbons. To a plant, carbon is carbon is carbon. So plants take in carbon-14 alongside carbon-12, and incorporate the two kinds of carbon atom in sugars, in the same proportion as they exist in the atmosphere. The carbon that is incorporated from the atmosphere (complete with the same proportion of carbon-14 atoms) is rapidly (compared to the half-life of carbon-14) spread through the food chain, as plants are eaten by herbivores, herbivores by carnivores and so on. All living creatures, whether plants or animals, have approximately the same ratio of carbon-12 to carbon-14, which is the same ratio as you'll find in the atmosphere.

So, when is the clock zeroed? At the moment when a living creature, whether animal or plant, dies. At that moment, it is severed from the food chain, and detached from the inflow of fresh carbon-14, via plants, from the atmosphere. As the centuries go by, the carbon-14 in the corpse, or lump of wood, or piece of cloth, or whatever it is, steadily decays to nitrogen-14. The ratio of carbon-14 to carbon-12 in the specimen therefore gradually drops further and further below the standard ratio that living creatures share with the atmosphere. Eventually it will be all carbon-12 – or, more strictly, the carbon-14 content will become too small to measure. And the ratio of carbon-12 to carbon-14 can be used to calculate the time that has elapsed since the death of the creature cut it off from the food chain and its interchange with the atmosphere.

That's all very well, but it only works because there is a continuously replenished supply of carbon-14 in the atmosphere. Without that, the carbon-14 with its short half-life would long since have disappeared from the Earth, along with all other naturally occurring isotopes with short half-lives. Carbon-14 is special because it is continually being made in a 2-step process by cosmic rays ejecting neutrons from atoms in the atmosphere. Nitrogen is the commonest gas in the atmosphere and its mass number is 14, the same as carbon-14's. The difference is that carbon-14 has 6 protons and 8 neutrons, while nitrogen-14 has 7 protons and 7 neutrons (neutrons, remember, have near-enough the same mass as protons). Some of the ejected neutrons strike nitrogen nuclei and convert one of their protons into a neutron. When this happens, the atom becomes carbon-14, carbon being one lower than nitrogen in the periodic table. The rate at which this conversion occurs is approximately constant from century to century, which is why carbon dating works. Actually the rate is not exactly constant, and ideally we need to compensate for this. Fortunately we have an accurate calibration of the fluctuating supply of carbon-14 in the atmosphere and can take this into account to refine our dating calculations. Remember that, over roughly the same age range as is covered by carbon dating, we have an alternative method of dating wood – dendrochronology – which is completely accurate to the nearest year. By looking at the carbon-dated ages of wood samples whose age is independently known from tree-ring dating, we can calibrate the fluctuating errors in carbon-dating. Then we can use these calibration measurements when we go back to organic samples for which we don't have tree-ring data (the majority).

Carbon dating is a comparatively recent invention, going back only to the 1940s. In its early years, substantial quantities of organic material were needed for the dating procedure. Then, in the 1970s, a technique called mass spectrometry was adapted to carbon dating, with the result that only a tiny quantity of organic material is now needed. This has revolutionized archaeological dating. The most celebrated example is the Shroud of Turin. Since this notorious piece of cloth seems mysteriously to have imprinted on it the image of a bearded, crucified man, many people hoped it might hail from the time of Jesus. It first turns up in the historical record in the mid-fourteenth century in France, and nobody knows where it was before that. It has been housed in Turin since 1578, under the custody of the Vatican since 1983. When mass spectrometry made it possible to date a tiny sample of the shroud, rather than the substantial swathes that would have been needed before, the Vatican allowed a small strip to be cut off. The strip was divided into three parts and sent to three leading laboratories specializing in carbon dating, in Oxford, Arizona and Zurich. Working under conditions of scrupulous independence – not comparing notes – the three laboratories reported their verdicts on the date when the flax from which the cloth had been woven died.

Oxford said AD 1200, Arizona 1304 and Zurich 1274. These dates are all – within normal margins of error – compatible with each other and with the date in the 1350s at which the shroud is first mentioned in history. The dating of the shroud remains controversial, but not for reasons that cast doubt on the carbon-dating technique itself. For example, the carbon in the shroud might have been contaminated by a fire, which is known to have occurred in 1532. I won't pursue the matter further, because the shroud is of historical, not evolutionary, interest. It is a nice example, however, to illustrate the method, and the fact that, unlike dendrochronology, it is not accurate to the nearest year, only to the nearest century or so.

I have repeatedly emphasized that there are lots of different clocks that the modern evolutionary detective can use, and also that they work best on different, but overlapping timescales. Radioactive clocks can be used to give independent estimates of the age of one piece of rock, bearing in mind that all the clocks were zeroed simultaneously when this very same piece of rock solidified. When such comparisons have been made, the different clocks agree with each other – within the expected margins of error. This gives great confidence in the correctness of the clocks. Thus mutually calibrated and verified on known rocks, these clocks can be carried with confidence to interesting dating problems, such as the age of the Earth itself. The currently agreed age of 4.6 billion years is the estimate upon which several different clocks converge. Such agreement is not surprising, but unfortunately we need to emphasize it because, astonishingly, as I pointed out in the Introduction (and have documented in the Appendix), some 40 per cent of the American population, and a somewhat smaller percentage of the British population, claim to believe that the age of the Earth, far from being measured in billions of years, is less than 10,000 years. Lamentably, especially in America and over much of the Islamic world, some of these history-deniers wield power over schools and their syllabuses.

Now, a history-denier could claim, say, that there is something wrong with the potassium argon clock. What if the present very slow rate of decay of potassium-40 has only been in operation since Noah's flood? If, before that, the half-life of potassium-40 was radically different, only a few centuries, say, rather than 1.26 billion years? The special pleading in such claims is glaring. Why on Earth should the laws of physics change, just like that, so massively and so conveniently? And it glares even more when you have to make mutually adjusted special pleading claims for each one of the clocks separately. At present, the applicable isotopes all agree with each other in placing the origin of the Earth at between four and five billion years ago. And they do so on the assumption that their half-lives have always been the same as we can measure today – as the known laws of physics, indeed, strongly suggest they should. The history-deniers would have to fiddle the half-lives of all the isotopes in their separate proportions, so that they all end up agreeing that the Earth began 6,000 years ago. Now that's what I call special pleading! And I haven't even mentioned various other dating methods which also produce the same result, for example 'fission track dating'. Bear in mind the huge differences in timescales of the different clocks, and think of the amount of contrived and complicated fiddling with the laws of physics that would be needed in order to make all the clocks agree with each other, across the orders of magnitude, that the Earth is 6,000 years old and not 4.6 billion! Given that the sole motive for such fiddling is the desire to uphold the origin myth of a particular set of Bronze Age desert tribesmen, it is surprising, to say the least, that anyone is fooled by it.

There is one more type of evolutionary clock, the molecular clock, but I shall postpone discussing it until Chapter 10, after introducing some other ideas of molecular genetics.

CHAPTER 5

BEFORE OUR VERY EYES

I HAVE used the metaphor of a detective, coming on the scene of a crime after it is all over and reconstructing from the surviving clues what must have happened. But perhaps I was too ready to concede the impossibility of viewing evolution as an eye witness. Although the vast majority of evolutionary change took place before any human being was born, some examples are so fast that we can see evolution happening with our own eyes during one human lifetime.

There's a plausible indication that this may have happened even with elephants, which Darwin himself picked out as one of the slowest-reproducing animals, with one of the longest generation turnovers. One of the main causes of mortality among African elephants is humans with guns hunting ivory, either for trophies or to sell for carving. Naturally the hunters tend to pick on the individuals with the largest tusks. This means that, at least in theory, smaller-tusked individuals will be at a selective advantage. As ever with evolution, there will be conflicting selection pressures, and what we see evolving will be a compromise. Larger tuskers doubtless have an advantage when it comes to competition with other elephants, and this will be balanced against their disadvantage when they encounter men with guns. Any increase in hunting activity, whether in the form of illegal poaching or legal hunting, will tend to shift the balance of advantage towards smaller tusks. All other things being equal, we might expect an evolutionary trend towards smaller tusks as a result of human hunting, but we'd probably expect it to take millennia to be detectable. We would not expect to see it within one human lifetime. Now let's look at some figures.

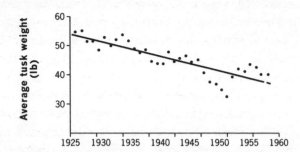

**Tusk weight
in Ugandan
elephants**

The graph above shows data from the Uganda Game Department, published in 1962. Referring only to elephants legally shot by licensed hunters, it shows mean tusk weight in pounds (that dates it) from year to year between 1925 and 1958 (during which time Uganda was a British protectorate). The dots are annual figures. The line through the dots is drawn not by eye but by a statistical technique called linear regression.* You can see that there is a decreasing trend over the thirty-three years. And the trend is highly statistically significant, which means that it is almost certainly a real trend, not a random chance effect.

The fact that there is a statistically significant trend towards shrinking tusks doesn't necessarily mean it is an evolutionary trend. If you were to plot a graph of mean height of 20-year-old men, from year to year during the twentieth century, you'd see in many countries a significant trend towards getting taller. This is normally reckoned to be not an evolutionary trend, but rather an effect of improved nutrition. Nevertheless, in the case of the elephants we have good

---

* Think of it like this. Imagine all possible straight lines. For each line, calculate how closely it fits the dots, by measuring the distance of each dot from the line, and adding up all the distances (after squaring them, for a good mathematical reason which would take us too far afield). Of all possible straight lines, the one that *minimizes* the sum of the squared dot-to-line distances, summed over all the dots, is the fitted regression line. It shows us the trend, without our eyes being distracted by all the mess of the individual points. There are separate calculations that statisticians do to calculate how *reliable* the line is as an indicator of a trend. These are called tests of statistical significance. They make use of the breadth of scatter about the line.

reason to suspect the existence of strong selection against large tusks. Reflect that, although the graph refers to tusks obtained from licensed kills, the selection pressure that produced the trend could well have come mostly from poaching. We must seriously entertain the possibility that it is a true evolutionary trend, in which case it is a remarkably rapid one. We must be cautious before concluding too much. It could be that we are observing strong natural selection, which is highly likely to result in changes in gene frequencies in the population, but such genetic effects have not so far been demonstrated. It could be that the difference between large-tusked and small-tusked elephants is a non-genetic difference. Nevertheless, I am inclined to take seriously the possibility that it is a true evolutionary trend.

More to the point, my colleague Dr Iain Douglas-Hamilton, who is the world authority on wild African elephant populations, takes it seriously and believes, surely rightly, that it needs looking into more closely. He suspects that the trend started long before 1925 and has continued after 1958. He has reason to think that the same cause, operating in the past, underlies the tusklessness of many local populations of Asian elephants. We seem to have here a prima facie case of rapid evolution taking place before our very eyes, one that would repay further research.

Let me turn, now, to another case, one in which there is some intriguing recent research: a study of lizards on Adriatic islands.

THE LIZARDS OF POD MRCARU

There are two small islets off the Croatian coast called Pod Kopiste and Pod Mrcaru. In 1971 a population of common Mediterranean lizards, *Podarcis sicula*, which mainly eat insects, was present on Pod Kopiste but there were none on Pod Mrcaru. In that year experimenters transported five pairs of *Podarcis sicula* from Pod Kopiste and released them on Pod Mrcaru. Then, in 2008, another group of mainly Belgian scientists, associated with Anthony Herrel, visited the islands to see what had happened. They found a flourishing population of lizards on Pod Mrcaru, which DNA analysis confirmed were indeed *Podarcis sicula*. These are presumed to have descended from the original five pairs that were transported. Herrel and his colleagues made observations on the descendants of the transported lizards, and compared them with lizards living on the original ancestral island. There were marked differences. The scientists made the probably justified assumption that the lizards on the ancestral island, Pod Kopiste, were unchanged representatives of the ancestral lizards of thirty-six years before. In other words, they presumed they were comparing the evolved lizards of Pod Mrcaru with their unevolved 'ancestors' (meaning their contemporaries but of ancestral type) on Pod Kopiste. Even if this presumption is wrong – even if, for example, the lizards of Pod Kopiste have been evolving just as fast as the lizards of Pod Mrcaru – we are still observing evolutionary divergence in nature, over a timescale of decades: the sort of timescale that humans can observe within one lifetime.

And what were the differences between the two island populations, differences that had taken a mere thirty-seven years or so to evolve?* Well, the Pod Mrcaru lizards – the 'evolved' population – had significantly larger heads than the 'original' Pod Kopiste population: longer, wider and taller heads. This translates into a markedly greater bite force. A change of this kind typically goes with a shift to a more vegetarian diet and, sure enough, the lizards of Pod Mrcaru eat significantly more plant material than the 'ancestral' type on Pod Kopiste. From the almost exclusive diet of insects (arthropods, in the terms of the graph opposite) still enjoyed by the modern Pod Kopiste population, the lizards on Pod Mrcaru had shifted to a largely vegetarian diet, especially in summer.

Why would an animal need a stronger bite when shifting to a

---

* Up to twice as long if the Pod Kopiste lizards have been evolving at the same rate since the shared ancestor of thirty-seven years ago.

Summer diet of lizards on two Adriatic islands

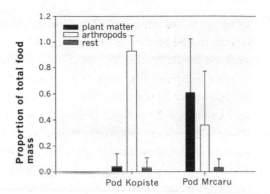

vegetarian diet? Because plant, but not animal, cell walls are stiffened by cellulose. Herbivorous mammals like horses, cattle and elephants have great millstone-like teeth for grinding cellulose, quite different from the shearing teeth of carnivores and the needly teeth of insectivores. And they have massive jaw muscles, and correspondingly robust skulls for the muscle attachments (think of the stout midline crest along the top of a gorilla's skull).* Vegetarians also have characteristic peculiarities of the gut. Animals generally can't digest cellulose without the aid of bacteria or other micro-organisms, and many vertebrates set aside a blind alley in the gut called the caecum, which houses such bacteria and acts as a fermentation chamber (our appendix is a vestige of the larger caecum in our more vegetarian ancestors). The caecum, and other parts of the gut, can become quite elaborate in specialist herbivores. Carnivores usually have simpler guts than herbivores, and smaller too. Among the complications that become inserted in herbivore guts are things called caecal valves. Valves are incomplete partitions, sometimes muscular, which can serve to regulate or slow down the flow of material through the gut, or simply increase the surface area of the interior of the

---

* The same gorilla-like features in the skull and teeth of our robust cousin *Paranthropus boisei* ('nutcracker man', also nicknamed 'Zinj' and 'Dear Boy') indicate that it was almost certainly vegetarian.

Caecal valve

caecum. The picture on the left shows the caecum cut open in a related species of lizard which eats a lot of plant material. The valve is indicated by the arrow. Now, the fascinating thing is that, although caecal valves don't normally occur in *Podarcis sicula* and are rare in the family to which it belongs, those valves have actually started to evolve in the population of *P. sicula* on Pod Mrcaru, the population that has, for only the past thirty-seven years, been evolving towards herbivory. The investigators discovered other evolutionary changes in the lizards of Pod Mrcaru. The population density increased, and the lizards ceased to defend territories in the way that the 'ancestral' population on Pod Kopiste did. I should repeat that the only thing that is really exceptional about this whole story, and the reason I am telling it here, is that it all happened so extremely rapidly, in a matter of a few decades: evolution before our very eyes.

FORTY-FIVE THOUSAND GENERATIONS OF EVOLUTION IN THE LAB

The average generation turnover of those lizards is about two years, so the evolutionary change observed on Pod Mrcaru represents only about eighteen or nineteen generations. Just think what you might see in three or four decades if you followed the evolution of bacteria, whose generations are measured in hours or even minutes, rather than years! Bacteria offer another priceless gift to the evolutionist. In some cases you can freeze them for an indefinite length of time and then bring them back to life again, whereupon they resume reproduction as if nothing had happened. This means that experimenters can lay down their own 'living fossil record', a snapshot of the exact point the evolutionary process had reached at any desired time. Imagine if we could bring Lucy, the magnificent pre-human fossil discovered by Don Johanson, back to life from a deep-freeze and set her kind evolving anew! All this has been achieved with the bacterium *Escherichia coli*, in a spectacular long-term experiment by the bacteriologist Richard Lenski and his colleagues at Michigan State University. Scientific research nowadays is often a team effort. In what follows, I may sometimes use the name 'Lenski' for brevity, but you should read it as 'Lenski and the colleagues and students in his lab.' As we shall see, the Lenski experiments are distressing to creationists, and for a very good reason. They are a beautiful demonstration of evolution in action, something it is hard to laugh off even when your motivation to do so is very strong. And the motivation for dyed-in-the-wool creationists is very strong indeed. I'll return to this at the end of the section.

E. coli is a common bacterium. Very common. There are about a hundred billion billion of them around the world at any one time, of which about a billion, by Lenski's calculation, are in your large intestine at this very moment. Most of them are harmless or even beneficial, but nasty strains occasionally hit the headlines. Such periodic evolutionary innovation is not surprising if you do the sums, even though mutations are rare events. If we assume that the probability of a gene mutating during any one act of bacterial reproduction is as low as one in a billion, the numbers of bacteria are so colossal that just about every gene in the genome will have mutated somewhere in the world, every day. As Richard Lenski says, 'That's a lot of opportunity for evolution.'

Lenski and his colleagues exploited that opportunity, in a controlled way, in the lab. Their work is extremely thorough and careful in every detail. The details really contribute to the impact of the evidence for evolution that these experiments provide, and I am therefore not going to stint in explaining them. This means that the next few pages are inevitably somewhat intricate – not difficult, just intricately detailed. It would probably be best not to read this section of the book when tired, at the end of a long day. What makes it easier to follow is that every detail makes sense: none of it leaves us scratching our heads and wondering what that was all about. So, please come with me, step by step, through this splendidly constructed and elegantly executed set of experiments.

These bacteria reproduce asexually – by simple cell division – so it is easy to clone up a huge population of genetically identical individuals in a short time. In 1988, Lenski took one such population and infected twelve identical flasks, all of which contained the same nutrient broth, including glucose as the vital food source. The twelve flasks, each with its founding population of bacteria, were then placed in a 'shaking incubator' where they were kept nice and warm, and shaken to keep the bacteria well distributed throughout the liquid. These twelve flasks founded twelve lines of evolution that were destined to be kept separate from one another for two decades and counting: sort of like the twelve tribes of Israel, except that in the case of the tribes of Israel there was no law against their mixing.

The twelve tribes of bacteria were not kept in the same twelve flasks for all that time. On the contrary, each tribe had a new flask every day. Imagine twelve *lines* of flasks, stretching away into the distance, each line more than 7,000 flasks long! Every day, for each of the twelve tribes, a new virgin flask was infected with liquid from the previous day's flask. A small sample, exactly one-hundredth of the volume of the old flask, was drawn out and squirted into the new flask, which contained a fresh supply of glucose-rich broth. The population of bacteria in the flask then started to skyrocket; but it always levelled off by the next day as the supply of food gave out and starvation set in. In other words, the population in every flask multiplied itself hugely, then reached a plateau, at which point a new infective sample was drawn and the cycle renewed the next day. Thousands of times through their high-speed equivalent of geological time, therefore, these bacteria went through the same daily repeated cycles of bonanza expansion, followed by starvation, from which a lucky hundredth were rescued and carried, in a glass Noah's Ark, to a fresh – but again temporary – glucose bonanza: perfect perfect perfect conditions for evolution, and, what is more, the experiment was done in twelve separate lines in parallel.

Lenski and his team have continued this daily routine for more than twenty years so far. This means about 7,000 'flask generations' and 45,000 bacterial generations – averaging between six and seven bacterial generations per day. To put that into perspective, if we were to go back 45,000 human generations, that would be about a million years, back to the time of *Homo erectus*, which is not very long ago. So, whatever evolutionary change Lenski may have clocked up in the equivalent of a million years of bacterial generations, think how much more evolution might happen in, say, 100 million years of mammal evolution. And even 100 million years is comparatively recent, by geological standards.

In addition to the main evolution experiment, the Lenski group used the bacteria for various illuminating spin-off experiments, for example replacing glucose with another sugar, maltose, after 2,000 generations, but I shall concentrate on the central experiment, which used glucose throughout. They sampled the twelve tribes at intervals throughout the twenty years, to see how evolution was progressing. They also froze samples of each of the tribes as a source of resuscitatable 'fossils' representing strategic points along the evolutionary way. It is hard to exaggerate how brilliantly conceived this series of experiments is.

Here's a little example of the excellent forward planning. You remember I said that the twelve founding flasks were all seeded from the same clone and therefore started out genetically identical. But that wasn't quite true – for an interesting and cunning reason. The Lenski lab had earlier exploited a gene called *ara* which comes in two forms, Ara+ and Ara–. You can't tell the difference until you take a sample of the bacteria and 'plate them out' on an agar plate that contains a nutritious broth plus the sugar arabinose and a chemical dye called tetrazolium. 'Plating out' is one of the things bacteriologists do. It means putting a drop of liquid, containing bacteria, on a plate covered with a thin sheet of agar gel and then incubating the plate. Colonies of bacteria grow out as expanding circles – miniature fairy rings* – from the drops, feeding on the nutrients mixed in with the agar. If

---

* That's no idle metaphor, for fairy rings of mushrooms achieve their circular form for exactly the same reason.

the mixture contains arabinose and the indicator dye, the difference between Ara+ and Ara– is revealed, as if by heating invisible ink: they show up as white and red colonies, respectively. The Lenski team find this colour distinction useful for labelling purposes, as we shall see, and they anticipated this usefulness by setting up six of their twelve tribes as Ara+ and the other six as Ara–. Just to give one example of how they exploited the colour coding of the bacteria, they used it as a check on their own laboratory procedures. When performing their daily ritual of infecting new flasks, they took care to handle Ara+ and Ara– flasks alternately. That way, if they ever made a mistake – splashed a transfer pipette with liquid or something like that – it would show up when they later subjected samples to the red/white test. Ingenious? Yes. And scrupulous. Really good scientists have to be both.

But forget about Ara+ and Ara– for the moment. In all other respects, the founding populations of the twelve tribes started out identical. No other differences between Ara– and Ara+ have been detected, so they really could be treated as convenient colour markers, as ornithologists put colour rings on birds' legs.

Right, then. We have our twelve tribes, marching through their own highly speeded-up equivalent of geological time, in parallel, under the same conditions of repeated boom and bust. The interesting question was, would they stay the same as their ancestors? Or would they evolve? And if they evolved, would all twelve tribes evolve in the same way, or would they diverge from one another?

The broth, as I have said, contained glucose. It was not the only food there, but it was the limiting resource. This means that running out of glucose was the key factor that caused the population size, in every flask every day, to stop climbing and reach a plateau. To put it another way, if the experimenters had put more glucose in the daily flasks, the population plateau at the end of the day would have been higher. Or, if they had added a second dollop of glucose after the plateau was reached, they would have witnessed a second spurt of population growth, up to a new plateau.

In these conditions, the Darwinian expectation was that, if any mutation arose that assisted an individual bacterium to exploit glucose more efficiently, natural selection would favour it, and it would spread through the flask as mutant individuals out-reproduced non-mutant individuals. Its type would then disproportionately infect the next flask in the lineage and, as flask took over from flask, pretty soon the mutant would have a monopoly of its tribe. Well, this is exactly what happened in all twelve tribes. As the 'flask generations' went by, all twelve lines improved over their ancestors: got better at exploiting glucose as a food source. But, fascinatingly, they got better in different ways – that is, different tribes developed different sets of mutations.

How did the scientists know this? They could tell by sampling the lineages as they evolved, and comparing the 'fitness' of each sample against 'fossils' sampled from the original founding population. Remember that 'fossils' are frozen samples of bacteria which, when unfrozen, carry on living and reproducing normally. And how did Lenski and his colleagues make this comparison of 'fitness'? How did they compare 'modern' bacteria with their 'fossil' ancestors? With great ingenuity. They took a sample of the putatively evolved population and put it in a virgin flask. And they put a same-sized sample of the unfrozen ancestral population in the same flask. Needless to say, these experimentally mixed flasks were from then on entirely removed from contact with the continuing lineages of the twelve tribes in the long-term evolution experiment. This side experiment was done with samples that played no further part in the main experiment.

So, we have a new experimental flask containing two competing strains, 'modern' and 'living fossil', and we want to know which of the two strains will out-populate the other. But they are all mixed up, so how do you tell? How do you distinguish the two strains, when they are mixed together in the 'competition flask'? I told you it was ingenious. You remember the colour coding, with the 'reds' (Ara−) and the 'whites' (Ara+)? Now, if you wanted to compare the fitness of, say, Tribe 5 with the ancestral fossil population, what would you do? Let's suppose that Tribe 5 was Ara+. Well then, you'd make sure that the 'ancestral fossils' to which you now compared Tribe 5 were Ara−. And if Tribe 6 happens to be Ara−, the 'fossils' that you'd choose to unfreeze and mix *them* with would all be Ara+. The Ara+ and Ara− genes themselves, as the Lenski team already knew from previous work, have no effect on fitness. So they could use the colour markers to assay the competitive abilities of each of the evolving tribes, using fossilized 'ancestors' as the competitive standard, in every case. All they had to do was simply plate out samples from the mixed flasks and see how many of the bacteria growing on the agar were white and how many red.

As I say, in all twelve tribes the average fitness increased as the thousands of generations went by. All twelve lines got better at surviving in these glucose-limited conditions. The fitness increase could be attributed to several changes. Populations grew faster in successive flasks, and the average body size of the bacteria grew, in all twelve lines. The top graph opposite plots the average bacterial body size for one of the tribes, which was typical. The blobs represent real data points. The curve drawn is a mathematical approximation. It gives the best fit to the observed data for this particular kind of curve, which is called a hyperbola.* It is always possible that a more complicated mathematical function than a hyperbola would give an even closer fit to the data, but this hyperbola is pretty good, so it hardly seems worth bothering to try. Biologists often

---

* You remember the straight line that was the best fit to the data on the decline of elephant tusk size from 1925 to 1958? I explained the method as equivalent to trying all possible straight lines and finding the one that minimized the sum of squares of distances of dots on the graph from the line. But you can do the same thing while not limiting yourself to straight lines. You can look at all possible curves of a certain type defined by mathematicians. The hyperbola is one such curve. In this case, we look at all possible hyperbolas in turn, measure the distance to the line of every point on the graph, then add up the sum of the squared distances over all points. Do the same for all hyperbolas, and then choose the hyperbola that minimizes that sum. Lenski did a sort of short-cut equivalent to that exhaustive operation to arrive at the best-fit hyperbola, which is the one you see drawn in.

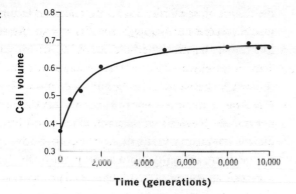

Lenski experiment: bacterial body size in one tribe

fit mathematical curves to observed data, but, unlike physicists, biologists are not accustomed to seeing such a close fit. Usually our data are too messy. In biology, as opposed to physical sciences, we only expect to get smooth curves when we have a very large quantity of data gathered under scrupulously controlled conditions. Lenski's research is a class act.

You can see that most of the increase in body size occurred in the first 2,000 or so generations. The next interesting question is this. Given that all twelve tribes increased in body size over evolutionary time, did they all increase in the same way, by the same genetic

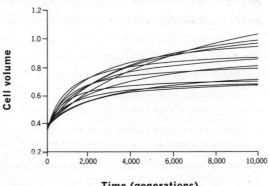

Lenski experiment: bacterial body size in twelve tribes

route? No, they didn't, and that's the second interesting result. The graph at the top of page 123 is for one of the twelve tribes. Now look at the equivalent hyperbolic best fits for all twelve (graph at the foot of page 123). Look how spread out they are. They all seem to be approaching a plateau, but the highest of the twelve plateaus is almost twice as high as the lowest. And the curves have different shapes: the curve that reaches the highest value by generation 10,000 starts by growing more slowly than some of the others, and then overtakes them before generation 7,000. Don't confuse these plateaus, by the way, with the daily plateaus of population size within each flask. We are now looking at curves in evolutionary time, measured in flask generations, not individual bacterial time, measured in hours within one flask.

What this evolutionary change suggests is that becoming larger is, for some reason, a good idea when you are struggling to survive in this alternating glucose-rich/glucose-poor environment. I won't speculate on why increasing body size might be an advantage – there are many possibilities – but it looks as though it must have been so, because all twelve tribes did it. But there are lots of different ways to become larger – different sets of mutations – and it looks as though different ways have been discovered by different evolutionary lineages in this experiment. That's pretty interesting. But perhaps even more interesting is that sometimes a pair of tribes seem to have independently discovered the *same* way of getting bigger. Lenski and a different set of colleagues investigated this phenomenon by taking two of the tribes, called Ara+1 and Ara−1, which seemed, over 20,000 generations, to have followed the same evolutionary trajectory, and looking at their DNA. The astonishing result they found was that 59 genes had changed their levels of expression in both tribes, and *all 59 had changed in the same direction*. Were it not for natural selection, such independent parallelism, in 59 genes independently, would completely beggar belief. The odds against its happening by chance are stupefyingly large. This is exactly the kind of thing creationists

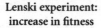

Lenski experiment: increase in fitness

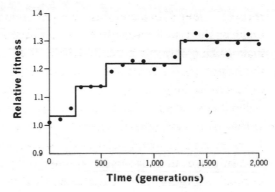

çay cannot happen, because they think it is too improbable to have happened by chance. Yet it actually happened. And the explanation, of course, is that it did *not* happen by chance, but because gradual, step-by-step, cumulative natural selection favoured the same – literally the same – beneficial changes in both lines independently.

The smooth curve in the graph of increasing cell size as the generations go by gives support to the idea that the improvement is gradual. But perhaps it is too gradual? Wouldn't you expect to see actual *steps*, as the population 'waits' for the next improving mutation to turn up? Not necessarily. It depends on factors such as the number of mutations involved, the magnitude of each mutation's effect, the variation in cell size that is caused by influences other than genes, and how often the bacteria were sampled. And interestingly, if we look at the graph of the increase in fitness, as opposed to cell size, we do see what could at least be interpreted as a more overtly stepped picture (above). You remember, when I introduced the hyperbola, I said it might be possible to find a more complicated mathematical function that would fit the data better. Mathematicians call it a 'model'. You could fit a hyperbolic model to these points, as in the previous graph, but you get an even better fit with a 'step model', as used in this picture. It is not such a close fit as the cell size graph's fit to a hyperbola. In neither case can it be proved that the data exactly fit the model, nor can that ever be done. But the data are at least compatible with the idea that the evolutionary change that we observe represents the stepwise accumulation of mutations.*

We have so far seen a beautiful demonstration of evolution in action: evolution before our very eyes, documented by comparing twelve independent lines, and also by comparing each line with 'living fossils', which literally, instead of only metaphorically, come from the past.

Now we are ready to move on to an even more interesting result. So far, I've implied that all twelve tribes evolved their improved fitness in the same general kind of way, differing only in detail – some being a bit faster, some a bit slower than others. However, the long-term experiment threw up one dramatic exception. Shortly after generation 33,000 something utterly remarkable happened. One out of the twelve lineages, called Ara–3, suddenly went berserk. Look at the graph opposite. The vertical axis, labelled OD, which stands for optical density or 'cloudiness', is a measure of population size in the flask. The liquid becomes cloudy because of the sheer numbers of bacteria; the thickness of the cloud can be measured as a number, and that number is our index of population density. You can see that up to about generation 33,000, the average population density of Tribe Ara–3 was coasting along at an OD of about 0.04, which was not very different from all the other tribes. Then, just after generation 33,100, the OD score of Tribe Ara–3 (and of that tribe alone among the twelve) went into vertical take-off. It shot up

---

* A stepwise pattern of evolution is to be expected in creatures such as bacteria, which (most of the time) don't reproduce sexually. In animals like us, who reproduce only sexually, evolutionary change doesn't usually 'hang about' while it 'waits' for a key mutation to turn up (this is a common mistake made by opponents of evolution with some pretensions to sophistication). Instead, sexually reproducing populations usually have a ready supply of genetic variation from which to select. Although originally generated by past mutation, a large number of genetic variants are often present in a gene pool at any one time, introduced by mutation a while back and now shuffled about by sexual recombination. Natural selection often acts to shift the balance of existing variation, rather than waiting for key mutations to turn up. In bacteria without sexual reproduction, the very idea of a gene pool doesn't properly apply. That is why we can realistically hope to see discrete steps, where we might not in a population of birds, mammals or fish.

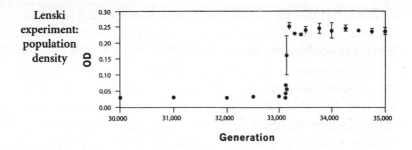

Lenski experiment: population density

OD

Generation

sixfold, to an OD value of about 0.25. The populations of successive flasks of this tribe soared. After only a few days the typical plateau at which flasks of this tribe stabilized had an OD number about six times greater than it had been, and than the other tribes were still showing. This higher plateau was then reached in all subsequent generations, in this tribe but no other. It was as though a large dose of extra glucose had been injected into every flask of Tribe Ara–3, but given to no other tribe. But that didn't happen. The same glucose ration was scrupulously administered to all the flasks equally.

What was going on? What was it that suddenly happened to Tribe Ara–3? Lenski and two colleagues investigated further, and worked it out. It is a fascinating story. You remember I said that glucose was the limiting resource, and any mutant that 'discovered' how to deal more efficiently with glucose would have an advantage. That indeed is what happened in the evolution of all twelve tribes. But I also told you that glucose was not the only nutrient in the broth. Another one was citrate (related to the substance that makes lemons sour). The broth contained plenty of citrate, but E. coli normally can't use it, at least not where there is oxygen in the water, as there was in Lenski's flasks. But if only a mutant could 'discover' how to deal with citrate, a bonanza would open up for it. This is exactly what happened with Ara–3. This tribe, and this tribe alone, suddenly acquired the ability to eat citrate as well as glucose, rather than only glucose. The amount of available food in each successive flask in the lineage therefore shot up. And so did the plateau at which the population in each successive flask daily stabilized.

Having discovered what was special about the Ara–3 tribe, Lenski and his colleagues went on to ask an even more interesting question. Was this sudden improvement in ability to draw nourishment all due to a single dramatic mutation, one so rare that only one of the twelve lineages was fortunate enough to undergo it? Was it, in other words, just another mutational step, like the ones that seemed to be demonstrated in the small steps of the fitness graph on page 125? This seemed to Lenski unlikely, for an interesting reason. Knowing the average mutation rate of each gene in the genome of these bacteria, he calculated that 30,000 generations was long enough for every gene to have mutated at least once in each of the twelve lines. So it seemed unlikely that it was the rarity of the mutation that singled Ara–3 out. It should have been 'discovered' by several other tribes.

There was another theoretical possibility, and an extremely tantalizing one. This is where the story starts to get quite complicated so, if it is late at night, it might be an idea to resume reading tomorrow . . .

What if the necessary biochemical wizardry to feed on citrate requires not just one mutation but two (or three)? We are not now talking about two mutations that build on each other in a simple additive way. If we were, it would be enough to get the two mutations in any order. Either one, on its own, would take you halfway (say) to the goal; and either one on its own would confer an ability to get some nourishment from citrate, but not as much as both mutations together would. That would be on a par with the mutations we have already discussed for increasing body size. But such a circumstance would not be rare enough to account for the dramatic uniqueness of Tribe Ara–3. No, the rarity of citrate metabolism suggests that we are looking for something more like the 'irreducible complexity' of creationist propaganda. This might be a biochemical pathway in which the product of one chemical reaction feeds into a second chemical reaction, and *neither can make any inroads at all without the other*. This would require two mutations, call them A and B, to catalyse the two reactions. On this hypothesis, you really would need *both* mutations *before there is any improvement whatsoever*, and that really would be improbable enough to account for the observed result that only one out of the twelve tribes achieved the feat.

That's all hypothetical. Could the Lenski group find out by experiment what was actually going on? Well, they could take great strides in that direction, making brilliant use of the frozen 'fossils', which are such a continual boon in this research. The hypothesis, to repeat, is that, at some time unknown, Tribe Ara–3 chanced to undergo a mutation, mutation A. This had no detectable effect because the other necessary mutation, B, was still lacking. Mutation B is equally likely to crop up in any one of the twelve tribes. Indeed, it probably did. But B is no use – has absolutely no beneficial effect at all – unless the tribe happens to be primed by the previous occurrence of mutation A. And only tribe Ara–3, as it happened, was so primed.

Lenski could even have phrased his hypothesis in the form of a testable prediction – and it is interesting to put it like this because it really is a prediction even though, in a sense, it is about the past. Here's how I would have put the prediction, if I had been Lenski:

> I shall thaw out fossils from Tribe Ara-3, dating from various points, strategically chosen, going back in time. Each of these 'Lazarus clones' will then be allowed to evolve further, on a similar regimen to the main evolution experiment, from which, of course, they will be completely isolated. And now, here's my prediction. Some of these Lazarus clones will 'discover' how to deal with citrate, but *only* if they were thawed out of the fossil record *after* a particular, critical generation in the original evolution experiment. We don't know – yet – when that magic generation was but we shall identify it, with hindsight, as the moment when, according to our hypothesis, mutation A entered the tribe.

You will be delighted to hear that this is exactly what Lenski's student Zachary Blount found, when he ran a gruelling set of experiments involving some forty trillion – 40,000,000,000,000 – E. coli cells from across the generations. The magic moment turned out to be approximately generation 20,000. Thawed-out clones of Ara–3 that dated from after generation 20,000 in the 'fossil record' showed increased probability of *subsequently* evolving citrate capability. No clones that dated from before generation 20,000 did. According to the hypothesis, after generation 20,000 the clones were now 'primed' to take advantage of mutation B whenever it came along. And there was no subsequent change in likelihood, in either direction, once the fossils' 'resurrection day' was later than the magic date of generation 20,000: whichever generation after 20,000 Blount sampled, the increased likelihood of those thawed fossils subsequently acquiring citrate capability remained the same. But thawed fossils from before generation 20,000 had no increased likelihood of developing citrate capability at all. Tribe Ara–3, before generation 20,000, was just like all the other tribes. Although its members belonged to Tribe Ara–3, they did not possess mutation A. But after generation 20,000, Tribe Ara–3 were 'primed'. Only they were able to take advantage of 'mutation B' when it turned up – as it probably did in several of the other tribes, but to no good effect. There are moments of great joy in scientific research, and this must surely have been one of them.

Lenski's research shows, in microcosm and in the lab, massively speeded up so that it happened before our very eyes, many of the essential components of evolution by natural selection: random mutation followed by non-random natural selection; adaptation to the same environment by separate routes independently; the way successive mutations build on their predecessors to produce evolutionary change; the way some genes rely, for their effects, on the presence of other genes. Yet it all happened in a tiny fraction of the time evolution normally takes.

There is a comic sequel to this triumphant tale of scientific endeavour. Creationists hate it. Not only does it show evolution in action; not only does it show new information entering genomes without the intervention of a designer, which is something they have all been told to deny is possible ('told to' because most of them don't understand what 'information' means); not only does it demonstrate the power of natural selection to put together combinations of genes that, by the naïve calculations so beloved of creationists, should be tantamount to impossible; it also undermines their central dogma of 'irreducible complexity'. So it is no wonder they are disconcerted by the Lenski research, and eager to find fault with it.

Andrew Schlafly, creationist editor of 'Conservapedia', the notoriously misleading imitation of Wikipedia, wrote to Dr Lenski demanding access to his original data, presumably implying that there was some doubt as to their veracity. Lenski had absolutely no obligation even to reply to this impertinent suggestion but, in a very gentlemanly way, he did so, mildly suggesting that Schlafly might make the effort to read his paper before criticizing it. Lenski went on to make the telling point that his best data are stored in the form of frozen bacterial cultures, which anybody could, in principle, examine to verify his conclusions. He would be happy to send samples to any bacteriologist qualified to handle them, pointing out that in unqualified hands they might be quite dangerous. Lenski listed these qualifications in merciless detail, and one can almost hear the relish with which he did so, knowing full well that Schlafly – a *lawyer*, if you please, not a scientist at all – would hardly be able to spell his way through the words, let alone qualify as a bacteriologist competent to carry out advanced and safe laboratory procedures, followed by statistical analysis of the results. The whole matter was trenchantly summed up by the celebrated scientific blogwit PZ Myers, in a passage beginning, 'Once again, Richard Lenski has replied to the goons and fools at Conservapedia, and boy, does he ever outclass them.'

Lenski's experiments, especially with the ingenious 'fossilization' technique, show the power of natural selection to wreak evolutionary change on a timescale that we can appreciate in a human lifetime, before our very eyes. But bacteria provide other impressive, if less clearly worked-out, examples. Many bacterial strains have evolved resistance to antibiotics in spectacularly short periods. After all, the first antibiotic, penicillin, was developed, heroically, by Florey and Chain as recently as the Second World War. New antibiotics have been coming out at frequent intervals since then, and bacteria have evolved resistance to just about every one of them. Nowadays, the most ominous example is MRSA (methycillin-resistant *Staphylococcus aureus*), which has succeeded in making many hospitals positively dangerous places to visit. Another menace is '*C. diff.*' (*Clostridium difficile*). Here again, we have natural selection favouring strains that are resistant to antibiotics; but the effect is overlain by another one. Prolonged use of antibiotics tends to kill 'good' bacteria in the gut, along with the bad ones. *C. diff.*, being resistant to most antibiotics, is greatly helped by the *absence* of other species of bacteria with which it would normally compete. It is the principle of 'my enemy's enemy is my friend'.

I was mildly irritated to read a pamphlet in my doctor's waiting room warning of the danger of failing to finish a course of antibiotic pills. Nothing wrong with that warning; but it was the reason given that worried me. The pamphlet explained that bacteria are 'clever'; they 'learn' to cope with antibiotics. Presumably the authors thought the phenomenon of antibiotic resistance would be easier to grasp if they called it learning rather than natural selection. But to talk of bacteria being clever, and of learning, is downright confusing, and above all it doesn't help the patient to make sense of the instruction to carry on taking the pills until they are finished. Any fool can see that it is not plausible to describe a bacterium as clever. Even if there were clever bacteria, why would stopping prematurely make any difference to the learning prowess of a clever bacterium? But as soon as you start thinking in terms of natural selection, it makes perfect sense.

Like any poison, antibiotics are likely to be dosage dependent. A sufficiently high dose will kill all the bacteria. A sufficiently low dose will kill none. An intermediate dose will kill some, but not all. If there is genetic variation among bacteria, such that some are more susceptible to the antibiotic than others, an intermediate dose will be tailor-made to select in favour of genes for resistance. When the doctor tells you to finish taking the pills, it is to increase the chances of killing *all* the bacteria and avoid leaving behind resistant, or semi-resistant, mutants. With hindsight we might say that if only we had all been better educated in Darwinian thinking, we would have woken up sooner to the dangers of resistant strains being selected. Pamphlets like the one in my doctor's waiting room don't help with that education – and what a sadly missed opportunity to teach something of the wondrous power of natural selection.

GUPPIES

My colleague Dr John Endler, recently moved from North America to the University of Exeter, told me the following marvellous – well, also depressing – story. He was travelling on a domestic flight in the United States, and the passenger in the next seat made conversation by asking him what he did. Endler replied that he was a professor of biology, doing research on wild guppy populations in Trinidad. The man became increasingly interested in the research and asked many questions. Intrigued by the elegance of the theory that seemed to underlie the experiments, he asked Endler what that theory was, and who originated it. Only then did Dr Endler drop what he correctly guessed would be his bombshell: 'It's called Darwin's theory of evolution by natural selection!' The man's whole demeanour instantly changed. His face went red; abruptly, he turned away, refused to speak further and terminated what had hitherto been an amiable conversation. More than amiable, indeed: Dr Endler writes to me that the man had 'asked some excellent questions before this, indicating that he was enthusiastically and intellectually following the argument. This is really tragic.'

The experiments that John Endler recounted to his closed-minded fellow passenger are elegant and simple, and they serve beautifully to illustrate the speed with which natural selection can go to work. It is fitting that I should use Endler's own research here, because he is also the author of *Natural Selection in the Wild*, the leading book in which examples of such studies have been collected, and their methods laid out.

Guppies are popular freshwater aquarium fish. As with the pheasants we met in Chapter 3, the males are more brightly coloured than the females, and aquarists have bred them to become even brighter. Endler studied wild guppies (*Poecilia reticulata*) living in mountain streams in Trinidad, Tobago and Venezuela. He noticed that local populations were strikingly different from each other. In some populations the adult males were rainbow-coloured, almost as bright as those bred in aquarium tanks. He surmised that their ancestors had been selected for their bright colours by female guppies, in the same manner as cock pheasants are selected by hens. In other areas the males were much drabber, although they were still brighter than the females. Like the females, though less so, they were well camouflaged against the gravelly bottoms of the streams in which they live. Endler showed, by elegant quantitative comparisons between many locations in Venezuela and Trinidad, that the streams where the males were less bright were also the streams where predation was heavy. In streams with only weak predation, males were more brightly coloured, with larger, gaudier spots, and more of them: here the males were free to evolve bright colours to appeal to females. The pressure from females on males to evolve bright colours was there all the time, in all the various separate populations, whether the local predators were pushing in the other direction strongly or weakly. As ever, evolution finds a compromise between selection pressures. What was interesting about the guppies is that Endler could actually see how the compromise varied in different streams. But he did much better than that. He went on to do experiments.

Suppose you wanted to set up the ideal experiment to demonstrate the evolution of camouflage: what would you do? Camouflaged animals resemble the background on which they are seen. Could you set up an experiment in which animals actually evolve, before your very eyes, to resemble a background that you have experimentally

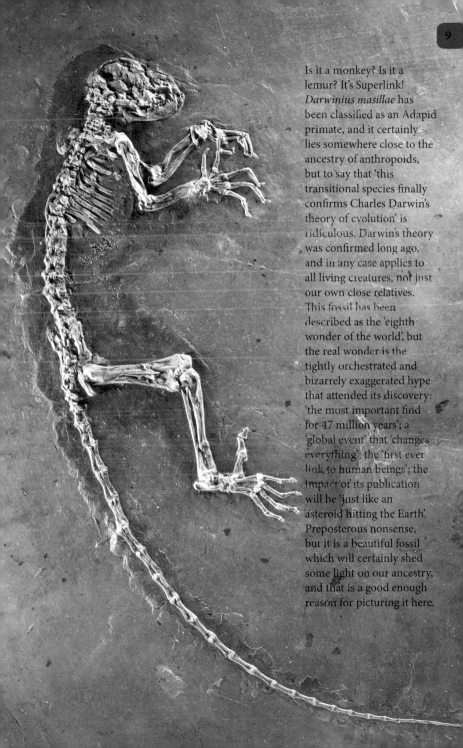

Is it a monkey? Is it a lemur? It's Superlink! *Darwinius masillae* has been classified as an Adapid primate, and it certainly lies somewhere close to the ancestry of anthropoids, but to say that 'this transitional species finally confirms Charles Darwin's theory of evolution' is ridiculous. Darwin's theory was confirmed long ago, and in any case applies to all living creatures, not just our own close relatives. This fossil has been described as the 'eighth wonder of the world', but the real wonder is the tightly orchestrated and bizarrely exaggerated hype that attended its discovery: 'the most important find for 47 million years'; a 'global event' that 'changes everything'; the 'first ever link to human beings'; the impact of its publication will be 'just like an asteroid hitting the Earth'. Preposterous nonsense, but it is a beautiful fossil which will certainly shed some light on our ancestry, and that is a good enough reason for picturing it here.

(a) The Devonian era when the land waited, full of promise, for the exodus of the fishes from the water. Embodying this huge transition was Canada's prize discovery *Tiktaalik* (b) and (c) – like all 'missing' links, just waiting to be found. But not all the animals who discovered the land stayed there: manatees (d, with babies) and dugongs (e) – together called sirenians, because of their alleged resemblance to mermaids, as perceived by frustrated sailors – returned to the water. Some groups – as suggested by the fascinating *Odontochelys semitestacea*, the proto-turtle without a top shell (f) – having returned to the water, may even have made another return to the land later.

a

b

c

d

e

f

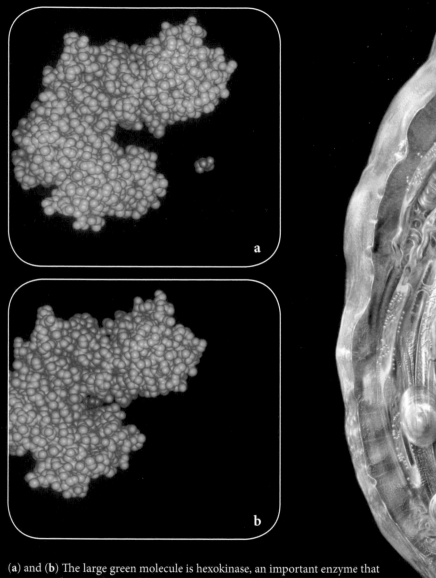

(a) and (b) The large green molecule is hexokinase, an important enzyme that processes glucose (the small brown molecule) by adding phosphate to it. The open 'jaws' (the 'active site' of the enzyme) in (a) clamp shut on the glucose (b), hold it while phosphate is added, then let it go. (c) Even a single cell is breathtakingly complicated. Far from being just a bag full of juice, the cell is packed with elaborate membranous machines and molecular conveyor belts. The key to understanding how such complexity is put together is that it is all done locally, by small entities obeying *local* rules.

c

14

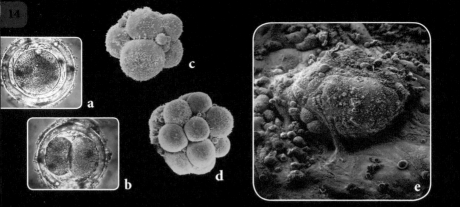

Stages in human development. The fertilized
egg cell or zygote (a) splits into two (b), then
four, then eight (c), then sixteen (d), all without
any increase in total size. At ten days, the
embryo implants into the wall of the uterus (e).
At twenty-two days the neural tube begins to
form (f). At twenty-four days (g) the embryo
resembles a tiny fish. At twenty-five days (h) the
face is forming. The little holes near the back of
the head are the embryonic ears.

g

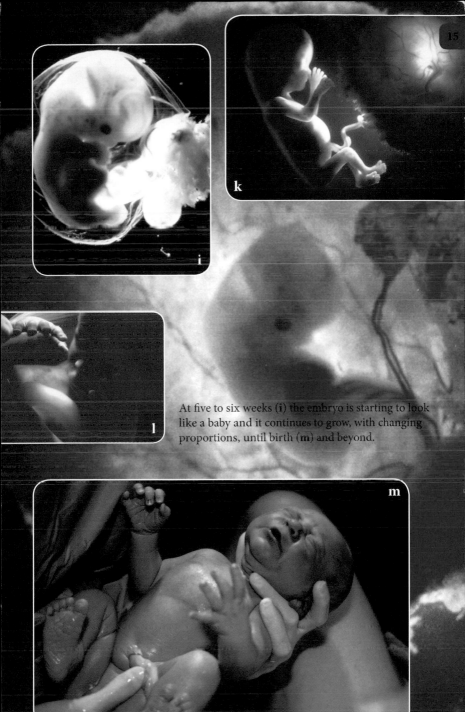

At five to six weeks (**i**) the embryo is starting to look like a baby and it continues to grow, with changing proportions, until birth (**m**) and beyond.

Surely one of the wonders of the world.
Starlings flocking in winter over Otmoor,
near Oxford. Group mind? No, local units
obeying local rules.

provided for them? Preferably two backgrounds, with a different population on each? The aim is to do something like the selection of two lines of maize plants for high and low oil content that we saw in Chapter 3. But in these experiments the selection will be done not by humans but by predators and by female guppies. The only thing that will separate the two experimental lines is the different backgrounds that we shall supply.

Take some animals of a camouflaged species, perhaps a species of insect, and assign them randomly to different cages (or enclosures, or ponds, whatever is suitable) which have differently coloured, or differently patterned, backgrounds. For example, you might give half the enclosures a green foresty background and the other half a reddish-brown, deserty background. Having put your animals in their green or brown enclosures, you'd then leave them to live and breed for as many generations as you have time for, after which you'd come back to see whether they had evolved to resemble their backgrounds, green or brown respectively. Of course, you only expect this result if you put predators in the enclosure too. So, let's put, say, a chameleon in. In *all* the enclosures? No, of course not. This is an experiment, remember; so you'd put a predator in half the green enclosures and half the red enclosures. The experiment would be to test the prediction that, in enclosures with a predator, the insects would evolve to become either green or brown – to become more similar to their background. But in the enclosures without a predator, they might if anything evolve to become more different from their background, to be conspicuous to females.

I have long nursed an ambition to do exactly this experiment with fruit flies (because their reproductive turnover time is so short) but, alas, I never got around to it. So I am especially delighted to say that this is exactly what John Endler did, not with insects but with guppies. Obviously he didn't use chameleons for predators, but instead chose a fish called the pike cichlid (pronounced 'sick lid'), *Crenicichla alta*, which is a dangerous predator of these guppies in the wild. Nor did he use green versus brown backgrounds – he opted for something more interesting than that. He noticed that guppies derive much of their camouflage from their spots, often quite large ones, whose patterning resembles the patterning of the gravelly bottoms of their native streams. Some streams have coarser, more pebbly gravel, others finer, more sandy gravel. Those were the two backgrounds he used, and you'll agree that the camouflage he was seeking was subtler and more interesting than my green versus brown.

Endler got a large greenhouse, to simulate the tropical world of the guppies, and set up ten ponds inside it. He put gravel on the bottom of all ten ponds, but five of them had coarse, pebbly gravel and the other five had finer, sandy gravel. You can see where this is going. The prediction is that, when exposed to strong predation, the guppies on the two backgrounds will diverge from each other over evolutionary time, each in the direction of matching its own background. Where predation is weak or non-existent, the prediction is that the males should tend in the direction of becoming more conspicuous, to appeal to females.

Instead of putting predators in half the ponds and no predators in the other half, again Endler did something more subtle. He had three levels of predation. Two ponds (one fine and one coarse gravel) had no predators at all. Four ponds (two fine and two coarse gravel) had the dangerous pike cichlid. In the remaining four ponds, Endler introduced another species of fish, *Rivulus hartii*, which, despite its English name, 'killifish' (actually that's quite irrelevant since it is named after a Mr Kille), is relatively harmless to guppies. It is a 'weak predator', whereas the pike cichlid is a strong predator. The 'weak predator' situation is a better control condition than no predators at all. This is because, as Endler explains, he was trying to simulate two natural conditions, and he knows of no natural streams that are totally free of predators: thus the comparison between strong and weak predation is a more natural comparison.

So, here's the set-up: guppies were assigned randomly to ten ponds, five with coarse gravel and five with fine gravel. All ten colonies of guppies were allowed to breed freely for six months with no predators. At this point the experiment proper began. Endler put one 'dangerous predator' into each of two coarse gravel ponds and two fine gravel ponds. He put six 'weak predators' (six rather than one, to give a closer approximation to the relative densities of the two kinds of fish in the wild) into each of two coarse gravel ponds and two fine gravel ponds. And the remaining two ponds just carried on as before, with no predators at all.

After the experiment had been running for five months, Endler took a census of all the ponds, and counted and measured the spots on all the guppies in all the ponds. Nine months later, that is, after fourteen months in all, he took another census, counting and measuring in the same way. And what of the results? They were spectacular, even after so short a time. Endler used various measures of the fishes' colour patterns, one of which was 'spots per fish'. When the guppies were first put into their ponds, before the predators were introduced, there was a very large range of spot numbers, because the fish had been gathered from a wide variety of streams, of widely varying predator content. During the six months before any predators were introduced, the mean number of spots per fish shot up. Presumably this was in response to selection by females. Then, at the point when the predators were introduced, there was a dramatic change. In the four ponds that had the dangerous predator, the mean number of spots plummeted. The difference was fully apparent at the five-month census, and the number of spots had declined even further by the fourteen-month census. But in the two ponds with no predators, and the four ponds with weak predation, the number of spots continued to increase. It reached a plateau as early as the five-month census, and stayed high for the fourteen-month census. With respect to spot number, weak predation seems to be pretty much the same as no predation, over-ruled by sexual selection by females who prefer lots of spots.

So much for spot number. Spot size tells an equally interesting story. In the presence of predators, whether weak or strong, coarse gravel promoted relatively larger spots, while fine gravel favoured relatively smaller spots. This is easily interpreted as spot size mimicking stone size. Fascinatingly, however, in the ponds where there were no predators at all, Endler found exactly the reverse. Fine gravel favoured large spots on male guppies, and coarse gravel favoured small spots. They are more conspicuous if they do *not* mimic the stones on their respective backgrounds, and that is good for attracting females. Neat!

Yes, neat. But that was in the lab. Could Endler get similar results in the wild? Yes. He went to a natural stream that contained the dangerous pike cichlids, in which the male guppies were all relatively inconspicuous. He caught guppies of both sexes and transplanted them to a tributary of the same stream that contained no guppies and no dangerous predators, although the weak predator killifish were present. He left them there to get on with living and breeding, and went away. Twenty-three months later, he returned and re-examined the guppies to see what had happened. Amazingly, after less than two years, the males had shifted noticeably in the direction of being more brightly coloured – pulled by females, no doubt, and freed to go there by the absence of dangerous predators.

One of the nice things about science is that it is a public activity. Scientists publish their methods as well as their conclusions, which means that anybody else, anywhere in the world, can repeat their work. If they don't get the same results, we want to know the reason why. Usually they don't just repeat previous work but extend it: carry it further. John Endler's brilliant research on guppies was just begging to be continued and extended. Among those who have taken it up is David Reznick of the University of California at Riverside.

Nine years after Endler sampled his experimental stream with such spectacular results, Reznick and his colleagues revisited the place and sampled the descendants of Endler's experimental population yet again. The males were now very brightly coloured. The female-driven trend that Endler observed had continued, with a vengeance. And that wasn't all. You remember the silver foxes of Chapter 3, and how artificial selection for one characteristic (tameness) pulled along in its wake a whole cluster of others: changes in breeding season, in ears, tail, coat colour and other things? Well, a similar thing happened with the guppies, under natural selection.

Reznick and Endler had already noticed that when you compare guppies in predator-infested streams with guppies in streams with only weak predation, colour differences are only the tip of the iceberg. There is a whole cluster of other differences. Guppies from low-predation streams reach sexual maturity later than those from high-predation streams, and they are larger when they reach adulthood; they produce litters of young less frequently, and their litters are smaller, with larger offspring. When Reznick examined the descendants of Endler's guppies, his findings were almost too good to be true. The ones that had been freed to follow female-driven sexual selection rather than predator-driven selection for individual survival had not only become more brightly coloured: in all the other respects I have just listed, these fish had evolved the full cluster of other changes, to match those normally found in wild populations free from predators. The guppies matured at a later age than in predator-infested streams, they were larger, and they produced fewer and larger offspring. The balance had shifted towards the norm for predator-free pools, where sexual attractiveness takes priority. And it all happened staggeringly fast, by evolutionary standards. Later in the book we shall see that the evolutionary change witnessed by Endler and Reznick, driven purely by natural selection (strictly including sexual selection), raced ahead at a speed comparable to that achieved by artificial selection of domestic animals. It is a spectacular example of evolution before our very eyes.

One of the surprising things we have learned about evolution is that it can be both very fast – as we have seen in this chapter – and, under other circumstances, as we know from the fossil record, very slow. Slowest of all are those living creatures that we call 'living fossils'. They are not literally brought back from the dead like Lenski's frozen bacteria. But they are creatures that have changed so little since their remote ancestors that it is almost as though they were fossils.

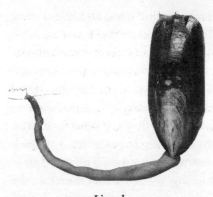

Lingula

My favourite living fossil is the brachiopod *Lingula*. You don't need to know what a brachiopod is. They would surely have been staples on the menu, had seafood restaurants flourished before the great Permian extinction a quarter of a billion years ago – the most catastrophic extinction of all time. A superficial glance might confuse them with bivalve molluscs – mussels and their kind – but they are really very different. Their two shells are top and bottom, where mussels' shells are left and right. In evolutionary history bivalves and brachiopods were, as Stephen Jay Gould memorably put it, ships that pass in the night. A few brachiopods survived 'the Great Dying' (Gould's phrase again), and modern *Lingula* (above) is so similar to *Lingulella*, the fossil below, that the fossil was originally given the same generic name, *Lingula*. This particular specimen of *Lingulella* goes back to the Ordovician era, 450 million years ago. But there are fossils, also originally named *Lingula* and now known as *Lingulella*, going back more than half a billion years to the Cambrian era. I should admit, however, that a fossilized shell is not a lot to go on, and some zoologists dispute *Lingula*'s claim to be an almost wholly unchanged 'living fossil'.

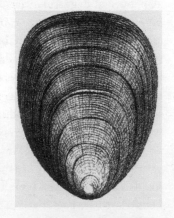

Lingulella – almost identical to its modern relatives

Many of the problems that we meet in evolutionary argumentation arise only because animals are inconsiderate enough to evolve at different rates, and might even be inconsiderate enough not to evolve at all. If there were a law of nature dictating that quantity of evolutionary change must always be obligingly proportional to elapsed time, degree of resemblance would faithfully reflect closeness of cousinship. In the real world, however, we have to contend with evolutionary sprinters like birds, who leave their reptile origins standing in the Mesozoic dust – helped, in our perception of their uniqueness, by the happenstance that their neighbours in the evolutionary tree were all killed by a celestial catastrophe. At the other extreme, we have to contend with 'living fossils' like *Lingula* which, in extreme cases, have changed so little that they might almost interbreed with their remote ancestors, if only a matchmaking time-machine could procure them a date.

Lingula is not the only famous example of a living fossil. Others include *Limulus*, the horseshoe 'crab', and coelacanths, which we shall meet in the next chapter.

CHAPTER 6

MISSING LINK? WHAT DO YOU MEAN, 'MISSING'?

CREATIONISTS are deeply enamoured of the fossil record, because they have been taught (by each other) to repeat, over and over, the mantra that it is full of 'gaps': 'Show me your "intermediates"!' They fondly (very fondly) imagine that these 'gaps' are an embarrassment to evolutionists. Actually, we are lucky to have any fossils at all, let alone the massive numbers that we now do have to document evolutionary history – large numbers of which, by any standards, constitute beautiful 'intermediates'. I shall emphasize in Chapters 9 and 10 that we don't need fossils in order to demonstrate that evolution is a fact. The evidence for evolution would be entirely secure, even if not a single corpse had ever fossilized. It is a bonus that we do actually have rich seams of fossils to mine, and more are discovered every day. The fossil evidence for evolution in many major animal groups is wonderfully strong. Nevertheless there are, of course, gaps, and creationists love them obsessively.

Let's again make use of our analogy of the detective coming to the scene of a crime to which there were no eye witnesses. The baronet has been shot. Fingerprints, footprints, DNA from a sweat stain on the pistol, and a strong motive all point towards the butler. It's pretty much an open and shut case, and the jury and everybody in the court is convinced that the butler did it. But a last-minute piece of evidence is discovered, in the nick of time before the jury retires to consider what had seemed to be their inevitable verdict of guilty: somebody remembers that the baronet had installed spy cameras against burglars. With bated breath, the court watches the films. One of them shows the butler in the act of opening the drawer in his pantry, taking out a pistol, loading it, and creeping stealthily out of the room with a malevolent gleam in his eye. You might think that this solidifies the case against the butler even further. Mark the sequel, however. The butler's defence lawyer astutely points out that there was no spy camera in the library where the murder took place, and no spy camera in the corridor leading from the butler's pantry. He wags his finger, in that compelling way that lawyers have made their own. 'There's a *gap* in the video record! We don't know what happened after the butler left the pantry. There is clearly insufficient evidence to convict my client.'

In vain the prosecution lawyer points out that there was a second camera in the billiard room, and this shows, through the open door, the butler, gun at the ready, creeping on tiptoe along the passage towards the library. Surely this plugs the gap in the video record? Surely the case against the butler is now unassailable? But no. Triumphantly the defence lawyer plays his ace. 'We don't know what happened before or after the butler passed the open door of the billiard room. There are now *two* gaps in the video record. Ladies and gentlemen of the jury, my case rests. There is now even less evidence against my client than there was before.'

The fossil record, like the spy camera in the murder story, is a *bonus*, something that we had no right to expect as a matter of entitlement. There is already more than enough evidence to convict the butler without the spy camera, and the jury were about to deliver a guilty verdict before the spy camera was discovered. Similarly, there is more than enough evidence for the fact of evolution in the comparative study of modern species (Chapter 10) and their geographical distribution (Chapter 9). We don't *need* fossils – the case for evolution is watertight without them; so it is paradoxical to use *gaps* in the fossil record as though they were evidence against evolution. We are, as I say, lucky to have fossils at all.

What *would* be evidence against evolution, and very strong evidence at that, would be the discovery of even a single fossil in the wrong geological stratum. I have already made this point in Chapter 4. J. B. S. Haldane famously retorted, when asked to name an observation that would disprove the theory of evolution, 'Fossil rabbits in the Precambrian!' No such rabbits, no authentically anachronistic fossils of any kind, have ever been found. All the fossils that we have, and there are very very many indeed, occur, without a single authenticated exception, in the right temporal sequence. Yes, there are gaps, where there are no fossils at all, and that is only to be expected. But not a single solitary fossil has ever been found *before* it could have evolved. That is a very telling fact (and there is no reason why we should expect it on the creationist theory). As I briefly mentioned in Chapter 4, a good theory, a scientific theory, is one that is vulnerable to disproof, yet is not disproved. Evolution could so easily be disproved if just a single fossil turned up in the wrong date order. Evolution has passed this test with flying colours. Sceptics of evolution who wish to prove their case should be diligently scrabbling around in the rocks, desperately trying to find anachronistic fossils. Maybe they'll find one. Want a bet?

The biggest gap, and the one the creationists like best of all, is the one that preceded the so-called Cambrian Explosion. A little more than half a billion years ago, in the Cambrian era, most of the great animal phyla – the main divisions within the animal world – 'suddenly' appear in the fossil record. Suddenly, that is, in the sense that no fossils of these animal groups are known in rocks older than the Cambrian, not suddenly in the sense of instantaneously: the period we are talking about covers about 20 million years. Twenty million years feels short when it is half a billion years ago. But of course it represents exactly the same amount of time for evolution as 20 million years today! Anyway, it is still quite sudden, and, as I wrote in a previous book, the Cambrian shows us a substantial number of major animal phyla

> already in an advanced state of evolution, the very first time
> they appear. It is as though they were just planted there, without

any evolutionary history. Needless to say, this appearance of
sudden planting has delighted creationists.

That last sentence shows that I was savvy enough to realize that
creationists would like the Cambrian Explosion. I was not (back
in 1986) savvy enough to realize that they'd gleefully quote my
lines back at me in their own favour, over and over again, carefully
omitting my careful words of explanation. On a whim, I just searched
the World Wide Web for 'It is as though they were just planted there,
without any evolutionary history' and obtained no fewer than 1,250
hits. As a crude control test of the hypothesis that the majority of
these hits represent creationist quote-minings, I tried searching,
as a comparison, for the clause that immediately follows the above
quotation in *The Blind Watchmaker*: 'Evolutionists of all stripes
believe, however, that this really does represent a very large gap in
the fossil record'. I obtained a grand total of 63 hits, compared to the
1,250 hits for the previous sentence. The ratio of 1250 to 63 is 19.8.
We might call this ratio the Quote Mining Index.

I have dealt with the Cambrian Explosion at length, especially in
Unweaving the Rainbow. Here I'll add just one new point, illustrated
by the flatworms, Platyhelminthes. This great phylum of worms
includes the parasitic flukes and tapeworms, which are of great medical
importance. My favourites, however, are the free-living turbellarian
worms, of which there are more than four thousand species: that's
about as numerous as all the mammal species put together. Some of
these turbellarians are creatures of great beauty, as the two pictured
opposite show. They are common, both in water and on land, and
presumably have been common for a very long time. You'd expect,
therefore, to see a rich fossil history. Unfortunately, there is almost
nothing. Apart from a handful of ambiguous trace fossils, not a
single fossil flatworm has ever been found. The Platyhelminthes,
to a worm, are 'already in an advanced state of evolution, the very
first time they appear. It is as though they were just planted there,
without any evolutionary history.' But in this case, 'the very first

Turbellarians – no fossil record, but they must have been there all along

time they appear' is not the Cambrian but today. Do you see what this means, or at least ought to mean for creationists? Creationists believe that flatworms were created in the same week as all other creatures. They have therefore had exactly the same time in which to fossilize as all other animals. During all the centuries when all those bony or shelly animals were depositing their fossils by the thousands, the flatworms must have been living happily alongside them, but without leaving any significant trace of their presence in the rocks. What, then, is so special about gaps in the record of those animals that *do* fossilize, given that the past history of the flatworms amounts to *one big gap*: even though the flatworms, by the creationists' own account, have been living for the same length of time? If the gap before the Cambrian Explosion is used as evidence that most animals suddenly sprang into existence in the Cambrian, exactly the same 'logic' should be used to prove that the flatworms sprang into existence yesterday. Yet this contradicts the creationist's belief that flatworms were created during the same creative week as everything else. You cannot have it both ways. This argument, at a stroke, completely destroys the creationist case that the Precambrian gap in the fossil record weakens the evidence for evolution.

Why, on the evolutionary view, are there so few fossils before the Cambrian era? Well, presumably, whatever factors applied to the flatworms throughout geological time to this day, those same factors applied to the rest of the animal kingdom before the Cambrian. Probably, most animals before the Cambrian were soft-bodied like modern flatworms, probably also rather small like modern turbellarians – just not good fossil material. Then something happened half a billion years ago to allow animals to fossilize freely – the arising of hard, mineralized skeletons, for example.

An earlier name for 'gap in the fossil record' was 'missing link'. The phrase enjoyed a vogue in late Victorian England, and persisted into the twentieth century. Inspired by a misunderstanding of Darwin's theory, it was used as an insult in roughly the same way as 'neanderthal' is colloquially (and unjustly) used today. Among the *Oxford English Dictionary*'s list of representative quotations is a 1930 one in which D. H. Lawrence tells of a woman who wrote to say his name 'stank' and went on, 'You, who are a mixture of the missing-link and the chimpanzee.'

The original meaning, a confused one as I shall show, implied that the Darwinian theory lacked a vital link between humans and other primates. Another of the dictionary's illustrative quotations, a Victorian one, uses it like this: 'I've heard talk o' some missing link, atween men and puggies' ('puggie' was a Scottish dialect word for monkey). History-deniers, to this day, are very fond of saying, in what they imagine is a taunting tone of voice: 'But you still haven't found the missing link,' and they often throw in a jibe about Piltdown Man, for good measure. Nobody knows who perpetrated the Piltdown hoax, but he has a lot to answer for.* The fact that one of the first candidates for a man-ape fossil to be discovered turned out to be a hoax provided an excuse for history-deniers to ignore the very numerous fossils that are not hoaxes; and they still haven't stopped crowing about it. If only they would look at the facts, they'd soon discover that we now have a rich supply of intermediate fossils linking modern humans to the common ancestor that we share with chimpanzees. On the human side of the divide, that is. Interestingly, there are as yet no fossils

---

* Majority opinion suspects the amateur palaeontologist Charles Dawson, but Stephen Jay Gould intriguingly floated the alternative theory that it might have been Pierre Teilhard de Chardin. You may recognize Teilhard's name as the Jesuit theologian whose later book, *The Phenomenon of Man*, was to receive the greatest negative book review of all time, from the matchless Peter Medawar (reprinted in *The Art of the Soluble* and *Pluto's Republic*).

linking that ancestor (which was neither chimpanzee nor human) to modern chimpanzees. Perhaps this is because chimpanzees live in forests, which don't provide good fossilizing conditions. If anything it is chimpanzees, not humans, who today have a right to complain of missing links!

That, then, is one meaning of 'missing link'. It is the alleged gap between humans and the rest of the animal kingdom. The missing link in that sense is, to put it mildly, no longer missing. I shall return to this in the next chapter, which is specifically about human fossils.

Another meaning concerns the alleged paucity of so-called 'transitional forms' between major groups: between reptiles and birds, for example, or between fish and amphibians. 'Produce your intermediates!' Evolutionists often respond to this challenge from history-deniers by throwing them the bones of *Archaeopteryx*, the famous 'intermediate' between 'reptiles' and birds. This is a mistake, as I shall show. *Archaeopteryx* is not the answer to a challenge, because there is no challenge worth answering. To put up a single famous fossil like *Archaeopteryx* panders to a fallacy. In fact, for a large number of fossils, a good case can be made that every one of them is an intermediate between something and something else. The alleged challenge that seems to be answered by *Archaeopteryx* is based on an outdated conception, the one that used to be known as the Great Chain of Being; and that is the title under which I shall deal with it later in this chapter.

The silliest of all these 'missing link' challenges are the following two (or variants of them, of which there are many). First, 'If people came from monkeys via frogs and fish, then why does the fossil record not contain a "fronkey"?' I have seen an Islamic creationist ask truculently why there are no crocoducks. And, second, 'I'll believe in evolution when I see a monkey give birth to a human baby.' This last one makes the same mistake as all the others, plus the additional one of thinking that major evolutionary change happens overnight.

As it happens, two of these fallacies crop up next door to each other in the long list of comments that follow an article in the *Sunday*

Times (London) about a television documentary on Darwin that I presented:

> Dawkins opinion on religion is absurd since Evolution is nothing more than a religion itself – you have to believe we all came from a single cell . . . and that a snail can become a monkey etc. Ha Ha – that's the most laughable religion yet!!
>
> *Joyce, Warwickshire, UK*
>
> Dawkins should explain why science has failed to find the missing links. Faith in unfounded science is more fairy tale stuff than faith in God.
>
> *Bob, Las Vegas, USA*

This chapter will deal with all these related fallacies, beginning with the silliest of all, since the answer to it will serve as an introduction to the others.

'SHOW ME YOUR CROCODUCK!'

'Why doesn't the fossil record contain a fronkey?' Well, of course, monkeys are not descended from frogs. No sane evolutionist ever said they were, or that ducks are descended from crocodiles or vice versa. Monkeys and frogs share an ancestor, which certainly looked nothing like a frog and nothing like a monkey. Maybe it looked a bit like a salamander, and we do indeed have salamander-like fossils dating from the right time. But that is not the point. Every one of the millions of species of animals shares an ancestor with every other one. If your understanding of evolution is so warped that you think we should expect to see a fronkey and a crocoduck, you should also wax sarcastic about the absence of a doggypotamus and an elephanzee. Indeed, why limit yourself to mammals? Why not a kangaroach (intermediate between kangaroo and cockroach), or an octopard (intermediate between octopus and leopard)? There's an infinite number of animal names you can string together in that way.* Of course hippopotamuses are not descended from dogs, or vice versa. Chimpanzees are not descended from elephants or vice versa, just as monkeys are not descended from frogs. No modern species is descended from any other modern species (if we leave out very recent splits). Just as you can find fossils that approximate to the common ancestor of a frog and a monkey, so you can find fossils that approximate to the common ancestor of elephants and chimpanzees. Here is one called *Eomaia*, which lived in the early Cretaceous period, a little more than 100 million years ago.

As you can see, *Eomaia* was nothing like a chimpanzee and nothing like an elephant. Vaguely like a shrew, it probably was pretty similar to their common ancestor, with which it was roughly contemporary, and you can see that a lot of evolutionary change has taken place along both pathways from an *Eomaia*-like ancestor to an elephant descendant, and from the same *Eomaia*-like ancestor to a chimpanzee descendant. But it is not in any sense an elephanzee. If it were, it would also have to

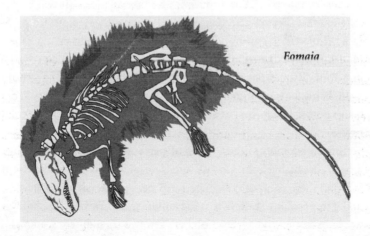

Eomaia

* I'm using 'infinite' in the common, often abused, rhetorical sense of very very large. The actual number is the number of pairwise combinations of every species with every other – and that's as near infinite as makes no practical difference!

be a dogatee, for whatever is the common ancestor of a chimpanzee and an elephant is also the common ancestor of a dog and a manatee. And it would also have to be an aardvapotamus, for the same ancestor is also the common ancestor of an aardvark and a hippopotamus. The very idea of a dogatee (or an elephanzee, or an aardvapotamus or a kangaroceros or a buffalion) is deeply unevolutionary and ridiculous. So is a fronkey, and it is a disgrace that the perpetrator of that little witlessism, the Australian itinerant preacher John Mackay, has been touring British schools in 2008 and 2009, masquerading as a 'geologist', teaching innocent children that if evolution were true the fossil record should contain 'fronkeys'.

An equally ludicrous example is to be found in the Muslim apologist Harun Yahya's enormous, lavishly produced, glossily illustrated and fatuously ignorant book *Atlas of Creation*. This book obviously cost a fortune to produce, which makes it all the more astounding that it was distributed free to tens of thousands of science teachers, including me. Notwithstanding the prodigious sums of money spent on this book, the errors in it have become legendary. In the service of illustrating the falsehood that most ancient fossils are indistinguishable from their modern counterparts, Yahya shows a sea snake as an 'eel' (two animals so different that they are placed in different classes of vertebrates), a starfish as a 'brittlestar' (actually different classes of echinoderms), a sabellid (annelid) worm as a crinoid 'sea lily' (an echinoderm: this pair come not just from different phyla but from different sub-kingdoms, so that they could hardly be more distant from each other if they tried, while still both being animals) and – best of all – a fishing lure as a 'caddis fly' (see colour page 8).

But in addition to these gems of partisan risibility, the book has a section on missing links. One picture is seriously offered to illustrate the fact that there is no intermediate form between a fish and a starfish. I find it impossible to believe that the author seriously thinks evolutionists would expect to find a transition between two such differing animals as a starfish and a fish. I therefore cannot help suspecting that he knows his audience all too well, and is deliberately and cynically exploiting their ignorance.

'I'LL BELIEVE IN EVOLUTION WHEN A MONKEY GIVES BIRTH TO A HUMAN BABY'

Once again, humans are not descended from monkeys. We share a common ancestor with monkeys. As it happens, the common ancestor would have looked a lot more like a monkey than a man, and we would indeed probably have called it a monkey if we had met it, some 25 million years ago. But even though humans evolved from an ancestor that we could sensibly call a monkey, no animal gives birth to an instant new species, or at least not one as different from itself as a man is from a monkey, or even from a chimpanzee. That isn't what evolution is about. Evolution not only is a gradual process as a matter of fact; it *has* to be gradual if it is to do any explanatory work. Huge leaps in a single generation – which is what a monkey giving birth to a human would be – are almost as unlikely as divine creation, and are ruled out for the same reason: too statistically improbable. It would be so nice if those who oppose evolution would take a tiny bit of trouble to learn the merest rudiments of what it is that they are opposing.

THE PERNICIOUS LEGACY OF THE GREAT CHAIN OF BEING

Underlying much of the fallacious demand for 'missing links' is a medieval myth, which occupied men's minds right up to the age of Darwin and stubbornly confused them after it. This is the myth of the Great Chain of Being, according to which everything in the universe sat on a ladder, with God at the top, then archangels, then various ranks of angels, then human beings, then animals, then plants, then down to stones and other inanimate creations. Given that this goes way back to a time when racism was second nature, I hardly need add that human beings were not all sitting on the same rung. Oh no. And of course males were a healthy rung above females of their kind (which was why I let myself get away with 'occupied *men's* minds' in the opening sentence of this section). But it was the alleged hierarchy within the animal kingdom that had the greatest capacity to muddy the waters when the idea of evolution burst upon the scene. It seemed natural to suppose that 'lower' animals evolved into 'higher' animals. And if this were so, we should expect to see 'links' between them, all the way up and down the 'ladder'. A ladder with lots of missing rungs lacks conviction. It is this image of the rungless ladder that lurks behind much of the scepticism about 'missing links'. But the entire ladder myth is deeply misconceived and un-evolutionary, as I shall now show.

So glibly do the phrases 'higher animals' and 'lower animals' trip off our tongues that it comes as a shock to realize that, far from effortlessly slotting into evolutionary thinking as one might suppose, they were – and are – deeply antithetical to it. We think we know that chimpanzees are higher animals and earthworms are lower, we think we've always known what that means, and we think evolution makes it even clearer. But it doesn't. It is by no means clear that it means anything at all. Or if it means anything, it means so many different things as to be misleading, even pernicious.

Here is a list of the more or less distinctly confusing things you might mean when you say, for example, that a monkey is 'higher' than an earthworm.

1   *'Monkeys evolved from earthworms.'* This is false, just as it is false that humans evolved from chimpanzees. Monkeys and earthworms share a common ancestor.

2   *'The common ancestor of monkeys and earthworms was more like an earthworm than like a monkey.'* Well, that potentially makes more sense. You can even use the word 'primitive' in a semi-precise way, if you define it as 'resembling ancestors', and it is obviously true that some modern animals are more primitive in this sense than others. What that exactly means, if you think about it, is that the more primitive of a pair of species has changed less since the common ancestor (*all* species, without exception, share a common ancestor if you go back far enough). If neither

species has changed dramatically more than the other, the word 'primitive' should not be used in comparing them.

It's worth pausing here to make a related point. It is hard to measure degrees of resemblance. And there is in any case no *necessary* reason why the common ancestor of two modern animals should be more like one than the other. If you take two animals, say a herring and a squid, it is *possible* that one of them resembles the common ancestor more than the other, but it doesn't follow that this *has* to be the case. There has been an exactly equal amount of time for both to have diverged from the ancestor, so the prior expectation of an evolutionist might be, if anything, that no modern animal should be more primitive than any other. We might expect both of them to have changed to the same extent, but in different directions, since the time of the shared ancestor. This expectation, as it happens, is often violated (as in the case of monkey and earthworm), but there is no necessary reason why we should expect it to be. Moreover, the different parts of animals don't all have to evolve at the same rate. An animal might be primitive from the waist down but highly evolved from the waist up. Less facetiously, one of them might be more primitive in its nervous system, the other more primitive in its skeleton. Notice especially that 'primitive' in the sense of 'resembling ancestors' does not have to go with 'simple' (meaning less complex). A horse's foot is simpler than a human foot (it has only a single digit instead of five, for example), but the human foot is more primitive (the ancestor that we share with horses had five digits, as we do, so the horse has changed more). This leads us on to the next item in our list.

3   '*Monkeys are cleverer [or prettier, have larger genomes, more complicated body plans, etc. etc.] than earthworms.*' This kind of zoological snobbery is a mess when you start trying to apply it scientifically. I mention it only because it is so readily confused with the other meanings, and the best way to sort out confusion is to expose it. You could imagine a large number of scales along

which you might rank animals – not just the four scales I have mentioned. Animals that are high on one of these ladders may or may not be high on another. Mammals certainly have larger brains than salamanders, but they have smaller genomes than some salamanders.

4 *'Monkeys are more like humans than earthworms are.'* This is undeniable for the particular example of monkeys and earthworms. But so what? Why should we choose humans as the standard against which we judge other organisms? An indignant leech might point out that earthworms have the great virtue of being more like leeches than humans are. Despite the Great Chain of Being's traditional ranking of humans between animals and angels, there is no evolutionary justification for the common assumption that evolution is somehow 'aimed' at humans, or that humans are 'evolution's last word'. It is remarkable how commonly this vainglorious assumption thrusts itself forward. At its crudest level, you meet it in the ubiquitously querulous, 'If chimps evolved into us, how come there are still chimps around?' I've already mentioned this, and I'm not joking. I meet this question again and again and again, sometimes from apparently well-educated people.*

5 *'Monkeys [and other 'higher' animals] are better at surviving than earthworms [and other 'lower' animals].'* This doesn't even begin to be sensible, or even true. All living species have survived at least into the present. Some monkeys, such as the exquisite golden tamarin, are in danger of going extinct. They are much less good at surviving than earthworms are. Rats and

---

* 'Well-educated' reminds me of Peter Medawar's wickedly astute observation that 'the spread of secondary and latterly of tertiary education has created a large population of people, often with well-developed literary and scholarly tastes, who have been educated far beyond their capacity to undertake analytical thought'. Isn't that priceless? It is the kind of writing that makes me want to rush out into the street to share with somebody – anybody – because it is too good to keep to oneself.

cockroaches flourish, despite being regarded by many people as 'lower' than gorillas and orang-utans, which are perilously close to extinction.

I hope I've said enough to show what nonsense it is to rank modern species on a ladder, as though it were obvious what you meant by 'higher' and 'lower', and to show how thoroughly unevolutionary it is. You can imagine lots and lots of ladders; it might sometimes be sensible to rank animals on at least some of the ladders separately, but the ladders are not well correlated with each other, and none of them has any right to be called an 'evolutionary scale'. We have seen the historical temptation to crude errors such as 'Why aren't there any fronkeys?' But the pernicious legacy of the Great Chain of Being also feeds the challenge 'Where are the intermediates between major animal groups?' and, nearly as discreditably, underlies the tendency of evolutionists to *answer* such a challenge by trotting out particular fossils, such as *Archaeopteryx*, the celebrated 'intermediate between reptiles and birds'. Nevertheless, there is something else going on underneath the *Archaeopteryx* fallacy, and it is of general importance; so I shall give it a couple of paragraphs, using *Archaeopteryx* as a particular example of a general case.

Zoologists have traditionally divided the vertebrates into classes: major divisions with names like mammals, birds, reptiles and amphibians. Some zoologists, called 'cladists',* insist that a proper class must consist of animals all of whom share a common ancestor which belonged to that class and which has no descendants outside that group. The birds would be an example of a good class.† All birds

---

* From the term 'clade', meaning a group of organisms believed to comprise all the evolutionary descendants of a common ancestor.

† At least according to a consensus of zoologists, and I shall continue to use the birds, for the sake of argument, as an example of a good class. Recent fossil research is showing up a number of feathered dinosaurs, and it is open to somebody to claim that some of the modern animals we call birds are descended from a different group of feathered dinosaurs than others. If the most recent common ancestor of all modern birds turns out to be an animal that would not be classified as a bird, I would have to revise my statement that the birds constitute a good class.

are descended from a single ancestor that would also have been called a bird and would have shared with modern birds the key diagnostic characters – feathers, wings, a beak, etc. The animals commonly called reptiles are not a good class in this sense. This is because, at least in conventional taxonomies, the category explicitly *excludes* birds (they constitute their own class) and yet some 'reptiles' as conventionally recognized (e.g. crocodiles and dinosaurs) are closer cousins to birds than they are to other 'reptiles' (e.g. lizards and turtles). Indeed, some dinosaurs are closer cousins to birds than they are to other dinosaurs. 'Reptiles', then, is an artificial class, because birds are artificially *excluded*. In a strict sense, if we were to make reptiles a truly natural class, we should have to include birds as reptiles. Cladistically inclined zoologists avoid the word 'reptiles' altogether, splitting them into Archosaurs (crocodiles, dinosaurs and birds), Lepidosaurs (snakes, lizards and the rare *Sphenodon* of New Zealand) and Testudines (turtles and tortoises). Non-cladistically inclined zoologists are happy to use a word like 'reptile' because they find it descriptively useful, even if it does artificially exclude the birds.

But what is it about the birds that tempts us to hive them off from the reptiles? What is it that seems to justify bestowing on birds the accolade of 'class', when they are, evolutionarily speaking, just one branch within reptiles? It is the fact that the immediately surrounding reptiles, birds' close neighbours on the tree of life, happen to be extinct, while the birds, alone of their kind, marched on. The closest relatives of birds are all to be found among the long-extinct dinosaurs. If a wide variety of dinosaur lineages had survived, birds would not stand out: they would not have been elevated to the status of their own class of vertebrates, and we would not be asking any such question as 'Where are the missing links between reptiles and birds?' *Archaeopteryx* would still be a nice fossil to have in your museum, but it would not play its present starring role as the stock answer to (what we can now see is) an empty challenge: 'Produce your intermediates.' If the cards of extinction had fallen differently, there would just be lots of dinosaurs running about, including some feathered, flying, beaked dinosaurs called birds. And indeed, fossilized feathered dinosaurs are now increasingly being discovered, so it is becoming vividly clear that there really is no major 'Produce your missing link!' challenge to which *Archaeopteryx* is the reply.

Let's proceed, now, to some of the major transitions in evolution, where 'links' have been alleged to be 'missing'.

UP FROM THE SEA

Short of rocketing into space, it is hard to imagine a bolder or more life-changing step than leaving the water for dry land. The two life-zones are different in so many ways that moving from one to the other demands a radical shift in almost all parts of the body. Gills that are good at extracting oxygen from water are all but useless in air, and lungs are useless in water. Methods of propulsion that are speedy, graceful and efficient in water are dangerously clumsy on land, and vice versa. No wonder 'fish out of water' and 'like a drowning man' have both become proverbial phrases. And no wonder 'missing links' in this region of the fossil record are of more than ordinary interest.

If you go back far enough, everything lived in the sea – watery, salty *alma mater* of all life. At various points in evolutionary history, enterprising individuals from many different animal groups moved out on to the land, sometimes eventually to the most parched deserts, taking their own private sea water with them in blood and cellular fluids. In addition to the reptiles, birds, mammals and insects we see all around us, other groups that have succeeded in making the great trek out of life's watery womb include scorpions, snails, crustaceans such as woodlice and land crabs, millipedes and centipedes, spiders and their kin, and at least three phyla of worms. And we mustn't forget the plants, onlie begetters of usable carbon, without whose prior invasion of the land none of the other migrations could have happened.

Fortunately the transitional stages of our exodus, as fish emerged on to the land, are beautifully documented in the fossil record. So are the transitional stages going the other way much later, as the ancestors of whales and dugongs forsook their hard-won home on dry land and returned to their ancestral seas. In both cases, links that once were missing now abound and grace our museums.

When we say that 'fish' emerged on to the land, we have to remember that 'fish', like 'reptiles', do not constitute a natural group. Fish are defined by exclusion. Fish are all the vertebrates except those that moved on to the land. Because all the early evolution of vertebrates took place in water, it is not surprising that most of the surviving branches of the vertebrate tree are still in the sea. And we still call them 'fish' even when they are only distantly related to other 'fish'. Trout and tuna are closer cousins to humans than they are to sharks, but we call them all 'fish'. And lungfish and coelacanths are closer cousins to humans than they are to trout and tuna (and of course sharks) but, again, we call them 'fish'. Even sharks are closer cousins to humans than they are to lampreys and hagfish (the only modern survivors of the once thriving and diverse group of jawless fishes) but again, we call them all fish. Vertebrates whose ancestors never ventured on to land all look like 'fish', they all swim like fish (unlike dolphins, which swim with an up-and-down bending of the spine instead of side to side like a fish), and they all, I suspect, taste like fish.

To an evolutionist, as we just saw in the example of reptiles and birds, a 'natural' group of animals is a group all of whose members are closer cousins to each other than they are to all non-members of the group. 'Birds', as we saw, are a natural group, since they share a most recent common ancestor that is not shared by any non-bird. By the same definition, 'fish' and 'reptiles' are not natural groups. The most recent common ancestor of all 'fish' is shared by many non-fish too. If we push our distant cousins the sharks to one side, we mammals belong to a natural group that includes all modern bony fish (bony as opposed to cartilaginous sharks). If we then push to one side the bony 'ray-finned fishes' (salmon, trout, tuna, angel fish: just about all the fish you are likely to see that are not sharks), the natural group to which we belong includes all land vertebrates plus the so-called lobe-finned fishes. It is from the ranks of the lobe-finned fishes that we sprang, and we must now pay special attention to the lobefins.

Lobefins today have dwindled to the lungfishes and the coelacanths ('dwindled' as 'fish', that is, but mightily expanded on land: we land vertebrates are aberrant lungfish). They are 'lobefins' because their fins are like legs rather than the ray fins of familiar fishes. Indeed, *Old Fourlegs* was the title of a popular book on coelacanths written by J. L. B. Smith, the South African biologist most responsible for bringing them to the world's attention after the first live one was dramatically discovered in 1938 in the catch of a South African trawler. 'I would not have been more surprised if I had seen a dinosaur walking down the street.' Coelacanths had been known before, as fossils, but they had been thought extinct since the time of the dinosaurs. Smith movingly wrote of the moment when he first cast eyes on this astonishing find, to which he had been summoned by its discoverer, Margaret Latimer (he later named it *Latimeria*), to give his expert opinion:

> We went straight to the Museum. Miss Latimer was out for the moment, the caretaker ushered us into the inner room and there was the – Coelacanth, yes, God! Although I had come prepared, that first sight hit me like a white-hot blast and made me feel shaky and queer, my body tingled. I stood as if stricken to stone. Yes, there was not a shadow of doubt, scale by scale, bone by bone, fin by fin, it was a true Coelacanth. It could have been one of those creatures of 200 million years ago come alive again. I forgot everything else and just looked and looked, and then almost fearfully went close up and touched and stroked, while my wife watched in silence. Miss Latimer came in and greeted us warmly. It was only then that speech came back, the exact words I have forgotten, but it was to tell them that it was true, it was really true, it was unquestionably a Coelacanth. Not even I could doubt any more.

Coelacanths are closer cousins to us than they are to most fish. They have changed somewhat since the time of our shared ancestor, but not enough to be moved out of the category of animals that, colloquially and to a fisherman, would be classified as fish. But they, and lungfish, are definitely closer cousins to us than to trout, tuna and the majority of fish. Coelacanths and lungfish are examples of 'living fossils'.

Nevertheless, we are not descended from lungfish, or from coelacanths. We share an ancestor with lungfish, which looked more like a lungfish than it looked like us. But it didn't look much like either. Lungfish may be living fossils, but they are still not very like our ancestors. In the quest for those, we must instead seek real fossils in the rocks. And in particular we are interested in fossils from the Devonian era that capture the transition between water-dwelling fish and the first vertebrates to live on land. Even among real fossils, we would be too optimistic if we hoped literally to find our ancestors. We can, however, hope to find cousins of our ancestors that are sufficiently close to tell us approximately what they were like.

One of the most famous gaps in the fossil record – conspicuous enough to have been given a name, 'Romer's Gap' (A. S. Romer was a famous American palaeontologist), stretches from about 360 million years ago, at the end of the Devonian period, to about 340 million years ago, in the early part of the Carboniferous, the 'Coal Measures'. After Romer's Gap, we find unequivocal amphibians crawling through the swamps, a rich radiation of salamander-like animals, some of them as large as crocodiles, which they superficially resembled. It seems to have been an age of giants, for there were dragonflies with a wing span as long as my arm, the largest insects that ever lived.* Starting about

---

* It's been suggested, by the way, that this gigantism was made possible by the higher oxygen content in the atmosphere at that time. Insects lack lungs, and they breathe by means of tiny air tubes that pipe air throughout the body. Air tubes can't mount such an intricately comprehensive distribution system as blood tubes can, and it is plausible that this limits body size. That limit would have been higher in an atmosphere with 35% oxygen, instead of the mere 21% that we breathe today. This provides a satisfying explanation for the giant dragonflies, but it may not necessarily be the right one. Incidentally, I'm puzzled why, with so much oxygen about, things didn't burst into flames all the time. Perhaps they did. Forest fires must have been more common than today, and the fossils indicate a high incidence of fire-resistant plant species. It is not certain why the oxygen content of the atmosphere peaked during the Carboniferous and Permian. It may be associated with the sequestering of so much carbon under the ground, as coal.

340 million years ago, we might almost call the Carboniferous the amphibian equivalent of the age of dinosaurs. Before that, however, was Romer's Gap. And before his gap, Romer could see only fish, lobe-finned fish, living in water. Where were the intermediates, and what led them to venture out on to the land?

My undergraduate imagination at Oxford was fired by the lectures of the prodigiously knowledgeable Harold Pusey who, despite his dry and prolonged delivery, had a gift for seeing beyond dry bones to the flesh-and blood animals that had to make a living in some departed world.* His evocation of what drove some lobe-finned fish to develop lungs and legs, which was derived from Romer himself, made memorable sense to my student ears, and it still makes sense to me even though it is less fashionable among modern palaeontologists than it was in Romer's time. Romer, and Pusey, envisaged annual droughts during which lakes and ponds and streams dried up, only to flood again the following year. Fishes that made their living in water could benefit from a temporary ability to survive on land, while they dragged themselves from a shallow lake or pond that was threatened with imminent desiccation to a deeper one in which they could survive until the next wet season. On this view, our ancestors didn't so much emerge on to the dry land as use the dry land as a temporary bridge to escape back into the water. Many modern animals do the same.

Rather unfortunately, Romer introduced his theory with a preamble whose purpose was to show that the Devonian era was a time of drought. Consequently, when more recent evidence undermined this assumption, it seemed to undermine the whole Romer theory. He'd have done better to omit the preamble, which was, in any case, overkill. As I argued in *The Ancestor's Tale*, the

---

* An Oxford don of the old school, who believed he was there to teach undergraduates, he would not have survived in today's research-assessment culture. With scarcely a single published article to his name, his legacy rests in the generations of grateful pupils to whom he imparted his wisdom and at least some of his immense learning.

theory still works, even if the Devonian was less drought-ridden than Romer originally thought.

Let us, in any case, return to the fossils themselves. They trickle sparsely through the late Devonian, the period immediately preceding the Carboniferous: tantalizing traces of 'missing links', animals that went some way towards bridging the gap between the lobe-finned fishes that were so abundant in Devonian seas, and the amphibians that later slithered through the Carboniferous swamps. On the fish side of the gap, *Eusthenopteron* was discovered in 1881 in a collection of fossils from Canada. It seems to have been a surface-hunting fish and probably didn't ever come on land, notwithstanding some early imaginative reconstructions. Nevertheless, it did have several anatomical similarities to the amphibians of 50 million years later, including its skull bones, its teeth and, above all, its fins. Although they were probably used for swimming and not walking, the bones followed the typical pattern of a tetrapod (the name given to all land vertebrates). In the forelimb, a single humerus was joined to two bones, the radius and ulna, joined to lots of little bones, which we tetrapods would call carpals, metacarpals and fingers. And the hind limb shows a similar tetrapod-like pattern.

Then, near the amphibian side of the gap, some 20 million years later, at the border between the Devonian and Carboniferous, great excitement was caused by the 1932 discovery in Greenland of *Ichthyostega*. Don't be misled by thoughts of cold and ice, by the way. Greenland in the days of *Ichthyostega* was on the equator. *Ichthyostega* was first reconstructed by the Swedish palaeontologist Erik Jarvik

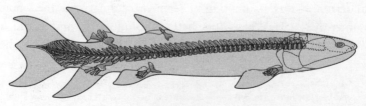

Eusthenopteron

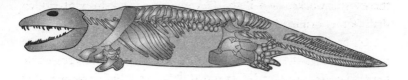

Ichthyostega

in 1955, and again he portrayed it as closer to a land-dweller than modern experts do. The most recent reconstruction, by Per Ahlberg at Jarvik's old university of Uppsala, places *Ichthyostega* mostly in the water, although it probably made occasional forays on to the land. Nevertheless, it looked more like a giant salamander than a fish, and it had the flat head that is so characteristic of amphibians. Unlike all modern tetrapods, which have five fingers and toes (at least in the embryo, although they may lose some in the adult), *Ichthyostega* had seven toes. It seems that the early tetrapods enjoyed more freedom to 'experiment' with varying numbers of digits than we have today. Presumably at some point the embryological processes fixed upon five, and a step was taken that was hard to reverse. Although, admittedly, not as hard as all that. There are individual cats, and indeed humans, who have six toes. These extra toes probably arise through a duplication error in embryology.

Another exciting discovery, also from tropical Greenland and also dating from the boundary between the Devonian and the Carboniferous, was *Acanthostega*. *Acanthostega*, too, had a flat, amphibian skull and tetrapod-like limbs; but it too departed, and

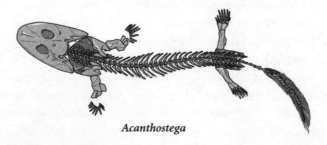

Acanthostega

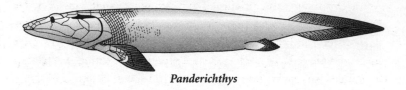

Panderichthys

even further than *Ichthyostega*, from what we now think of as the five-finger standard. It had eight digits. The scientists most responsible for our knowledge of it, Jenny Clack and Michael Coates of Cambridge University, believe that, like *Ichthyostega*, *Acanthostega* was largely a water-dweller, but it had lungs and its limbs strongly suggest that it could cope with land as well as water if it had to. Again, it looked pretty much like a giant salamander. Moving back now to the fish side of the divide, *Panderichthys*, also from the late Devonian, is also slightly more amphibian-like, and slightly less fish-like, than *Eusthenopteron*. But if you saw it you would surely want to call it a fish rather than a salamander.

So, we are left with a gap between *Panderichthys*, the amphibian-like fish, and *Acanthostega*, the fish-like amphibian. Where is the 'missing link' between them? A team of scientists from the University of Pennsylvania, including Neil Shubin and Edward Daeschler, set out to find it. Shubin made their quest the basis for a delightful series of reflections on human evolution in his book *Your Inner Fish*. They deliberately thought about where might be the best place to look, and carefully chose a rocky area of exactly the right late Devonian age in the Canadian Arctic. There they went – and struck zoological gold. *Tiktaalik!* A name never to be forgotten. It comes from an Inuit word for a large freshwater fish. As for the specific name, *roseae*, let me tell a cautionary tale against myself. When I first heard the name, and saw photographs like the one reproduced on colour page 10, my mind immediately leapt to the Devonian, the 'Old Red Sandstone', the colour of the eponymous county of Devon, the colour of Petra ('A rose-red city, half as old as time'). Alas, I was quite wrong. The photograph exaggerates the rosy glow. The name was chosen in honour of a benefactor who helped finance the expedition to the

Arctic Devonian. I was privileged to be shown *Tiktaalik roseae* by Dr Daeschler when I had lunch with him in Philadelphia, shortly after its discovery, and the lifelong zoologist in me – or perhaps my inner fish – was moved to speechlessness. Through rose-tinted spectacles I imagined I was gazing upon the face of my direct ancestor. Unrealistic as that was, this not-so-rose-red fossil was probably as close as I was going to get to meeting a real dead ancestor half as old as time.

If you were to meet a real live *Tiktaalik*, snout to snout, you might start back as if threatened by a crocodile, for that is what its face resembled. A crocodile's head on a salamander's trunk, attached to a fish's rear end and tail. Unlike any fish, *Tiktaalik* had a neck. It could turn its head. In almost every particular, *Tiktaalik* is the perfect missing link – perfect, because it almost exactly splits the difference between fish and amphibian, and perfect because it is missing no longer. We have the fossil. You can see it, touch it, try to appreciate the age of it – and fail.

I MUST GO DOWN TO THE SEA AGAIN[*]

The move from water to land launched a major redesign of every aspect of life, from breathing to reproduction: it was a great trek through biological space. Nevertheless, with what seems almost wanton perversity, a good number of thoroughgoing land animals later turned around, abandoned their hard-earned terrestrial retooling, and trooped back into the water again. Seals and sea lions have only gone part-way back. They show us what the intermediates might have been like, on the way to extreme cases such as whales and dugongs. Whales (including the small whales we call dolphins), and dugongs with their close cousins the manatees, ceased to be

---

[*] This seems to be correct. *The Oxford Dictionary of Quotations* suggests that the commonly quoted 'seas' stems from a misprint in Masefield's original 1902 edition: a nice example of a successful mutant meme.

land creatures altogether and reverted to the full marine habits of their remote ancestors. They don't even come ashore to breed. They do, however, still breathe air, having never developed anything equivalent to the gills of their earlier marine progenitors. Other animals that have returned from land to water, at least some of the time, are pond snails, water spiders, water beetles, crocodiles, otters, sea snakes, water shrews, Galapagos flightless cormorants, Galapagos marine iguanas, yapoks (aquatic marsupials from South America), platypuses, penguins and turtles.

Whales were long an enigma, but recently our knowledge of whale evolution has become rather rich. Molecular genetic evidence (see Chapter 10 for the nature of this kind of evidence) shows that the closest living cousins of whales are hippos, then pigs, then ruminants. Even more surprisingly, the molecular evidence shows that hippos are more closely related to whales than they are to the cloven-hoofed animals (such as pigs and ruminants) which look much more like them. This is another example of the mismatch that can sometimes arise between closeness of cousinship and degree of physical resemblance. We noted it above in connection with fish that are closer cousins to us than they are to other fish. In that case, the anomaly arose because our lineage left the water for the land, and consequently surged away in evolution, leaving our close fish cousins, the lungfish and coelacanths, resembling our more distant fish cousins because they all stayed in the water. Now we meet the same phenomenon again, but in reverse. Hippos stayed, at least partly, on land, and so still resemble their more distant land-dwelling cousins, the ruminants, while their closer cousins, the whales, took off into the sea and changed so drastically that their affinities with hippos escaped all biologists except molecular geneticists. As when their remote fishy ancestors originally went in the other direction, it was a bit like taking off into space, or at least like launching a balloon, as the ancestors of whales floated free of the constraining burden of gravity and severed their moorings to dry land.

At the same time, the once rather scanty fossil record of whale evolution has been convincingly filled out, mostly by a new trove from Pakistan. However, the story of fossil whales has been so well treated in other recent books, for example Donald Prothero's *Evolution: What the Fossils Say and Why it Matters*, and, more recently, Jerry Coyne's *Why Evolution is True*, that I have decided not to cover the same details here. Instead, I have confined myself to one diagram (below), taken from Prothero's book, showing a sequence of fossils ordered in time. Note the careful way the picture is drawn. It is tempting – and older books used to do this – to draw sequences of fossils with arrows from older to younger ones. But nobody can say, for example, that *Ambulocetus* was descended from *Pakicetus*. Or that *Basilosaurus* was descended from *Rodhocetus*. Instead, the diagram follows the more cautious policy of suggesting that, for example, whales are descended from a contemporary cousin of *Ambulocetus*

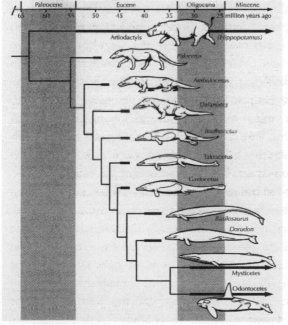

Fossil whales

Figure 14.16. Evolution of whales from land creatures, showing the numerous transitional fossils now documented from the Eocene beds of Africa and Pakistan. (Drawing by Carl Buell)

which was probably rather like *Ambulocetus* (and might even have been *Ambulocetus*). The fossils shown are representative of various stages of whale evolution. The gradual disappearance of the hind limbs, the transformation of the front limbs from walking legs to swimming fins, and the flattening of the tail into flukes, are among the changes that emerged in elegant cascade.

That's all I'm going to say about the fossil history of whales, because it has been so well treated in the books I mentioned. The other, less numerous and diverse but just as thoroughly aquatic group of marine mammals, the sirenians – dugongs and manatees – are not so well documented in the fossil record, but one outstandingly beautiful 'missing link' has recently been discovered. Roughly contemporary with *Ambulocetus*, the Eocene 'walking whale', is *Pezosiren*, the 'walking manatee' fossil from Jamaica. It looks pretty much like a manatee or dugong, except that it has proper walking legs both front and rear, where they have flippers in the front and no limbs at all in the rear. The picture opposite shows a modern dugong skeleton above, *Pezosiren* below.

Just as whales are related to hippos, so sirenians are related to elephants, as a great deal of evidence, including the all-important molecular evidence, attests. *Pezosiren*, however, probably lived like a hippo, spending much of its time in water and using its legs to walk on the bottom as well as swim. The skull is unmistakably sirenian. *Pezosiren* may or may not be the actual ancestor of modern manatees and dugongs, but it is certainly excellent casting for the role.

This book was about to go to the printer when exciting news came in, from the journal *Nature*, of a new fossil from the Canadian Arctic, plugging a gap in the ancestry of modern seals, sea lions and walruses (collectively 'pinnipeds'). A single skeleton, about 65 per cent complete, *Puijila darwini* dates from the early Miocene epoch (about 20 million years ago). That's recent enough that the map of the world was almost the same as today. So this early seal/sea lion (they had not diverged yet) was an Arctic animal, a denizen of cold water. Evidence suggests that it lived and fished in fresh water (like

Modern dugong

Pezosiren – ancient dugong

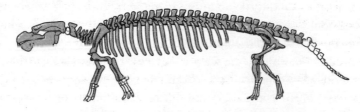

most otters except the famous sea otters of California), rather than in the sea (like most modern seals except the famous Lake Baikal seal). *Puijila* did not have flippers, but webbed feet. It probably ran like a dog on land (very unlike a modern pinniped) but spent much of its time in water, where it swam like a dog, unlike either of the two styles adopted respectively by modern seals and sea lions. *Puijila* neatly straddles the gap between land and water in the ancestry of pinnipeds. It is yet another delightful addition to our growing list of 'links' that are no longer missing.

I now want to turn to another group of animals that returned from the land to the water: a particularly intriguing example because some of them later reversed the process and returned to the land a second time! Sea turtles are, in one important respect, less fully given back to the water than whales or dugongs, for they still lay their eggs on beaches. Like all vertebrate returners to the water, turtles haven't given up breathing air, but in this department some of them go one better than whales. These turtles extract additional oxygen from the water through a pair of chambers at their rear end that are richly supplied with blood vessels. One Australian river turtle, indeed, gets the majority of its oxygen by breathing (as an Australian would not hesitate to say) through its arse.

Before going any further, I can't escape a tiresome point of terminology, and a regrettable vindication of George Bernard Shaw's observation that 'England and America are two countries divided by a common language.' In Britain, turtles live in the sea, tortoises live on land and terrapins live in fresh or brackish water. In America all these animals are 'turtles', whether they live on land or in water. 'Land turtle' sounds odd to me, but not to an American, for whom tortoises are the subset of turtles that live on land. Some Americans use 'tortoise' in a strict taxonomic sense to refer to the Testudinidae, which is the scientific name for modern land tortoises. In Britain, we'd be inclined to call any land-dwelling chelonian a tortoise, whether it is a member of the Testudinidae or not (as we shall see, there are fossil 'tortoises' that lived on land but are not members of the Testudinidae). In what follows, I'll try to avoid confusion, making allowance for readers in Britain and America (and Australia, where the usage is different again), but it's hard. The terminology is a mess, to put it mildly. Zoologists use 'chelonians' for all these animals, turtles, tortoises and terrapins, whichever version of English we speak.

The most instantly noticeable feature of chelonians is their shell. How did it evolve, and what did the intermediates look like? Where are the missing links? What (a creationist zealot might ask) is the use of half a shell? Well, amazingly, a new fossil has just been described, which eloquently answers that question. It made its debut in the journal *Nature* in the nick of time before I had to hand this book over to the publishers. It was an aquatic turtle, found in late Triassic sediments in China, and its age is estimated at 220 million years. Its name is *Odontochelys semitestacea*, from which you may deduce that, unlike a modern turtle or tortoise, it had teeth, and it did indeed have half a shell. It also had a much longer tail than a modern turtle or tortoise. All three of these features mark it out as prime 'missing link' material. The belly was covered by a shell, the so-called plastron, in pretty much the same way as that of a modern sea turtle. But it almost completely lacked the dorsal portion of the shell, known as the carapace. Its back was presumably soft, like a lizard's, although there were some hard, bony bits along the middle above the backbone, as in a crocodile, and the ribs were flattened, as though 'trying' to form the evolutionary beginnings of a carapace.

And here we have an interesting controversy. The authors of the paper that introduced *Odontochelys* to the world, Li, Wu, Rieppel, Wang and Zhao (for brevity, I'll call them the Chinese authors, although Rieppel is not Chinese), think that their animal was indeed halfway towards acquiring a shell. Others dispute *Odontochelys's* claim to demonstrate that the shell evolved in water. *Nature* has the admirable custom of commissioning experts other than the authors to write a commentary on the week's more interesting articles, which they publish in a section called 'News and Views'. The 'News and Views' commentary on the *Odontochelys* paper is by two Canadian biologists, Robert Reisz and Jason Head, and they offer an alternative interpretation. Maybe the whole shell had already evolved on land, before *Odontochelys's* ancestors went back to the water. And maybe *Odontochelys* lost its carapace after returning to the water. Reisz and Head point out that some of today's sea turtles, for example the giant leatherback turtle, have lost or greatly reduced the carapace, so their theory is quite plausible.

I need to digress for a brief aside on the question, 'What is the use of half a shell?' In particular, why would *Odontochelys* be armoured below but not above? Perhaps because danger threatened from below, which would suggest that these creatures spent a lot of their time swimming near the surface – and of course they had to come to the surface to breathe, anyway. Sharks today often attack from below, sharks would have been a menacingly important part of the world of *Odontochelys*, and there's no reason to suppose that their hunting habits were different in those times. As a parallel example, one of the most surprising achievements of evolution, the extra pair of eyes in the fish *Bathylychnops* (see over), is probably aimed at detecting predatory attacks from below. The main eye looks outwards, as in

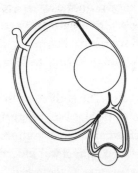

Bathylychnops' extra eye

any ordinary fish. But each of the two main eyes has an extra little eye, complete with lens and retina, tucked into its lower side. If *Bathylychnops* can go to the trouble (you know what I mean, don't be pedantic) of growing a whole extra pair of eyes, presumably to look out for attacks coming from below, it seems quite plausible that *Odontochelys* might grow armour aimed at fending off attacks from the same direction. The plastron makes sense. And if you want to say, yes, but why not have a carapace on top as well, just to be extra safe, the reply is easy. Shells are heavy and cumbersome, they are costly to grow and costly to carry around. There are always trade-offs in evolution. For land tortoises, the trade-off ends up favouring stout, heavy armour above as well as below. For many sea turtles, the trade-off favours a strong plastron underneath but lighter armour on top. And it is a plausible suggestion that *Odontochelys* just carried that trend a bit further.

If, on the other hand, the Chinese authors are right that *Odontochelys* was on its way to evolving a full shell, and that the shell evolved in water, it would seem to follow that modern land tortoises, which have well-developed shells, are descended from water turtles. This, as we shall see, is probably true. But it is remarkable, because it means that today's land tortoises represent a *second* migration from water to land. Nobody has ever claimed that any whales, or dugongs, *returned* to the land after invading the water. The alternative story for land tortoises is that they were on land all along and independently evolved the shell, in parallel to their aquatic cousins. This is by no means impossible; but, as it happens, we have good reason to believe that sea turtles did indeed return to the land for a second go at becoming land tortoises.

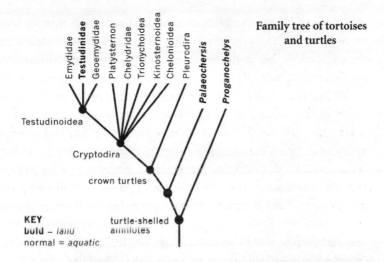

Family tree of tortoises and turtles

If you draw out the family tree of all modern turtles and tortoises, based on molecular and other comparisons, nearly all the branches are aquatic (normal type). Land tortoises are represented by bold type, and you can see that today's land tortoises constitute a single branch, the Testudinidae, deeply nested within rich branchings of otherwise aquatic chelonians. All their close cousins are aquatic. Modern land tortoises are a single twig on the bush of otherwise aquatic turtles. Their aquatic ancestors turned turtle and trooped back on to the land. This fact is compatible with the hypothesis that the shell evolved in water, in a creature like *Odontochelys*. But now we have another difficulty. If you look at the family tree, you'll notice that, in addition to the Testudinidae (all modern land tortoises) there are two fossil genera of fully shelled animals called *Proganochelys** and

---

* I'm advised that this doesn't make a lot of sense in Greek. If it were *Progonochelys* it would make perfect sense. It would mean something like 'ancestral tortoise' or 'primeval tortoise', and I can't help feeling that that may be what the original authors intended when they named it. Unfortunately, the rules of zoological nomenclature are strict, and even obvious mistakes can't be changed, once they are enshrined in a naming publication. The taxonomy is littered with such fossilized mistakes. My favourite is *Khaya*, African mahogany. Legend (which I long to believe) has it that in a local language it means 'I don't know', with the presumed subtext, 'And I don't care and why don't you stop asking stupid questions about plant names.'

Palaeochersis. These are drawn as land-dwellers, for reasons we shall come to in the next paragraph. They lie right outside the branches representing the water turtles. It would seem that these two genera are anciently terrestrial.

Before *Odontochelys* was discovered, these two fossils were the oldest known chelonians. Like *Odontochelys* they lived in the late Triassic, but about 15 million years later than *Odontochelys.* Some authorities have reconstructed them as living in fresh water, but recent evidence does indeed place them on land, as indicated by bold type on the diagram. You might wonder how we tell whether fossil animals, especially if only fragments are found, lived on land or in water. Sometimes it's pretty obvious. Ichthyosaurs were reptilian contemporaries of the dinosaurs, with fins and streamlined bodies. The fossils look like dolphins and they surely lived like dolphins, in the water. With turtles and tortoises it is a little less obvious. As you might expect, the biggest giveaway is their limbs. Paddles really are rather different from walking legs. Walter Joyce and Jacques Gauthier, of Yale University, took this common-sense intuition and provided the numbers to support it. They took three key measurements in the arm and hand bones of seventy-one species of living chelonians. I'll resist the temptation to explain their elegant calculations, but their conclusion was clear. These animals had had walking legs, not paddles. In British English, they were 'tortoises', not 'turtles'. They lived on land. They were only distant cousins, however, of modern land tortoises.

Now we seem to have a problem. If, as the authors of the paper describing *Odontochelys* believe, their half-shelled fossil shows that the shell evolved in water, how do we explain two genera of fully shelled 'tortoises' on land, 15 million years later? Until the discovery of *Odontochelys*, I would not have hesitated to say that *Proganochelys* and *Palaeochersis* were representative of the land-dwelling ancestral type *before* the return to water. The shell evolved on land. Some shelled tortoises returned to the sea, as seals, whales and dugongs were later to do. Others stayed on land, but went extinct. And then some sea turtles returned to the land, to give rise to all modern land tortoises. That's what I would have said – indeed what I did say in the earlier draft of this chapter that preceded the announcement of *Odontochelys*. But *Odontochelys* throws speculation back into the melting pot. We now have three possibilities, all equally intriguing.

1 *Proganochelys* and *Palaeochersis* might be survivors of the land-dwelling animals that had earlier sent some representatives to sea, including the ancestors of *Odontochelys*. This hypothesis would suggest that the shell evolved on land early, and *Odontochelys* lost the carapace in the water, retaining the ventral plastron.

2 The shell might have evolved in water, as the Chinese authors suggest, with the plastron over the belly evolving first, and the carapace over the back evolving later. In this case, what do we make of *Proganochelys* and *Palaeochersis*, who lived on land *after Odontochelys* lived, with its half shell, in water? *Proganochelys* and *Palaeochersis* might have evolved the shell independently. But there is another possibility:

3 *Proganochelys* and *Palaeochersis* might represent an earlier return from the water to the land. Isn't that a startlingly exciting thought?

We are already pretty confident of the remarkable fact that the turtles accomplished an evolutionary doubling back to the land: an early marque of land 'tortoises' went back to the watery environment of their even earlier fish ancestors, became sea turtles, then returned to the land yet again, as a new incarnation of land tortoises, the Testudinidae. That we know, or are nearly certain of. But now we are facing up to the additional suggestion that this doubling back happened *twice!* Not just to spawn the modern tortoises, but much longer ago, to give rise to *Proganochelys* and *Palaeochersis* in the Triassic.

In another book I described DNA as 'the Genetic Book of the

Dead'. Because of the way natural selection works, there is a sense in which the DNA of an animal is a textual description of the worlds in which its ancestors were naturally selected. For a fish, the genetic book of the dead describes ancestral seas. For us and most mammals, the early chapters of the book are all set in the sea and the later ones all out on land. For whales, dugongs, marine iguanas, penguins, seals, sea lions and turtles, there is a third section of the book which recounts their epic return to the proving grounds of their remote past, the sea. But for the land tortoises, perhaps twice independently on two widely separated occasions, there is yet a fourth section of the book devoted to a final – or is it? – re-emergence, yet again to the land. Can there be another animal for which the genetic book of the dead is such a palimpsest of multiple evolutionary U-turns? As a parting shot, I cannot help wondering about those freshwater and brackish water forms ('terrapins'), which are close cousins of the land tortoises. Did their ancestors move directly from the sea into brackish and then fresh water? Do they represent an intermediate stage on the way from the sea back to the land? Or is it possible that they constitute yet another doubling-back to the water from ancestors that were modern land tortoises? Have the chelonians been shuttling back and forth in evolutionary time between water and land? Could the palimpsest be even more densely over-written than I have so far suggested?

....

POSTSCRIPT

On 19 May 2009, as I was correcting the proofs of this book, a 'missing link' between lemur-like and monkey-like primates was announced in the online scientific journal *PLOS One*. Named *Darwinius masillae*, it lived 47 million years ago in rain forest in what is now Germany. It is claimed by the authors to be the most complete fossil primate ever found: not just bones but skin, hair, some internal organs and its last meal. Beautiful as *Darwinius masillae* undoubtedly is (see colour page 9), it comes trailing clouds of hype that obscure clear thinking. According to *Sky News* it is 'the eighth wonder of the world' which 'finally confirms Charles Darwin's theory of evolution'. Goodness me! The more-or-less nonsensical mystique of the 'missing link' seems to have lost none of its power.

CHAPTER 7

MISSING PERSONS? MISSING NO LONGER

D ARWIN'S treatment of human evolution in his most famous work, *On the Origin of Species*, is limited to twelve portentous words: 'Light will be thrown on the origin of man and his history.' That is the wording in the first edition, which is the edition I always cite unless otherwise stated. By the sixth (and last) edition, Darwin allowed himself to stretch a point, and the sentence became 'Much light will be thrown on the origin of man and his history.' I like to think of his pen, poised over the fifth edition, while the great man judiciously pondered whether to indulge himself in the luxury of 'Much'. Even with it, the sentence is a calculated understatement.

Darwin deliberately deferred his treatment of human evolution to another book, *The Descent of Man*. Perhaps it is not surprising that the two volumes of that later work devote more space to the topic of its subtitle, *Selection in Relation to Sex* (investigated largely in birds), than to human evolution. Not surprising because, at the time of Darwin's writing, there were no fossils at all linking us to our closest relatives among the apes. Darwin had only living apes to look at, and he used them well, arguing correctly (and almost alone) that our closest living relatives were all African (gorillas and chimpanzees – bonobos were not recognized as separate from chimpanzees in those days, but they are African too), and therefore predicting that, if ancestral human fossils were ever to be found, Africa was the place to search. Darwin regretted the paucity of fossils, but he maintained a robustly bullish attitude to it. Citing Lyell, his mentor and the great geologist of the time, he pointed out that 'in all the vertebrate classes the discovery of fossil remains has been an extremely slow and fortuitous process' and added, 'Nor should it be forgotten that those regions which are the most likely to afford remains connecting man with some extinct ape-like creature, have not as yet been searched by geologists.' He meant Africa, and the quest was not helped by the fact that his immediate successors largely ignored his advice and searched Asia instead.

It was indeed in Asia that the 'missing links' first began to become less missing. But those first fossils to be discovered were relatively recent, less than a million years old, dating from a time when hominids were pretty close to modern humans and had migrated out of Africa and reached the Far East. They were called 'Java Man' and 'Peking Man' after their discovery sites.* Java Man was discovered by the Dutch anthropologist Eugene Dubois in 1891. He named it *Pithecanthropus erectus*, signifying his belief that he had realized his life's ambition and found 'the missing link'. Disagreement came from two opposite sources, which rather proved his point: some said his fossil was purely human, others that it was a giant gibbon. Later in his rather embittered and cantankerous life, Dubois resented the suggestion that the more recently discovered Peking fossils were similar to his Java Man. Fiercely possessive about, not to say protective of, his fossil, Dubois believed that only Java Man was the true missing link. To emphasize the distinction from the various Peking Man fossils, he described them as far closer to modern man, and his own Java Man of Trinil as intermediate between man and ape.

....

* Predictably, the Peking fossil is now sometimes called Beijing Man. Why, since we are talking English rather than Chinese, do we go along with 'Beijing' at all, when referring to China's capital? There's a rather charming programme on British television called *Grumpy Old Men*, which is a genially edited collection of grouses and grizzles of just this kind. If I were on it, I would say something like the following. We don't dab on a splash of Eau de Köln to drown out the smell of Mumbai Duck, or go waltzing to the strains of 'The Blue Dunaj' or 'Tales from the Wien Woods'. We don't compare Neville Chamberlain, the Man of München, to Napoleon's retreat from Moskva. Nor yet (though give it time) do we take our snuffling little pet Beij for walkies. What's wrong with Peking, when it's the English language we are speaking? I was delighted recently to meet a member of the British diplomatic corps, fluent in Mandarin, who had played a leading role in our embassy in what he insisted on calling Peking.

Pithecanthropus [Java Man] was not a man, but a gigantic genus allied to the gibbons, however superior to the gibbons on account of its exceedingly large brain volume and distinguished at the same time by its faculty of assuming an erect attitude and gait. It had the double cephalization [ratio of brain size to body size] of the anthropoid apes in general and half that of man . . .

It was the surprising volume of the brain – which is very much too large for an anthropoid ape, and which is small compared with the average, though not smaller than the smallest human brain – that led to the now almost general view that the 'Ape Man' of Trinil, Java was really a primitive Man. Morphologically, however, the calvaria [skullcap] closely resembles that of anthropoid apes, especially the gibbon . . .

It can't have improved Dubois' temper that others took him to be saying that *Pithecanthropus* was just a giant gibbon, not intermediate between them and humans at all, and he was at pains to reassert his earlier stand: 'I still believe, now more firmly than ever, that the *Pithecanthropus* of Trinil is the real "missing link".'

Creationists have from time to time used as a political weapon the allegation that Dubois backed off from his claim that *Pithecanthropus* was an intermediate ape-man. The creationist organization Answers in Genesis has, however, added it to their list of discredited arguments which they now say should not be used. It is to their credit that they maintain such a list at all. As I said, both the Java and Peking specimens of *Pithecanthropus* have now been shown to be quite young, less than a million years old. They are now classified along with us in the genus *Homo*, retaining Dubois' specific name *erectus*: *Homo erectus*.

Dubois chose the wrong part of the world for his single-minded quest for the 'missing link'. It was natural for a Dutchman to head first for the Dutch East Indies, but a man of his dedication should have followed Darwin's advice and gone on to Africa: for Africa is where our ancestors evolved, as we shall see. So what were these *Homo erectus* specimens doing out of Africa? The phrase 'out of Africa' has been borrowed from Karen Blixen* to refer to the great exodus of our ancestors from Africa. But there were two exoduses and it is important not to confuse them. Relatively recently, maybe less than 100,000 years ago, roving bands of *Homo sapiens* looking pretty much like us left Africa and diversified into all the races that we see around the world today: Inuit, native Americans, native Australians, Chinese, and so on. It is to this recent exodus that the phrase 'out of Africa' is normally applied. But there was an earlier exodus from Africa, and these *erectus* pioneers left fossils in Asia and Europe, including the Java and Peking specimens. The oldest fossil known outside Africa was found in the central Asian country of Georgia and dubbed 'Georgian Man': a diminutive creature whose (rather well-preserved) skull is dated, by modern methods, to about 1.8 million years ago. It has been called *Homo georgicus* (by some taxonomists, although others don't recognize it as a separate species) to indicate that it seems rather more primitive than the rest of the early refugees from Africa, who are all classified as *Homo erectus*. Some stone tools slightly older than Georgian Man have just been discovered

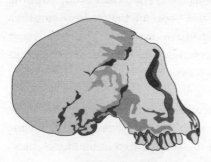

Homo georgicus

in Malaysia, sparking a new search for fossil bones in that peninsula. But in any case, all these early Asian fossils are pretty close to modern humans and all are nowadays classified in the genus *Homo*; for our earlier antecedents we must go to Africa. First, though, let's

---

* Pen name Isak Dinesen, but I like to use her real name because I spent my earliest childhood near Karen, the village 'at the foot of the Ngong Hills' which is still named after her.

pause to ask what we should
expect of a 'missing link'.

Suppose, for the sake of
argument, we take seriously
the original confused meaning
of the term 'missing link', and
seek an intermediate between
chimpanzees (see right) and
ourselves. We are not descended
from chimpanzees, but it is

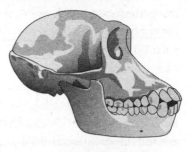

Chimpanzee

a fair bet that the common ancestor that we share with them was
more like a chimp than like us. In particular, it didn't have a huge
brain like ours, it probably didn't walk upright as we do, it probably
was a lot hairier than we are, and it surely didn't have such advanced
human features as language. So, even though we must remain
adamant, in the face of common misunderstanding, that we are not
descended from chimpanzees, there's still no harm in asking what
an intermediate between something like a chimpanzee and us would
look like.

Well, hair and language don't fossilize well, but we can get good
clues about brain size from the skull, and good clues about gait from
the whole skeleton (including the skull, for the *foramen magnum*,
the hole for the spinal cord, points downwards in bipeds, more
backwards in quadrupeds). Possible candidates for missing links
might have any of the following attributes:

1 Intermediate brain size and intermediate gait: perhaps a sort
   of stooping shamble rather than the proudly erect bearing
   favoured by sergeant majors and deportment mistresses.

2 Chimpanzee-sized brain with human upright gait.

3 Large, more human-like brain, walking on all fours like
   a chimp.

So, bearing these possibilities in mind, let's examine some of the many African fossils that are now available to us, but unfortunately were not available to Darwin.

I'M STILL MISCHIEVOUSLY HOPING . . .

Molecular evidence (which I shall come on to in Chapter 10) shows that the common ancestor we share with chimpanzees lived about six million years ago or a bit earlier, so let's split the difference and look at some three-million-year-old fossils. The most famous fossil of this vintage is 'Lucy', classified by her discoverer in Ethiopia, Donald Johanson, as *Australopithecus afarensis*. Unfortunately we have only fragments of Lucy's cranium, but her lower jaw is unusually well preserved. She was small by modern standards, although not as small as *Homo floresiensis*, the tiny creature the newspapers have irritatingly dubbed 'the Hobbit', which died out tantalizingly recently on the island of Flores in Indonesia. Lucy's skeleton is complete enough to suggest that she walked upright on the ground, but probably also sought refuge in trees, where she was an agile climber. There is good evidence that the bones attributed to Lucy really did all come from a single individual. The same is not true of the so-called 'First Family', a collection of bones from at least thirteen individuals, similar to Lucy and of approximately the same vintage, who somehow became buried together, also in Ethiopia. The fragments from Lucy and from the First Family give a good idea of what *Australopithecus afarensis* looked like, but it is hard to make an authentically complete reconstruction from pieces of many different individuals. Fortunately, a rather complete skull known as AL 444-2 (right) was discovered in 1992 in the same area of Ethiopia, and this confirmed the tentative reconstructions that had previously been made.

The conclusion from studies of Lucy and her kind is that they had brains about the same size as chimpanzees' but, unlike chimpanzees, they walked upright on their hind legs, as we do – the second of our three hypothetical scenarios. 'Lucys' were a bit like upright-walking

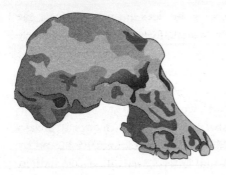

AL 444-2

chimps. Their bipedality is dramatically confirmed by the poignantly evocative set of footprints discovered by Mary Leakey in fossilized volcanic ash. These are further south, at Laetoli in Tanzania, and they are older than Lucy and AL 444-2: about 3.6 million years. They are usually attributed to a pair of *Australopithecus afarensis* walking together (hand in hand?) but what matters is that, by 3.6 million years ago, an erect ape walked the Earth, on two feet which were pretty much like ours although its brain was the size of a chimpanzee's.

It seems quite likely that the species we call *Australopithecus afarensis* – Lucy's species – included our ancestors of three million years ago. Other fossils have been placed in different species of the same genus, and it is virtually certain that our ancestors were members of that genus. The first Australopithecine to be discovered, and the type specimen of the genus, was the so-called Taung Child. At the age of three and a half the Taung Child was eaten by an eagle. The evidence is that damage marks to the eye sockets of the fossil are identical to marks made by modern eagles on modern monkeys as they rip out their eyes. Poor little Taung Child, shrieking on the wind as you were borne aloft by the aquiline fury, you would have found no comfort in your destined fame, two and a half million years on, as the type specimen of *Australopithecus africanus*. Poor Taung mother, weeping in the Pliocene.

The type specimen is the first individual of a new species to be named and officially given the virgin label in a museum. Theoretically, later finds are compared against the type specimen to see if they match. The Taung Child was discovered and given brand new genus and species names by the South African anthropologist Raymond Dart in 1924.

What's the difference between 'species' and 'genus'? Let's get the question swiftly out of the way, before proceeding. Genus is the more inclusive division. A species belongs within a genus, and often it shares the genus with other species. *Homo sapiens* and *Homo erectus* are two species within the genus *Homo. Australopithecus africanus* and *Australopithecus afarensis* are two species within the genus *Australopithecus*. The Latin name of an animal or plant always includes a generic name (with an initial capital letter) followed by a specific name (without a capital letter). Both names are written in italics. Sometimes there is an additional sub-specific name, which follows the specific name, as in, for example, *Homo sapiens neanderthalensis*. Taxonomists often dispute names. Many, for example, would speak of *Homo neanderthalensis* not *Homo sapiens neanderthalensis*, elevating Neanderthal man from sub-species to species status. Generic names and specific names are also often disputed, and often change with successive revisions in the scientific literature. *Paranthropus boisei* has been, in its time, *Zinjanthropus boisei* and *Australopithecus boisei*,* and is still often referred to, informally, as a robust Australopithecine – as opposed to the two 'gracile' (slender) species of *Australopithecus* mentioned above. One of the main messages of this chapter concerns the somewhat arbitrary nature of zoological classification.

Raymond Dart, then, gave the name *Australopithecus* to the Taung Child, the type specimen of the genus, and we have been stuck with this depressingly unimaginative name for our ancestor ever since. It simply means 'southern ape'. Nothing to do with Australia, which

---

* Unlike diseases, which are often named after their discoverers, new species are named *by* their discoverers but never *after* themselves. It is a nice opportunity for a biologist to honour the name of another, or, as in this case, a benefactor. Not surprisingly, my distinguished colleague the late W. D. Hamilton was several times honoured in this way. Arguably one of Darwin's greatest successors of the twentieth century, he had a lugubrious manner reminiscent of A. A. Milne's Eeyore (not the deplorable Walt Disney version, of course). Hamilton was once on a small boat on an expedition up the Amazon when he was stung by a wasp. Knowing what a great entomologist he was, his companion said, 'Bill, do you know the name of that wasp?' 'Yes,' Bill murmured gloomily in his most Eeyoreish voice. 'As a matter of fact it's named after me.'

just means 'southern country'. You'd think Dart might have thought of a more imaginative name for such an important genus. He might even have guessed that other members of the genus would later be discovered north of the equator.

Slightly older than the Taung Child, one of the most beautifully preserved skulls we have, although lacking a lower jaw, is called 'Mrs Ples'. Mrs Ples, who may actually have been a small male rather than a large female, obtained 'her' nickname because she was originally classified in the genus *Plesianthropus*. This means 'nearly human', which is a better name than 'southern ape'. One might have hoped that, when later taxonomists decided that Mrs Ples and her kind were really of the same genus as the Taung Child, *Plesianthropus* would have become the name for all of them. Unfortunately, the rules of zoological nomenclature are strict to the point of pedantry. Priority of naming takes precedence over sense and suitability. 'Southern ape' might be a lousy name but no matter: it predates the much more sensible *Plesianthropus* and we seem to be stuck with it, unless . . . I'm still mischievously hoping somebody will uncover, in a dusty drawer in a South African museum, a long-forgotten fossil, clearly the same kind as Mrs Ples and the Taung Child, but bearing the scrawled label, '*Hemianthropus* type specimen, 1920'. At a stroke, all the museums in the world would immediately have to relabel their *Australopithecus* specimens and casts, and all books and articles on hominid prehistory would have to follow suit. Word-processing programs across the world would work overtime sniffing out any occurrences of *Australopithecus* and replacing them with *Hemianthropus*. I can't think of any other case where international

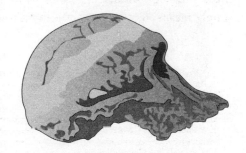

'Mrs Ples'

rules are potent enough to dictate a worldwide and backdated change of language overnight.

Now for my next important point about allegedly missing links and the arbitrariness of names. Obviously, when Mrs Ples's name was changed from *Plesianthropus* to *Australopithecus*, nothing changed in the real world at all. Presumably nobody would be tempted to think anything else. But consider a similar case where a fossil is re-examined and moved, for anatomical reasons, from one genus to another. Or where its generic status is disputed – and this very frequently happens – by rival anthropologists. It is, after all, essential to the logic of evolution that there must have existed individuals sitting exactly on the borderline between two genera, say *Australopithecus* and *Homo*.

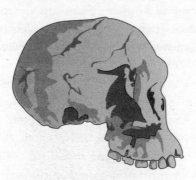

KNM ER 1813

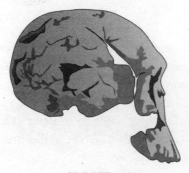

KNM ER 1470

It is easy to look at Mrs Ples and a modern *Homo sapiens* skull and say, yes, there is no doubt these two skulls belong in different genera. If we assume, as almost every anthropologist today accepts, that all members of the genus *Homo* are descended from ancestors belonging to the genus we call *Australopithecus*, it necessarily follows that, somewhere along the chain of descent from one species to the other, there must have been at least one individual who sat exactly on the borderline. This is an important point, so let me stay with it a little longer.

Bearing in mind the shape of Mrs Ples's skull as a representative of *Australopithecus africanus* 2.6 million years ago, have a look at the top skull opposite, called KNM ER 1813. Then look at the one underneath it, called KNM ER 1470. Both are dated at approximately 1.9 million years ago, and both are placed by most authorities in the genus *Homo*. Today, 1813 is classified as *Homo habilis*, but it wasn't always. Until recently, 1470 was too, but there is now a move afoot to reclassify it as *Homo rudolfensis*. Once again, see how fickle and transitory our names are. But no matter. both have an apparently agreed foothold in the genus *Homo*. The obvious difference from Mrs Ples and her kind is that she had a more forward-protruding face and a smaller brain-case. In both respects, 1813 and 1470 seem more human, Mrs Ples more 'ape-like'.

Now look at the skull below, called 'Twiggy'. Twiggy is also normally classified nowadays as *Homo habilis*. But her forward-pointing muzzle has more of a suggestion of Mrs Ples about it than of 1470 or 1813. You will perhaps not be surprised to be told that Twiggy has been placed by some anthropologists in the genus *Australopithecus* and by other anthropologists in *Homo*. In fact, each of these three fossils has been,

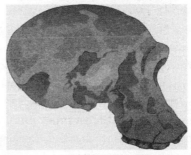

'Twiggy'

at various times, classified as *Homo habilis* and as *Australopithecus habilis*. As I have already noted, some authorities at some times have given 1470 a different specific name, changing *habilis* to *rudolfensis*. And, to cap it all, the specific name *rudolfensis* has been fastened to both generic names, *Australopithecus* and *Homo*. In summary, these three fossils have been variously called, by different authorities at different times, the following range of names:

KNM ER 1813:   *Australopithecus habilis, Homo habilis*
KNM ER 1470:   *Australopithecus habilis, Homo habilis,*
                 Australopithecus rudolfensis, Homo
                 rudolfensis
OH 24 ('Twiggy'): *Australopithecus habilis, Homo habilis*

Should such a confusion of names shake our confidence in evolutionary science? Quite the contrary. It is exactly what we should expect, given that these creatures are all evolutionary intermediates, links that were formerly missing but are missing no longer. We should be positively worried if there were no intermediates so close to borderlines as to be difficult to classify. Indeed, on the evolutionary view, the conferring of discrete names should actually become impossible if only the fossil record were more complete. In one way, it is fortunate that fossils are so rare. If we had a continuous and unbroken fossil record, the granting of distinct names to species and genera would become impossible, or at least very problematical. It is a fair conclusion that the predominant source of discord among palaeoanthropologists – whether such and such a fossil belongs in this species/genus or that – is deeply and interestingly futile.

Hold in your head the hypothetical notion that we might, by some fluke, have been blessed with a continuous fossil record of all evolutionary change, with no links missing at all. Now look at the four Latin names that have been applied to 1470. On the face of it, the change from *habilis* to *rudolfensis* would seem to be a smaller change than the one from *Australopithecus* to *Homo*. Two species within a genus are more like each other than two genera. Aren't they? Isn't that the whole basis for the distinction between the genus level (say *Homo* or *Pan* as alternative genera of African apes) and the species level (say *troglodytes* or *paniscus* within the chimpanzees) in the hierarchy of classification? Well, yes, that is right when we are classifying modern animals, which can be thought of as the tips of the twigs on the evolutionary tree, with their antecedents on the inside of the tree's crown all comfortably dead and out of the way.

Naturally, those twigs that join each other further back (further into the interior of the tree's crown) will tend to be less alike than those whose junction (more recent common ancestor) is nearer the tips. The system works, as long as we don't try to classify the dead antecedents. But as soon as we include our hypothetically complete fossil record, all the neat separations break down. Discrete names become, as a general rule, impossible to apply. We can easily see this if we walk steadily backwards through time, much as we did with the rabbits in Chapter 2.

As we trace the ancestry of modern *Homo sapiens* backwards, there must come a time when the difference from living people is sufficiently great to deserve a different specific name, say *Homo ergaster*. Yet, every step of the way, individuals were presumably sufficiently similar to their parents and their children to be placed in the same species. Now we go back further, tracing the ancestry of *Homo ergaster*, and there must come a time when we reach individuals who are sufficiently different from 'mainstream' *ergaster* to deserve a different specific name, say *Homo habilis*. And now we come to the point of this argument. As we go back further still, at some point we must start to hit individuals sufficiently different from modern *Homo sapiens* to deserve a different genus name: say *Australopithecus*. The trouble is, 'sufficiently different from modern *Homo sapiens*' is another matter entirely from 'sufficiently different from the earliest *Homo*', here designated *Homo habilis*. Think about the first specimen of *Homo habilis* to be born. Her parents were *Australopithecus*. She belonged to a different genus from her parents? That's just dopey! Yes it certainly is. But it is not reality that's at fault, it's our human insistence on shoving everything into a named category. In reality, there was no such creature as the first specimen of *Homo habilis*. There was no first specimen of any species or any genus or any order or any class or any phylum. Every creature that has ever been born would have been classified – had there been a zoologist around to do the classifying – as belonging to exactly the same species as its parents and its children. Yet, with the hindsight of modernity, and with the benefit – yes, in this one paradoxical sense *benefit* – of the fact that most of the links are missing, classification into distinct species, genera, families, orders, classes and phyla becomes possible.

I wish we really did have a complete and unbroken trail of fossils, a cinematic record of all evolutionary change as it happened. I wish it, not least because I'd love to see the egg all over the faces of those zoologists and anthropologists who engage in lifelong feuds with each other over whether such and such a fossil belongs to this species or that, this genus or that. Gentlemen – I wonder why it never seems to be ladies – you are arguing about words, not reality. As Darwin himself said, in *The Descent of Man*, 'In a series of forms graduating insensibly from some apelike creature to man as he now exists, it would be impossible to fix on any definite point where the term "man" ought to be used.'

Let's move on through the fossils, and look at some more recent links among those that are no longer missing, although they were missing in Darwin's time. What intermediates can we find between ourselves and the various creatures like 1470 and Twiggy, who are sometimes called *Homo* and sometimes called *Australopithecus*? We've already met some of them, as Java Man and Peking Man, normally classified as *Homo erectus*. But those two lived in Asia, and there's good evidence that most of our human evolution took place in Africa. Java Man and Peking Man and their kind were emigrants from the mother continent of Africa. Within Africa itself, their equivalents are nowadays usually classified as *Homo ergaster*, although for many years they were all called *Homo erectus* – yet another

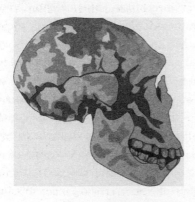

Homo erectus

illustration of the fickleness of our naming procedures. The most famous specimen of *Homo ergaster*, and one of the most complete pre-human fossils ever found, is the Turkana Boy, or Nariokotome Boy, discovered by Kamoya Kimeu, star fossil-finder of Richard Leakey's team of palaeontologists.

The Turkana Boy lived approximately 1.6 million years ago and died at the age of about eleven. There are indications that he would have grown to a height of 6 feet if he had lived to adulthood. His projected adult brain volume would have been about 900 cubic centimetres (cc). This was typical of *Homo ergaster/erectus* brains, which varied around 1,000 cc. It is significantly smaller than modern human brains, which vary around 1,300 or 1,400 cc, but larger than *Homo habilis* (around 600 cc) which in turn was larger than *Australopithecus* (around 400 cc) and chimpanzees (around the same). You'll remember we concluded that our ancestor of three million years ago had the brain of a chimpanzee but walked on its hind legs. From this we might presume that the second half of the story, from 3 million years ago to recent times, would be a tale of increasing brain size. And so, indeed, it proves.

Homo ergaster/erectus, of which we have many fossil specimens, is a very persuasive halfway link, no longer missing, between *Homo sapiens* today and *Homo habilis* two million years ago, which is in turn a beautiful link back to *Australopithecus* three million years ago, which, as we saw, could pretty well be described as an upright-walking chimpanzee. How many links do you need, before you concede that they are no longer 'missing'? And can we also bridge the gap between *Homo ergaster* and modern *Homo sapiens*? Yes: we have a rich lode of fossils, covering the last few hundred thousand years, which are intermediate between them. Some have been given species names, like *Homo heidelbergensis*, *Homo rhodesiensis* and *Homo neanderthalensis*. Others (and sometimes the same ones) are called 'archaic' *Homo sapiens*. But, as I keep repeating, names don't matter. What matters is that the links are no longer missing. Intermediates abound.

JUST GO AND LOOK

So, we have fine fossil documentation of gradual change, all the way from Lucy, the 'upright-walking chimp' of three million years ago, to ourselves today. How do history-deniers cope with this evidence? Some by literal denial. I encountered this in an interview I did for the Channel Four television documentary *The Genius of Charles Darwin* in 2008. I was interviewing Wendy Wright, President of 'Concerned Women for America'. Her opinion that 'The morning-after pill is a pedophile's best friend' gives a fair idea of her powers of reasoning, and she fully lived up to expectation during our interview. Only a very small part of the interview was used for the television documentary. What follows is a much fuller transcript, but obviously for the purposes of this chapter I have confined myself to those places where we discussed the fossil record of human ancestry.

> *Wendy*: What I go back to is the evolutionists are still lacking the science to back it up. But instead what happens is science that doesn't bolster the case for evolution gets censored out. Such as there is no evidence of evolution from going from one species to another species. If that, if evolution had occurred then surely whether it's going from birds to mammals or, or, even beyond that surely there'd be at least one evidence.

> *Richard*: There's a massive amount of evidence. I'm sorry but you people keep repeating that like a kind of mantra because you, you, just listen to each other. I mean, if only you would just open your eyes and look at the evidence.

> *Wendy*: Show it to me, show me the, show me the bones, show me the carcass, show me the evidence of the in-between stages from one species to another.

Richard: Every time a fossil is found which is in between one species and another you guys say, 'Ah, now we've got two gaps where there, where previously there was only one.' I mean almost every fossil you find is intermediate between something and something else.

Wendy [*laughs*]: If that were the case, the Smithsonian Natural History Museum would be filled with these examples but it isn't.

Richard: It is, it is . . . in the case of humans, since Darwin's time there's now an enormous amount of evidence about intermediates in human fossils and you've got various species of *Australopithecus* for example, and . . . then you've got *Homo habilis* – these are intermediates between *Australopithecus* which was an older species and *Homo sapiens* which is a younger species. I mean, why don't you see those as intermediates?

Wendy: . . . if evolution has had the actual evidence then it would be displayed in museums not just in illustrations.

Richard: I just told you about *Australopithecus, Homo habilis, Homo erectus, Homo sapiens* – archaic *Homo sapiens* and then modern *Homo sapiens* – that's a beautiful series of intermediates.

Wendy: You're still lacking the material evidence so . . .

Richard: The material evidence is there. Go to the museum and look at it . . . I don't have them here obviously, but you can go to any museum and you can see *Australopithecus*, you can see *Homo habilis*, you can see *Homo erectus*, you can see archaic *Homo sapiens* and modern *Homo sapiens*. A beautiful series of intermediates Why do you keep saying 'Present me with the evidence' when I've done so? Go to the museum and look.

Wendy: And I have. I have gone to the museums and there are so many of us who still are not convinced . . .

Richard: Have you seen, have you seen *Homo erectus*?

Wendy: And I think there's this effort, this rather aggressive effort to try and talk over us and to censor us. It seems to come out of a frustration that so many people still don't believe in evolution. Now if evolutionists were so confident in their beliefs there wouldn't be the effort to censor out information. It shows that evolution is still lacking and is questionable.

Richard: I am . . . I confess to being frustrated. It's not about suppression, it's about the fact that I have told you about four or five fossils . . . [*Wendy laughs*] . . . and you seem to simply be ignoring what I'm saying . . . Why don't you go and look at those fossils?

Wendy: . . . If they were in the museums which I've been to many times, then I would look at them objectively, but what I go back to is . . .

Richard: They are in the museum.

Wendy: What I go back to is that the philosophy of evolution can lead to ideologies that have been so destructive to the human race . . .

Richard: Yes, but wouldn't it be a good idea, instead of pointing to misperceptions of Darwinism, which have been hideously misused politically, if you tried to understand Darwinism, then you'd be in a position to counteract these horrible misunderstandings.

Wendy: Well actually we are so often forced by the aggressiveness of those who favour evolution. It's not as if we are hidden from this information that you keep presenting.

It's not as if it is unknown to us, because we can't get away from it. It's pushed on us all the time. But I think your frustration comes from the fact that so many of us who have seen your information still don't buy into your ideology.

Richard: Have you seen *Homo erectus*? Have you seen *Homo habilis*? Have you seen *Australopithecus*? I've asked you that question.

Wendy: What I've seen is that in the museums and in the textbooks whenever they claim to show the evolutionary differences from one species to another, it relies on illustrations and drawings . . . not any material evidence.

Richard: Well, you might have to go to the Nairobi Museum to see the original fossils but you can see casts of fossils – exact copies of these fossils in any major museum you care to look at.

Wendy: Well, let me ask you why you are so aggressive? Why is it so important to you that everyone believes like you believe?

Richard: I'm not talking about belief, I'm talking about facts. I've told you about certain fossils, and every time I ask you about them you evade the question and turn to something else.

Wendy: . . . There should be overwhelming tons of material evidence not just an isolated thing, but again, there is not evidence.

Richard: I happened to pick hominid fossils because I thought you'd be most interested in them, but you can find similar fossils from any vertebrate group you care to name.

Wendy: But I guess I go back to why is it so important to you that everyone believes in evolution . . .

Richard: I don't like the word belief. I prefer to just ask people to look at the evidence, and I'm asking you to look at the

evidence . . . I want you to go to museums and look at the
facts and don't believe what you have been told that there is
no evidence. Just go and look at the evidence.

Wendy [*laughs*]: And yes, and what I would say . . .

Richard: It's not funny. I mean, really go, go. I've told you
about hominid fossils, and you can go and see the evolution
of the horse, you can go and look at the evolution of the early
mammals, you can go and look at the evolution of fish, you
can go and look at the transition from fish to land-living
amphibians and reptiles. Any of those things you will find in
any good museum. Just open your eyes and look at the facts.

Wendy: And I would say open your eyes and see the
communities that have been built by those who believe in a
loving God who created each one of us . . .

It might seem, in that exchange, that I was being needlessly obstinate
in hammering home the request that she should go to a museum
and look, but I really meant it. These people have been coached to
say, 'There are no fossils, show me the evidence, show me just one
fossil . . .' and they say it so often that they come to believe it. So I
tried the experiment of mentioning three or four fossils to this
woman and not letting her get away with simply ignoring them.
The results are depressing, and a good example of the commonest
tactic used by history-deniers when confronted with the evidence of
history – namely, just ignore it and repeat the mantra: 'Show me the
fossils. Where are the fossils? There are no fossils. Just show me one
intermediate fossil, that's all I ask . . .'

Others befuddle themselves with names, and the inevitable
tendency names have to make false divisions where there are
none. Every fossil that might potentially be intermediate is always
classified as either *Homo* or *Australopithecus*. None is ever classified
as intermediate. Therefore there are no intermediates. But, as I have

explained above, this is an inevitable consequence of the conventions of zoological nomenclature, not a fact about the real world. The most perfect intermediate you could possibly imagine would *still* find itself shoehorned into either *Homo* or *Australopithecus*. In fact, it would probably be called *Homo* by half the palaeontologists and *Australopithecus* by the other half. And unfortunately, instead of getting together to agree that ambiguously intermediate fossils are exactly what we should *expect* on the evolution theory, the palaeontologists could probably be relied upon to give an entirely false impression by seeming almost to come to blows over their terminological disagreement.

It's a bit like the legal distinction between an adult and a minor. For legal purposes, and for deciding whether a young person is old enough to vote or join the army, it is necessary to make an absolute distinction. In 1969 the legal voting age in Britain was lowered from twenty-one to eighteen (in 1971 the same change was made in the USA). Now there's talk of lowering it to sixteen. But, whatever the legal voting age may be, nobody seriously thinks the stroke of midnight on the eighteenth (or twenty-first, or sixteenth) birthday actually turns you into a different kind of person. Nobody seriously believes there are two kinds of people, children and adults, with 'no intermediates'. Obviously we all understand that the whole period of growing up is one long exercise in intermediacy. Some of us, it might be said, have never really grown up. Similarly, human evolution, from something like *Australopithecus afarensis* to *Homo sapiens*, consisted of an unbroken series of parents giving birth to children who would certainly have been placed, by a contemporary taxonomist, in the same species as their parents. With hindsight, and for reasons that are not far from legalistic, modern taxonomists insist on tying a label around each fossil, which must say something like *Australopithecus* or *Homo*. Museum labels are positively not *allowed* to say 'halfway between *Australopithecus africanus* and *Homo habilis*'. History-deniers seize upon this naming convention as though it were *evidence* of a lack of intermediates in the real world. You might as well say there is no such thing as an adolescent because every single person you look at turns out to be either a voting adult (eighteen or over) or a non-voting child (under eighteen). It's tantamount to saying that the legal necessity for a voting age threshold proves that adolescents don't exist.

Back to the fossils again. If the creationist apologists are right, *Australopithecus* is 'just an ape', so its own predecessors are irrelevant to the search for 'missing links'. Nevertheless, we may as well look at them. There are a few, albeit rather fragmentary, traces. *Ardipithecus*, which lived 4–5 million years ago, is known mainly from teeth, but enough cranial and foot bones have been found to suggest, at least to most anatomists who have attended to it, that it walked upright. Much the same conclusion has been drawn by the respective discoverers of two even older fossils, *Orrorin* ('Millennium Man') and *Sahelanthropus* ('Toumai', below).

Sahelanthropus is remarkable in being very old (six million years, close in age to the common ancestor with chimpanzees) and in being found far west of the Rift Valley (in Chad, where its nickname, 'Toumai', means 'hope of life'). Other palaeoanthropologists are sceptical of the claims to bipedality that have been made on behalf of *Orrorin* and *Sahelanthropus* by their discoverers. And, as a cynic might note, for each such problematic fossil some of the doubters include the discoverers of the others!

Sahelanthropus

Palaeoanthropology, more than other fields of science, is notoriously plagued – or is it enlivened? – by rivalries. We have to admit that the fossil record connecting the upright-walking ape *Australopithecus* to the (presumably) quadrupedal ancestor that we share with chimpanzees is still poor. We don't know how our ancestors rose on to their hind legs. We need more fossils. But let's at least rejoice in the good fossil record that we – unlike Darwin – can enjoy, showing us the evolutionary transition from *Australopithecus*, with its chimpanzee-sized brain, to modern *Homo sapiens* with our balloon-like skull and big brain.

Throughout this section I have reproduced pictures of skulls and encouraged you to compare them. You perhaps noticed, for example, the protrusion of the muzzle in some fossils, or of the brow ridge. Sometimes the differences are quite subtle, which aids appreciation of the gradual transitions from one fossil to a later one. But now I want to introduce a complication, which will develop into an interesting point in its own right. The changes that take place within an individual's lifetime, as it grows up, are in any case much more dramatic than the changes we see as we compare adults in successive generations.

The skull below belongs to a chimpanzee shortly before birth. It is obviously completely different from the adult chimpanzee

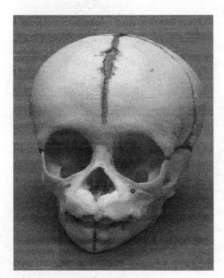

skull shown on page 187 and far more like a human (adult human as well as baby human). There's a much-reproduced picture (reproduced again overleaf) of an infant and an adult chimpanzee, which is often used to illustrate the interesting idea that in human evolution juvenile characteristics were retained into adulthood (or – which may not necessarily be quite the same thing – we become sexually mature when our bodies are still juvenile). I

**Chimpanzee shortly
before birth**

Lang's photos of infant and adult chimpanzees

thought the picture looked too good to be true, and I sent it to my colleague Desmond Morris for his expert opinion. Could it be a fake, I asked him? Had he ever seen a young chimp looking quite so human? Dr Morris is sceptical about the back and shoulders but is happy about the head itself. 'Chimps are characteristically hunched-up in posture and this one has a wonderfully upright human neck. But if you just take the head alone, the picture can be trusted.' Sheila Lee, the publishers' picture researcher for this book, tracked down the original source of this famous photograph, an expedition to the Congo in 1909–15 mounted by the American Museum of Natural History. The animals were dead when photographed, and she points out that the photographer, Herbert Lang, was also a taxidermist. It would be tempting to surmise that the oddly human posture of the baby chimpanzee is due to bad stuffing – were it not for the fact that, according to the museum, Lang photographed his specimens *before* stuffing them. Nevertheless, the posture of a dead chimp can be adjusted in the way that the posture of a live chimp cannot. Desmond Morris's conclusion seems to stand up. The human-like posture of the baby chimpanzee's shoulders may be suspect, but the head is reliable.

Taking the head at face value, even if the shoulders can't quite bear the burden of authenticity, you can immediately see how a comparison of adult fossil skulls might mislead us. Or, to put it more constructively, the dramatic difference between adult and juvenile heads shows us how easily a characteristic such as muzzle protrusion might change in just the right direction to become more – or indeed less – human. Chimpanzee embryology 'knows' how to make a human-like head, because it does it for every chimp as it passes through its infant years. It seems highly plausible that, as *Australopithecus* evolved through various intermediates to *Homo sapiens*, shortening the muzzle all along the way, it did so by the obvious route of retaining juvenile characteristics into adulthood (the process called neoteny, mentioned in Chapter 2). In any case, a great deal of evolutionary change consists of changes in the rate at which certain parts grow, relative to other parts. This is called heterochronic ('differently timed') growth. I suppose what I want to say is that evolutionary change is a doddle, once you accept the observed facts of embryological change. Embryos are shaped by differential growth – different bits of them grow at different rates. A baby chimpanzee's skull changes into an adult's skull via relatively fast growth of the bones of the jaws and muzzle compared to other bones of the skull. To repeat, every animal of every species changes, during its own embryological development, far more dramatically than the typical adult form changes from generation to generation as the geological ages go by. And this is my cue for a chapter on embryology and its relevance to evolution.

CHAPTER 8

YOU DID IT YOURSELF IN NINE MONTHS

Tʜᴀᴛ irascible genius J. B. S. Haldane, who did so much else besides being one of the three leading architects of neo-Darwinism, was once challenged by a lady after a public lecture. It's a word-of-mouth anecdote, and John Maynard Smith is sadly not available to confirm the exact words, but this is approximately how the exchange went:

> *Evolution sceptic:* Professor Haldane, even given the billions of years that you say were available for evolution, I simply cannot believe it is possible to go from a single cell to a complicated human body, with its trillions of cells organized into bones and muscles and nerves, a heart that pumps without ceasing for decades, miles and miles of blood vessels and kidney tubules, and a brain capable of thinking and talking and feeling.

> *JBS:* But madam, you did it yourself. And it only took you nine months.

The questioner was perhaps momentarily thrown off balance by the veering unexpectedness of Haldane's reply. Wind taken out of sails would have seemed an understatement. But maybe in one respect Haldane's retort left her unsatisfied. I don't know whether she asked a supplementary but, if so, it might have gone along these lines:

> *Evolution sceptic:* Ah yes, but the developing embryo follows genetic instructions. It is the *instructions* for how to build a

complicated body that you, Professor Haldane, claim evolved
by natural selection. And I still find it hard to believe, even
given a billion years for that evolution.

Perhaps she had a point. And even if a divine intelligence did prove
to be ultimately responsible for designing living complexity, it is
definitely not true that he *fashions* living bodies in anything like the
way that clay modellers, for example, or carpenters, potters, tailors
or car manufacturers go about their tasks. We may be 'wonderfully
developed' but we are not 'wonderfully *made*'. When children sing,
'He made their glowing colours / He made their tiny wings',* they
are uttering a childishly obvious falsehood. Whatever else God does,
he certainly doesn't *make* glowing colours and tiny wings. If he did

---

* I have been warned that 'All things bright and beautiful' will not necessarily strike my readers
as nostalgically as it does me. It is an Anglican hymn for children written by Mrs C. F. Alexander
in 1848, comfortably extolling the beauties of nature (and, in one verse, the political status quo)
with the refrain, 'The Lord God made them all'. It is the subject of a splendid parody written by
Eric Idle and sung by the Monty Python team:

All things dull and ugly
All creatures short and squat
All things rude and nasty
The Lord God made the lot.

Each little snake that poisons
Each little wasp that stings
He made their brutish venom
He made their horrid wings.

All things sick and cancerous
All evil great and small,
All things foul and dangerous
The Lord God made them all.

Each nasty little hornet
Each beastly little squid
Who made the spiky urchin?
Who made the sharks? He did!

All things scabbed and ulcerous
All pox both great and small
Putrid, foul and gangrenous
The Lord God made them all.

anything at all, it would be to supervise the embryonic development of things, for example by splicing together sequences of genes that direct a process of automated development. Wings are not made, they grow – progressively – from limb buds inside an egg.

God, to repeat this important point, which ought to be obvious but isn't, never made a tiny wing in his eternal life. If he made anything (he didn't in my view, but let it pass, that's not what I'm about here), what he made was an embryological *recipe*, or something like a computer program for controlling the embryonic development of a tiny wing (plus lots of other things too). Of course, God might claim that it is just as clever, just as breathtaking a feat of skill, to design a recipe or a program for a wing, as to make a wing. But for the moment, I just want to develop the distinction between *making* something like a wing, and what really happens in embryology.

NO CHOREOGRAPHER

The early history of embryology was riven between two opposing doctrines called preformationism and epigenesis. The distinction between them is not always clearly understood, so I shall spend a little time explaining these two terms. The preformationists believed that the egg (or sperm, for the preformationists were subdivided into 'ovists' versus 'spermists') contained a tiny miniature baby or 'homunculus'. All the parts of the baby were intricately in place, correctly disposed to each other, waiting only to be inflated like a compartmentalized balloon. This raises obvious problems. First, at least in its early naïve form, it requires what everybody knows to be false: that we inherit only from one parent – the mother for the ovists, the father for the spermists. Second, preformationists of this kind had to face a Russian-doll-style infinite regress of homunculi within homunculi   or if not infinite, at least long enough to take us back to Eve (Adam for the spermists). The only escape from the regress would be to construct the homunculus afresh in every generation by elaborately scanning the adult body of the previous generation. This 'inheritance of acquired characteristics' doesn't happen – otherwise Jewish boys would be born without foreskins, and gym-frequenting body-builders (but not their couch-potato twins) would conceive babies with rippling six-packs, pecs and glutes.

To be fair to the preformationists, they did face up, fairly and squarely, to the logical necessity of the regress, however absurd it seemed. At least some of them really did believe that the first woman (or man) contained miniaturized embryos of all her descendants, nested inside each other like Russian dolls. And there is a sense in which they had to believe that: a sense that is worth mentioning because it prefigures the nub of this chapter. If you believe Adam was 'made' rather than being born, you imply that Adam didn't have genes – or at least didn't need them in order to develop. Adam had no embryology but just sprang into existence. A related inference led the Victorian writer Philip Gosse (the father in Edmund Gosse's *Father and Son*) to write a book called *Omphalos* (Greek for 'navel') arguing that Adam must have had a navel, even though he was never born. A more sophisticated consequence of omphalogical reasoning would be that stars whose distance from us is more than a few thousand light years must have been created with ready-made light beams stretching almost all the way to us – otherwise we wouldn't be able to see them until the distant future! Making fun of omphalogy sounds frivolous, but there is a serious point here about embryology, which is the subject of this chapter. It is quite a difficult point to grasp – indeed, I am only in the process of grasping it myself – and I am approaching it from various directions.

For the reasons given, preformationism, at least in its original 'Russian doll' version, was always a non-starter. Is there a version of preformationism that could be sensibly revived in the DNA age? Well, perhaps, but I doubt it. Textbooks of biology repeat time and again that DNA is a 'blueprint' for building a body. It isn't. A true blueprint of, say, a car or a house embodies a one-to-one mapping from paper to finished product. It follows from this that a blueprint is reversible. It is as easy to go from house to blueprint as the other way around, precisely because it is a one-to-one mapping. Actually, it's easier, because you have to *build* the house, but you only have to take some measurements and then *draw* the blueprint. If you take an animal's body, no matter how many detailed measurements you take, you can't reconstruct its DNA. That's what makes it false to say that DNA is a blueprint.

It is theoretically possible to imagine – maybe that's the way things work on some alien planet – that DNA might have been a coded description of a body: a kind of three-dimensional map rendered into the linear code of DNA 'letters'. That really would be reversible. Scanning the body to make a genetic blueprint is not a totally ridiculous idea. If that is how DNA worked, we could represent it as a kind of neo-preformationism. It wouldn't raise the spectre of the Russian dolls. It isn't clear to me whether it would raise the spectre of inheritance from only one parent. DNA has a breathtakingly precise way of intersplicing half the paternal information with exactly half the maternal information, but how would it go about intersplicing half a scan of the mother's body with half a scan of the father's body? Let it pass: this is all so far from reality.

DNA, then, is emphatically not a blueprint. Unlike Adam, who was fashioned directly into his adult form, all real bodies develop and grow from a single cell through the intermediate stages of embryo, foetus, baby, child and adolescent. Maybe in some alien world living creatures assemble themselves from tip to toe as an ordered set of three-dimensional bio-pixels read out from a coded scan line. But that is not the way things work on our planet, and actually I think there are reasons – which I have dealt with elsewhere and so won't go into here – why it could never be so on any planet.*

---

* Note for professionals at the interface between biologists and computer scientists: Charles Simonyi, who speaks with the authority of a distinguished software designer, put it as follows, after reading an early draft of this chapter: '. . . the recipe (of the eye, brain, blood, etc.) is much much simpler than the blueprint for the same (in terms of bits or base-pairs) so evolution would be literally impossible (in less than 10^{100} years) especially because small variations in the (*cont.*)

The historical alternative to preformationism is epigenesis. If preformationism is all about blueprints, epigenesis is about something more like a recipe or a computer program. The *Shorter Oxford English Dictionary*'s definition is rather modern, and I'm not sure that Aristotle, who coined the word, would recognize it:

> **epigenesis**: A theory of the development of an organism by progressive differentiation of an initially undifferentiated whole.*

Principles of Development, by Lewis Wolpert and colleagues, describes epigenesis as the idea that new structures arise progressively. There is a sense in which epigenesis is self-evidently true, but details matter and the devil is in the cliché. How does the organism develop progressively? How does an initially undifferentiated whole 'know' how to differentiate progressively, if not by following a blueprint? The distinction I want to make in this chapter, which largely corresponds to the distinction between preformationism and epigenesis, is the distinction between planned architecture and *self-assembly*. The meaning of planned architecture is clear to us because we see it all around us in our buildings and other artefacts. Self-assembly is less familiar, and will need some attention from me. In the field of development, self-assembly occupies a position analogous to natural selection in evolution, although it is definitely not the same process. Both achieve, by automatic, non-deliberate, unplanned means, results that look, to a superficial gaze, as though they were meticulously planned.

(*cont.*) blueprint would be unlikely to have any positive effect, whereas a variation in the recipe would.' Alluding to my own 'computer biomorphs' and 'arthromorphs' (see Chapter 2), Dr Simonyi goes on: 'The artificial creatures that you [programmed for *The Blind Watchmaker* and *Climbing Mount Improbable*] are all described by recipes, not by blueprint – a blueprint would be just a jumble of vectors of black lines – can you imagine trying to play evolution on them by varying the endpoints of the black lines one at a time or even two at a time?' As you'd expect from one described by Bill Gates, no less, as 'one of the great programmers of all time', this is exactly right for the computer biomorphs, and it is surely right for living things too.

* There is a risk that 'epigenesis' will be confused with 'epigenetics', a modish buzz-word now enjoying its fifteen minutes of fame in the biological community. Whatever 'epigenetics' might mean (and its enthusiasts cannot seem to agree even with themselves, let alone with each other), all I intend to say about it here is that it is not the same thing as epigenesis.

J. B. S. Haldane spoke simple truth to his sceptical questioner, but he would not have denied that there is mystery, verging on the miraculous (but never quite getting there) in the very fact that a single cell gives rise to a human body in all its complexity. And the mystery is only somewhat mitigated by the feat's being achieved with the aid of DNA instructions. The reason the mystery remains is that we find it hard to imagine, even in principle, how we might set about writing the instructions for building a body in the way that the body is in fact built, namely by what I have just called 'self-assembly', which is related to what computer programmers sometimes call a 'bottom-up', as opposed to 'top-down', procedure.

An architect designs a great cathedral. Then, through a hierarchical chain of command, the building operation is broken down into separate departments, which break it down further into sub-departments, and so on until instructions are finally handed out to individual masons, carpenters and glaziers, who go to work until the cathedral is built, looking pretty much like the architect's original drawing. That's top-down design.

Bottom-up design works completely differently. I never believed this, but there used to be a myth that some of the finest medieval cathedrals in Europe had no architect. Nobody designed the cathedral. Each mason and carpenter would busy himself, in his own skilled way, with his own little corner of the building, paying scant attention to what the others were doing and no attention to any overall plan. Somehow, out of such anarchy, a cathedral would emerge. If that really happened, it would be bottom-up architecture. Notwithstanding the myth, it surely didn't happen like that for cathedrals.* But that pretty much *is* what happens in the building of a termite mound or an ant's nest – and in the development of an embryo. It is what makes embryology so remarkably different from anything we humans

---

* My medieval historian colleague Dr Christopher Tyerman confirms that this was indeed a myth that was invented in Victorian times for idealistic reasons, but that there was never a scintilla of truth in it.

are familiar with, in the way of construction or manufacture.

The same principle works for certain types of computer program, for certain types of animal behaviour, and – bringing the two together – for computer programs that are designed to simulate animal behaviour. Suppose we wanted to understand the flocking behaviour of starlings. There are some stunning films available on YouTube, from which the stills on colour page 16 are taken. These balletic manœuvres were photographed over Otmoor, near Oxford, by Dylan Winter. What is remarkable about the starlings' behaviour is that, despite all appearances, there is no choreographer and, as far as we know, no leader. Each individual bird is just following local rules.

The numbers of individual birds in these flocks can run into thousands, yet they almost literally never collide. That is just as well for, given the speed at which they fly, any such impact would severely injure them. Often the whole flock seems to behave as a single individual, wheeling and turning as one. It can look as though the separate flocks are moving through each other in opposite directions, maintaining their coherence intact as separate flocks. This makes it seem almost miraculous, but actually the flocks are at different distances from the camera and do not literally move through each other. It adds to the aesthetic pleasure that the edges of the flocks are so sharply defined. They don't peter off gradually, but come to an abrupt boundary. The density of the birds just inside the boundary is no less than in the middle of the flock, while it is zero outside the boundary. As soon as you think about it in that way, isn't it wondrously surprising?

The whole performance would make a more than usually elegant screensaver on a computer. You wouldn't want a real film of starlings because your screensaver would repeat the same identical balletic moves over and over, and therefore wouldn't exercise all the pixels equally. What you would want is a computer *simulation* of starling flocks; and, as any programmer will tell you, there's a right way and a wrong way to do it. Don't try to choreograph the whole ballet – that would be terribly bad programming style for this kind of task. I need to talk about the better way to do it because something like it is almost certainly how the birds themselves are programmed, in their brains. More to the point, it is a great analogy for how embryology works.

Here's how to program flocking behaviour in starlings. Devote almost all your effort to programming the behaviour of a single individual bird. Build into your robo-starling detailed rules for how to fly, and how to react to the presence of neighbouring starlings, depending on their distance and relative position. Build in rules for how much weight to give to the behaviour of neighbours, and how much weight to give to individual initiative in changing direction. These model rules would be informed by careful measurements of real birds in action. Endow your cyberbird with a certain tendency to vary its rules at random. Having written a complicated program to specify the behavioural rules of a single starling, now comes the definitive step that I am emphasizing in this chapter. *Don't* try to program the behaviour of a whole flock, as an earlier generation of computer programmers might have done. Instead, clone the single computer starling you have programmed. Make a thousand copies of your robo-bird, maybe all the same as each other, or maybe with some slight random variation among them in their rules. And now 'release' thousands of model starlings in your computer, so they are free to interact with each other, all obeying the same rules.

If you've got the behavioural rules right for a single starling, a thousand computer starlings, each one a dot on the screen, will behave like real starlings flocking in winter. If the flocking behaviour isn't quite right, you can go back and adjust the behaviour of the individual starling, perhaps in the light of further measurements of the behaviour of real starlings. Now clone up the new version a thousand times, in place of the thousand that didn't quite work. Keep iterating your reprogramming of the cloned-up single starling, until the flocking behaviour of thousands of them on the screen is a satisfyingly realistic screensaver. Calling it 'Boids', Craig

Reynolds wrote a program along these lines (not specifically for starlings) in 1986.

The key point is that there is no choreographer and no leader. Order, organization, structure – these all *emerge* as by-products of rules which are obeyed *locally* and many times over, not globally. And that is how embryology works. It is all done by local rules, at various levels but especially the level of the single cell. No choreographer. No conductor of the orchestra. No central planning. No architect. In the field of development, or manufacture, the equivalent of this kind of programming is *self-assembly*.

The body of a human, an eagle, a mole, a dolphin, a cheetah, a leopard frog, a swallow: these are so beautifully put together, it seems impossible to believe that the genes that program their development don't function as a blueprint, a design, a master plan. But no: as with the computer starlings, it is all done by individual cells obeying local rules. The beautifully 'designed' body *emerges* as a consequence of rules being *locally* obeyed by individual cells, with no reference to anything that could be called an overall global plan. The cells of a developing embryo wheel and dance around each other like starlings in gigantic flocks. There are differences, and they are important. Unlike starlings, cells are physically attached to each other in sheets and blocks: their 'flocks' are called 'tissues'. When they wheel and dance like miniature starlings, the consequence is that three-dimensional shapes are formed, as tissues invaginate in response to the movements of cells;* or swell or shrink due to local patterns of growth and cell death. The analogy I like for this is the paper-folding art of origami, suggested by the distinguished embryologist Lewis Wolpert in his book *The Triumph of the Embryo*; but before coming to that I need to clear out of the way some alternative analogies that might come to mind – analogies from among human crafts and manufacturing processes.

---

* Invaginate: 'fold inwards to form a hollow', 'turn or double back within itself' (*Shorter Oxford English Dictionary*).

ANALOGIES FOR DEVELOPMENT

It is surprisingly hard to find a good analogy for the development of living tissue, but you can find partial similarities to particular aspects of the process. A recipe captures something of the truth, and it is an analogy that I sometimes use, to explain why 'blueprint' is not appropriate. Unlike a blueprint, a recipe is irreversible. If you follow a cake recipe step by step, you'll end up with a cake. But you can't take a cake and reconstruct the recipe – certainly not the exact words of the recipe – whereas, as we have seen, you could take a house and reconstruct something close to the original blueprint. This is because of the one-to-one mapping between bits of house and bits of blueprint. With conspicuous exceptions such as the cherry on top, there is no one-to-one mapping between bits of cake and the words, say, or sentences of its recipe.

What other analogies to human manufacturing might there be? Sculpture is mostly way off the mark. A sculptor starts with a chunk of stone or wood and fashions it by subtraction, chipping away until the desired shape is all that remains. There is, admittedly, a somewhat sharp resemblance to one particular process in embryology called apoptosis. Apoptosis is programmed cell death, and it is involved, for example, in the development of fingers and toes. In the human embryo, the fingers and toes are all joined. In the womb, you and I had webbed feet and hands. The webbing disappeared (in most people: there are occasional exceptions) through programmed cell death. That is a bit reminiscent of the way a sculptor carves out a shape, but it is not common enough or important enough to capture how embryology normally works. Embryologists may briefly think 'sculptor's chisel', but they don't let the thought linger for long.

Some sculptors work not by subtractive carving but by taking a lump of clay, or soft wax, and kneading it into shape (which may subsequently be cast, in bronze for example). That again is not a good analogy for embryology. Nor is the craft of tailoring or dressmaking. Pre-existing cloth is cut, to shapes set out in a pre-planned pattern, then sewn together with other cut-out shapes. They are often then turned inside out to disguise the seams – and that bit, at least, is a good analogy to certain parts of embryology. But in general, embryology is no more like tailoring than it is like sculpture. Knitting might be better, in that the whole shape of a sweater, say, is built up from numerous individual stitches, like individual cells. But there are better analogies, as we shall see.

How about the assembly of a car, or other complicated machine, on a factory assembly line: is that a good analogy? Like sculpture and tailoring, assembly of pre-fabricated parts is an efficient way to make something. In a car factory, parts are pre-made, often by casting in moulds in a foundry (and there is, I think, nothing remotely like casting in embryology). Then the pre-made parts are brought together on an assembly line and screwed, riveted, welded or glued together, step by step according to a precisely drawn plan. Once again, embryology has nothing resembling a previously drawn plan. But there are resemblances to the ordered sticking together of pre-assembled parts, as when, in a car assembly plant, previously manufactured carburettors and distributor heads and fan belts and cylinder heads are brought together and joined in correct apposition.

Below are three kinds of virus. On the left is the tobacco mosaic virus (TMV), which parasitizes tobacco plants and other members of the family Solanaceae, such as tomatoes. In the middle is an

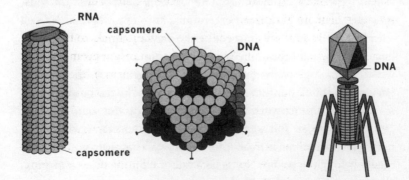

Three kinds of virus

adenovirus, which infects the respiratory system in many animals, including us. On the right is the T4 bacteriophage, which parasitizes bacteria. It looks like a lunar lander, and it behaves rather like one, 'landing' on the surface of a bacterium (which is very much larger) then lowering itself on its spidery 'legs', then thrusting a probe down the middle, through the bacterium's cell wall, and injecting its DNA inside. The viral DNA then hijacks the protein-making machinery of the bacterium, which is subverted into making new viruses. The other two viruses in the picture do something similar, although they don't look or behave like lunar landers. In all cases their genetic material hijacks the protein-making apparatus of the host cell and diverts its molecular production line to churning out viruses instead of its normal products.

Most of what you see in the pictures is the protein container for the genetic material, and in ('lunar lander') T4's case the machinery for infecting the host. What is interesting is the way in which this protein apparatus is put together. It really is self-assembled. Each virus is assembled from several previously made protein molecules. Each protein molecule, in a way that we shall see, has previously self-assembled into a characteristic 'tertiary structure' under the laws of chemistry given its particular sequence of amino acids. And then, in the virus, the protein molecules join up with each other to form a so-called 'quaternary structure', again by following local rules. There is no global plan, no blueprint.

The protein sub-units, which link up like Lego bricks to form the quaternary structure, are called capsomeres. Notice how geometrically perfect these little constructions are. The adenovirus in the middle has exactly 252 capsomeres, drawn here as little balls, arranged in an icosahedron. The icosahedron is that Platonic perfect solid that has 20 triangular faces. The capsomeres are arranged into an icosahedron not by any kind of master plan or blueprint but simply by each one of them obeying the laws of chemical attraction locally when it bumps into others like itself. This is how crystals are formed, and, indeed, the adenovirus could be described as a very small hollow crystal.

The 'crystallization' of viruses is an especially beautiful example of the 'self-assembly' that I am touting as a major principle by which living creatures are put together. The T4 'lunar lander' phage also has an icosahedron for its main DNA receptacle, but its self-assembled quaternary structure is more complex, incorporating additional protein units, assembled according to different local rules, in the injection apparatus and the 'legs' that are attached to the icosahedron.

Returning from viruses to the embryology of larger creatures, I come to my favourite analogy among human construction techniques: origami. Origami is the art of constructive paper-folding, developed to its most advanced level in Japan. The only origami creation I know how to make is the 'Chinese Junk'. I was taught it by my father, who learned it in a craze that swept through his boarding school during the 1920s.* One biologically realistic feature is that the 'embryology' of the Chinese junk passes through several intermediate 'larval' stages, which are in themselves pleasing creations, just as a caterpillar is a beautiful, working intermediate on the way to a butterfly, which it scarcely resembles at all. Starting with a simple square piece of paper, and simply folding it – never cutting it, never glueing it and never importing any other pieces – the procedure takes us through three recognizable 'larval stages': a 'catamaran', a 'box with two lids' and a 'picture in a frame', before culminating in the 'adult' Chinese junk itself. In favour of the origami analogy, when you first are taught how to make a Chinese junk, not only the junk itself but each of the three 'larval' stages – catamaran, cupboard, picture frame – comes as a surprise. Your hands may do the folding, but you are emphatically not following a blueprint for a Chinese junk, or for any of the larval stages. You are following a set of folding rules that seem to have no connection with the end product, until it finally emerges like a butterfly from its chrysalis. So the origami analogy captures

* The craze died out, but I reintroduced it to the same school in the 1950s, whereupon it spread just like a second epidemic of the same disease.

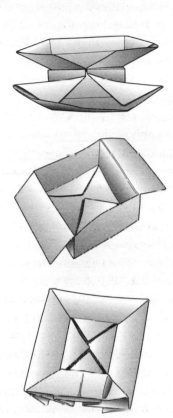

something of the importance of 'local rules' as opposed to a global plan.

Also in favour of the origami analogy, folding, invagination and turning inside out are some of the favourite tricks used by embryonic tissues when making a body. The analogy works especially well for the early embryonic stages. But it has its shortcomings, and here are two obvious ones. First, human hands are needed to do the folding. Second, the developing paper 'embryo' doesn't grow larger. It ends up weighing exactly as much as when it started. To acknowledge the difference, I shall sometimes refer to biological embryology as 'inflating origami', rather than just 'origami'.

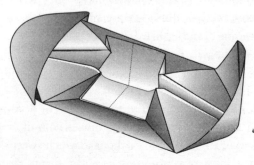

Chinese junk by
origami, with three
'larval stages': 'catamaran',
'box with two lids' and
'picture in a frame'

Actually, these two shortcomings kind of cancel each other out. The sheets of tissue that fold, invaginate and turn inside out in a developing embryo do indeed grow, and it is that very growth that provides part of the motive force which, in origami, is supplied by the human hand. If you wanted to make an origami model with a sheet of living tissue instead of dead paper, there is at least a sporting chance that, if the sheet were to grow in just the right way, not uniformly but faster in some parts of the sheet than in others, this might automatically cause the sheet to assume a certain shape – and even fold or invaginate or turn inside out in a certain way – without the need for hands to do the stretching and folding, and without the need for any global plan, but only local rules. And actually it's more than just a sporting chance, because it really happens. Let's call it 'auto-origami'. How does auto-origami work in practice, in embryology? It works because what happens in the real embryo, when a sheet of tissue grows, is that cells divide. And differential growth of the different parts of the sheet of tissue is achieved by the cells, in each part of the sheet, dividing at a rate determined by local rules. So, by a roundabout route, we return to the fundamental importance of bottom-up local rules as opposed to top-down global rules. It is a whole series of (far more complicated) versions of this simple principle that actually go on in the early stages of embryonic development.

Here's how the origami goes in the early stages of vertebrate development. The single fertilized egg cell divides to make two cells. Then the two divide to make four. And so on, with the number of cells rapidly doubling and redoubling. At this stage there is no growth, no inflation. The original volume of the fertilized egg is literally divided, as in slicing a cake, and we end up with a spherical ball of cells which is the same size as the original egg. It's not a solid ball but a hollow one, and it is called the blastula. The next stage, gastrulation, is the subject of a famous *bon mot* by Lewis Wolpert: 'It is not birth, marriage, or death, but gastrulation, which is truly the most important time in your life.'

Gastrulation is a kind of microcosmic earthquake which sweeps over the blastula's surface and revolutionizes its entire form. The tissues of the embryo become massively reorganized. Gastrulation typically involves a denting of the hollow ball that is the blastula, so that it becomes two-layered with an opening to the outside world (see the computer simulation on p. 231). The outer layer of this 'gastrula' is called the ectoderm, the inner layer is the endoderm, and there are also some cells thrown into the space between the ectoderm and endoderm, which are called mesoderm. Each of these three primordial layers is destined to make major parts of the body. For example, the outer skin and nervous system come from the ectoderm; the guts and other internal organs come from the endoderm; and the mesoderm furnishes muscle and bone.

The next stage in the embryo's origami is called neurulation. The diagram on the right shows a cross-section through the middle of the back of a neurulating amphibian embryo (it could be either a frog or a salamander). The black circle is the 'notochord', a stiffening rod that acts as a precursor of the backbone. The notochord is diagnostic of the phylum Chordata, to which we and all vertebrates belong (although we, like most modern vertebrates, have it only when we are embryos). In neurulation, as in gastrulation, invagination is much in evidence. You remember I said that the nervous system comes from ectoderm. Well, here's how. A section of ectoderm invaginates (progressively backwards along the body like a zip fastener), rolls itself up into a tube, and is pinched off where the sides of the tube 'zip up' so that it ends up running the

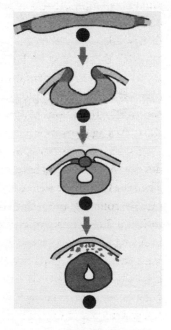

Neurulation

length of the body between the outer layer and the notochord. That tube is destined to become the spinal cord, the main nerve trunk of the body. The front end of it swells up and becomes the brain. And all the rest of the nerves are derived, by subsequent cell divisions, from this primordial tube.*

I don't want to get into the details of either gastrulation or neurulation, except to say that they are wonderful, and that the metaphor of origami holds up pretty well for both of them. I am concerned with the general principles by which embryos become more complicated through inflating origami. Below is one of the things that sheets of cells are observed to do during the course of embryonic development, for example during gastrulation. You can easily see how this invagination could be a useful move in inflating origami, and it does indeed play a major role in both gastrulation and neurulation.

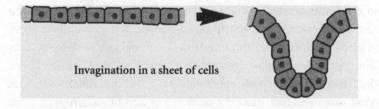

Invagination in a sheet of cells

Gastrulation and neurulation are accomplished early in development and they affect the whole shape of the embryo. Invagination and other 'inflating origami' manœuvres achieve these stages of early embryology, and they and similar tricks are involved later in development, when specialized organs like eyes and the heart

---

* I am sorry I am at a loss to explain why the notochord gets an 'h', like a musical or mathematical chord, while the spinal cord doesn't, like a bit of string. I have always found it mysterious, and have even wondered whether it might represent some long-forgotten but fossilized mistake. Admittedly, the *Oxford English Dictionary* lists 'chord' as an alternative spelling for the string kind of cord, but the difference does seem queer given that the spinal cord and the notochord run the length of the embryonic body, one above the other.

are made. But, given that there are no hands to do the folding, by what mechanical process are these dynamic movements achieved? Partly, as I have already said, by simple expansion itself. Cells multiply all through a sheet of tissue. Its area therefore increases and, having nowhere else to go, it has little choice but to buckle or invaginate. But the process is more controlled than that, and it has been deciphered by a group of scientists associated with the brilliant mathematical biologist George Oster, of the University of California at Berkeley.

MODELLING CELLS LIKE STARLINGS

Oster and his colleagues followed the same strategy we considered earlier in this chapter for a computer simulation of starlings flocking. Instead of programming the behaviour of a whole blastula, they programmed a single cell. Then they 'cloned up' lots of cells, all the same, and watched to see what happened when those cells got together in the computer. When I say they programmed the behaviour of a single cell, it would be better to say they programmed a mathematical model of a single cell, building into the model certain known facts about a single cell, but in simplified form. Specifically, it is known that the interiors of cells are criss-crossed by microfilaments: sort of miniature elastic bands, but with the additional property that they are capable of active contraction, like twitching muscle fibres. Indeed, the microfilaments use the same principle of contraction as muscle fibres.* The Oster model simplified the cell down to two dimensions for drawing on a computer screen, and with only half a dozen filaments, strategically placed in the cell, as you see in the

---

* And that's a fascinating story in its own right, by the way. It has gripped my imagination ever since the great Cambridge physiologist Joseph Needham (a polymath who became even better known as the leading expert on the history of Chinese science) came to my school to demonstrate it, at the invitation of his nephew who happened to be our student teacher at the time: a boon of nepotism for which I remain grateful. Under Dr Needham's guidance, we peered at muscle fibres down our microscopes and watched them shorten, as if by magic, when we gave them a drop of ATP, adenosine triphosphate, the universal energy currency of the body.

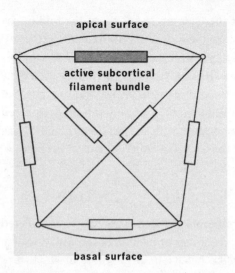

Microfilaments inside
Oster's model cell

diagram above. In the computer model, all the microfilaments were given certain quantitative properties with names that mean something to physicists: a 'viscous damping coefficient' and an 'elastic spring constant'. Never mind exactly what these mean: they are the kinds of things physicists like to measure in a spring. Although it is probable that in a real cell many filaments would be capable of contraction, Oster and his colleagues simplified matters by endowing only one of their six filaments with this capacity. If they could get realistic results even after throwing away some of the known properties of a cell, it would presumably be possible to get at least as good results with a more complicated model that kept those properties in. Rather than allowing the one contractile filament in their model to contract at will, they built into it a property which is common in certain kinds of muscle fibre: when stretched beyond a certain critical length, the fibre would respond by contracting to a much shorter length than the normal equilibrium length.

So, we have our model of a single cell: a greatly simplified model consisting of a two-dimensional outline in which are strung six elastic springs, one of which has the special property of responding to an externally imposed stretch by actively contracting. That is stage one of the modelling process. In stage two, Oster and his colleagues cloned up a few dozen of their model cells and arranged them in a circle, like a (two-dimensional) blastula. Then they took one cell and tweaked its contractile filament to provoke it into contracting. What happened next is almost too wonderful to bear. The model blastula gastrulated! Here are six screenshots showing what happened (a to f below). A wave of contraction spread sideways from the cell that was provoked, and the ball of cells spontaneously invaginated.

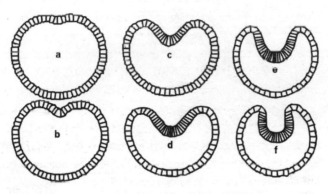

Oster's model blastula gastrulating

It gets even better. Oster and his colleagues tried the experiment, on their computer model, of lowering the 'firing threshold' of the contractile filaments. The result was a wave of invagination that went further, and actually pinched off a 'neural tube' (screenshots a to h, overleaf). It is important to understand what a model such as this really is. It is not an accurate representation of neurulation. Quite apart from the fact that it is two-dimensional and simplified in many other ways, the ball of cells that 'neurulated' (screenshot a) was not a two-layered 'gastrula' as it should have been. It was the same blastula-like starting point as we had for the model of gastrulation above. It doesn't matter: models are not supposed to be totally accurate in every detail. The model still shows how easy it is to mimic various

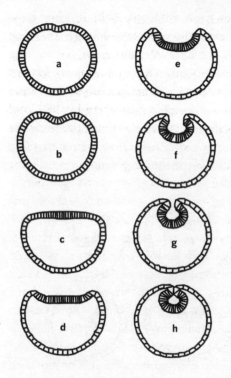

Formation of
'neural tube' in
Oster's model

aspects of the behaviour of cells in an early embryo. The fact that the two-dimensional 'ball' of cells responded spontaneously to the stimulus even though the model is simpler than the real situation makes this a more powerful piece of evidence. It reassures us that the evolution of the various procedures of early embryonic development need not have been all that difficult. Note that it is the model that is simple, not the phenomenon that it demonstrates. That is the hallmark of a good scientific model.

My purpose in expounding the Oster models has been to show the general *kind* of principle by which single cells can interact with each other to build a body, without any blueprint representing the whole body. Origami-like folding, Oster-style invagination and pinching off: these are just some of the simplest tricks for building embryos.

Other more elaborate ones come into play later in embryonic development. For example, ingenious experiments have shown that nerve cells, when they grow out from the spinal cord, or from the brain, find their way to their end organ not by following any kind of overall plan but by chemical attraction, rather as a dog sniffs around to find a bitch in season. An early classic experiment by the Nobel Prize-winning embryologist Roger Sperry illustrates the principle perfectly. Sperry and a colleague took a tadpole and removed a tiny square of skin from the back. They removed another square, the same size, from the belly. They then regrafted the two squares, but each in the other's place: the belly skin was grafted on the back, and the back skin on the belly. When the tadpole grew up into a frog, the result was rather pretty, as experiments in embryology often are: there was a neat postage stamp of white belly skin in the middle of the dark, mottled back, and another neat postage stamp of dark mottled skin in the middle of the white belly. And now for the point of the story. Normally, if you tickle a frog on its back with a bristle, the frog will wipe the place with a foot, as if deterring an irritating fly. But when Sperry tickled his experimental frog on the white 'postage stamp' on its back, it wiped its belly! And when Sperry tickled it on the dark postage stamp on its belly, the frog wiped its back.

What happens in normal embryonic development, according to Sperry's interpretation, is that axons (long 'wires', each one a narrow, tubular extension of a single nerve cell) grow questingly out from the spinal cord, sniffing like a dog for belly skin. Other axons grow out from the spinal cord, sniffing out back skin. And normally this gives the right result: tickles on the back feel as though they are on the back, while tickles on the belly feel as though they are on the belly. But in Sperry's experimental frog, some of the nerve cells sniffing out belly skin found the postage stamp of belly skin grafted on the back, presumably because it smelled right. And vice versa. People who believe in some sort of *tabula rasa* theory – whereby we are all born with a blank sheet for a mind, and fill it in by experience – must be surprised at Sperry's result. They would expect that frogs would learn from experience to feel their way around their own skin, associating the right sensations with the right places on the skin. Instead, it seems that each nerve cell in the spinal cord is labelled, say, a belly nerve cell or a back nerve cell, even before it makes contact with the appropriate skin. It will later find its designated target pixel of skin, wherever it may be. If a fly were to crawl up the length of its back, Sperry's frog would presumably experience the illusion that the fly suddenly leaped from back to belly, crawled a little further, then instantaneously leaped to the back again.

Experiments like this led Sperry to formulate his 'chemo-affinity' hypothesis, according to which the nervous system wires itself up not by following an overall blueprint but by each individual axon seeking out end organs with which it has a particular chemical affinity. Once again, we have small, local units following local rules. Cells in general bristle with 'labels', chemical badges that enable them to find their 'partners'. And we can go back to the origami analogy to find another place where the labelling principle comes in useful. Human paper origami doesn't use glue, but it could. And the origami of the embryo, whereby animal bodies put themselves together, does indeed use something equivalent to glue. Glues, rather, because there are lots of them, and this is where labelling comes triumphantly into its own. Cells have a complicated repertoire of 'adhesion molecules' on their surfaces, whereby they stick to other cells. This cellular glueing plays an important role in embryonic development in all parts of the body. There is a significant difference from the glues that we are familiar with, however. For us, glue is glue is glue. Some glues are stronger than others, and some glues set faster than others, and some glues are more suitable for wood, say, while others work better for metals or plastics. But that's about it for variety among glues.

Cell adhesion molecules are much more ingenious than that. More fussy, you could say. Unlike our artificial glues, which will stick to most surfaces, cell adhesion molecules bind only to particular other cell adhesion molecules of exactly the right kind. One class of adhesion molecules in vertebrates, the cadherins, come in about eighty currently known flavours. With some exceptions, each of these eighty or so cadherins will bind only to its own kind. Forget glue for a minute: a better analogy might be the children's party game where each child is assigned an animal, and they all have to mill about the room making noises like their own allotted animals. Each child knows that only one other child has been assigned the same animal as herself, and she has to find her partner by listening through the cacophony of farmyard imitations. Cadherins work like that. Perhaps, like me, you can dimly imagine how the judicious doping of cell surfaces with particular cadherins at strategic spots might refine and complicate the self-assembly principles of embryo origami. Note, once again, that this doesn't imply any kind of overall plan, but rather a piecemeal collection of local rules.

ENZYMES

Having seen how whole sheets of cells play the origami game in shaping the embryo, let's now dive inside a single cell, where we'll find the same principle of self-folding and self-crumpling, but on a much smaller scale, the scale of the single protein molecule. Proteins are immensely important, for reasons that I must take time to explain, beginning with a teasing speculation to celebrate the unique importance of proteins. I love speculating on how weirdly different we should expect life to be elsewhere in the universe, but one or two things I suspect are universal, wherever life might be found. All life will turn out to have evolved by a process related to Darwinian natural selection of genes. And it will rely heavily on proteins – or molecules which, like proteins, are capable of folding themselves up into a huge variety of shapes. Protein molecules are virtuosos of the auto-origamic arts, on a scale much smaller than that of the sheets of cells we have so far dealt with. Protein molecules are dazzling showcases of what can be achieved when local rules are obeyed on a local scale.

Proteins are chains of smaller molecules called amino acids, and these chains, like the sheets of cells we have been considering, also fold themselves, in highly determined ways but on a much smaller scale. In naturally occurring proteins (this is one fact that will presumably be different on alien worlds) there are only twenty kinds of amino acid, and all proteins are chains strung together from just this repertoire of twenty, drawn from a much larger set of possible amino acids. Now for the auto-origami. Protein molecules, simply following the laws of chemistry and thermodynamics, spontaneously and automatically twist themselves into precisely shaped three-dimensional configurations – I almost said 'knots' but, unlike hagfish (if I might impart a gratuitously inconsequential but engaging fact), proteins don't literally tie themselves in knots. The three-dimensional structure into which a protein chain folds and twists itself is the 'tertiary structure' that we briefly met when considering the self-assembly of viruses. Any given sequence of amino acids dictates a particular folding pattern. The amino-acid sequence, which itself is determined by the sequence of letters in the genetic code, determines the shape of the tertiary pattern.* The shape of the tertiary structure, in turn, has hugely important chemical consequences.

* This statement needs an important reservation. The determination of the amino-acid sequence by genes is indeed absolute. But the determination of the three-dimensional shape by the one-dimensional amino-acid sequence is not absolute, and it really matters. There are some sequences of amino acids that are capable of coiling up into two alternative 3D shapes. The proteins called prions, for example, have two stable shapes. These are discrete alternatives without stable intermediates, in the same way as a light switch is stable in the up position and in the down position but nowhere in between. Such 'switch proteins' can be disastrous or they can be useful. In the case of prions they are disastrous. In 'mad cow disease', a useful protein in the brain (it's a normal constituent of cell membranes) happens to have an alternative form – an alternative way to fold itself in auto-origami. The alternative form is normally never seen, but if it ever arises in one molecule it triggers neighbouring molecules to follow suit: they copy it and flip to the alternative form. Like a wave of falling dominoes, or like the irresponsible spreading of a rumour, the alternative prion form spreads through the brain, with disastrous results for the cow – or the person in the case of Creutzfeldt–Jakob disease, or the sheep in the case of scrapie. But sometimes molecules with the ability to auto-origami themselves into more than one alternative shape are useful. Without leaving the metaphor of the light switch we find a beautiful example. Rhodopsin, the protein in our eyes that is responsible for our sensitivity to light, has an embedded component called retinal (not itself a protein) which flips from its main stable configuration to its alternative configuration when hit by a photon of light. It then swiftly reverts, like a light switch on a cost-cutting timer, but meanwhile the flip has registered with the brain: 'Light detected at this pinpoint location here.' Jacques Monod's wonderful book, *Chance and Necessity*, is especially good on such bi-stable switch molecules.

The auto-origami by which protein chains fold and coil themselves is ruled by the laws of chemical attraction, and the laws determining the angles at which atoms bind to one another. Imagine a necklace of curiously shaped magnets. The necklace would not hang in a graceful catenary around a graceful neck. It would assume some other shape, becoming tangled up as the magnets latched on to each other and slotted into each other's nooks and crannies at various points along the length of the chain. Unlike the case of the protein chain, the exact shape of the tangle would not be predictable, because any magnet will attract any other. But it does suggest how chains of amino acids can spontaneously form a complicated knot-like structure, which may not look like a chain or a necklace.

The details of how the laws of chemistry determine the tertiary structure of a protein are not yet fully understood: chemists can't yet deduce, in all cases, how a given sequence of amino acids will coil up. Nevertheless, there is good evidence that the tertiary structure is *in principle* deducible from the sequence of amino acids. There's nothing mysterious about the phrase 'in principle'. Nobody can predict how a die will fall, but we all believe it is wholly determined by precise details of how it is thrown, plus some additional facts about wind resistance and so on. It is a demonstrated fact that a particular sequence of amino acids always does coil up into a particular shape, or one of a discrete set of alternative shapes (see the long footnote opposite). And – the important point for evolution – the sequence of amino acids is itself fully determined, through implementation of the rules of the genetic code, by the sequence of (triplets of) 'letters' in a gene. Even though it is not easy for human chemists to predict what change in protein shape will result from a particular genetic mutation, it is still a fact that once a mutation has occurred, the resulting change of protein shape will be in principle predict*able*. The same mutant gene will reliably produce the same altered protein shape (or discrete menu of alternative shapes). And that is all that matters for natural selection. Natural selection doesn't need to understand why a genetic change has a certain consequence.

It is sufficient that it does. If that consequence affects survival, the changed gene itself will stand or fall in the competition to dominate the gene pool, whether or not we understand the exact route by which the gene affects the protein.

Given that protein shape is immensely versatile, and given that it is determined by genes, why is it so supremely important? Partly because some proteins have a direct structural role to play in the body. Fibrous proteins, such as collagen, join together in stout ropes, which we call ligaments and tendons. But most proteins are not fibrous. Instead, they fold themselves up into their own characteristic globular shape, complete with subtle dents, and this shape determines the protein's characteristic role as an *enzyme*, which is a *catalyst*.

A catalyst is a chemical substance that speeds up, by as much as a billion or even a trillion times, a chemical reaction between other substances, while the catalyst itself emerges from the process unscathed and free to catalyse again. Enzymes, which are protein catalysts, are champions among catalysts because of their *specificity*: they are very fussy about precisely which chemical reactions they speed up. Or perhaps we could say: chemical reactions in living cells are very fussy about which enzymes speed them up. Many reactions in cell chemistry are so slow that, without the right enzyme, for practical purposes they don't occur at all. But with the right enzyme they happen very fast, and can churn out products in bulk.

Here's how I like to put it. A chemistry lab has hundreds of bottles and jars on its shelves, each containing a different pure substance: compounds and elements, solutions and powders. A chemist wishing to perform a certain chemical reaction selects two or three bottles, takes a sample from each, mixes them in a test tube or a flask, perhaps applies heat, and the reaction takes place. Other chemical reactions that could take place in the lab don't, because the glass walls of the bottles and jars prevent the ingredients meeting. If you want a different chemical reaction, you mix different ingredients in a different flask. Everywhere there are glass barriers keeping the pure substances separate from one another in bottles or jars, and keeping the reacting combinations separate from one another in test tubes or flasks or beakers.

The living cell, too, is a great chemistry lab, and it has a similarly large store of chemicals. But they aren't kept in separate bottles and jars on shelves. They are all mixed up together. It is as though a vandal, a chemical lord of misrule, entered the lab, seized all the bottles on all the shelves, and tipped them with anarchistic abandon into one great cauldron. Terrible thing to do? Well, it would be if they all reacted together, in all possible combinations. But they don't. Or if they do, the rate at which they react together is so slow that they might as well not be reacting at all. Except – and this is the whole point – if an enzyme is present. There is no need for glass bottles and jars to keep the substances apart because, to all intents and purposes, they are not going to react together anyway – *unless* the right enzyme is present. The equivalent of keeping the chemicals in stoppered bottles until you want to mix a particular pair, say A and B, is to mix all the hundreds of substances up in a great witch's brew, but supply only the right enzyme to catalyse the reaction between A and B and no other combination. Actually, the metaphor of the anarchically inclined bottle-tipper goes too far. Cells do contain an infrastructure of membranes between which, and within which, chemical reactions go on. To some extent, these membranes play the role of glass partitions between test tubes and flasks.

The point of this section of the chapter is that 'the right enzyme' achieves its 'rightness' largely through its physical shape (and that's important, because the physical shape is determined by genes, and it is genes whose variations are ultimately favoured or disfavoured by natural selection). Molecules aplenty are drifting and twisting and spinning through the soup that bathes the interior of a cell. A molecule of substance A might be happy to react with a molecule of substance B, but only if they happen to collide when facing in exactly the right direction, relative to each other. Crucially, that seldom happens – *unless* the right enzyme intervenes. The enzyme's precise shape, the shape into which it folded itself like a magnetic necklace, leaves it pitted with cavities and dents, each one of which itself has a precise shape. Each enzyme has a so-called 'active site', which is usually a particular dent or pocket, whose shape and chemical properties confer upon the enzyme its specificity. The word 'dent' doesn't adequately convey the specificity, the precision, of this mechanism. Perhaps a better comparison is with an electric socket. In what my friend the zoologist John Krebs calls 'the great plug conspiracy', different countries around the world have irritatingly adopted different arbitrary conventions for plugs and sockets. British plugs won't fit American sockets, or French sockets, and so on. The active sites on the surface of protein molecules are sockets into which only certain molecules will fit. But whereas the great plug conspiracy runs to only half a dozen separate shapes around the world (quite enough to constitute a continual annoyance to the traveller), the different kinds of sockets sported by enzymes are far more numerous.

Think of a particular enzyme, which catalyses the chemical combination of two molecules, P and Q, to make the compound PQ. One half of the active site 'socket' is just right for a molecule of type P to nestle into, like a jigsaw piece. The other half of the same socket is equally precisely shaped for a Q molecule to slot in – facing exactly the right way to combine chemically with the P molecule that is already there. Sharing a dent, firmly held at just the right angle to each other by the matchmaking enzyme molecule, P and Q unite. The new compound, PQ, now breaks away into the soup, leaving the active dent in the enzyme molecule free to bring together another P and another Q. A cell may be filled with swarms of identical enzyme molecules, all working away like robots in a car factory, churning out PQ in the cellular equivalent of industrial quantities. Put a different enzyme into the same cell, and it will churn out a different product, perhaps PR, or QS or YZ. The end product is different, even though the available raw materials are the same. Other types of enzymes are concerned not with constructing new compounds, but with breaking down old ones. Some of these enzymes are involved in digesting food, and they are exploited, too, in 'biological' washing powders.

But, since this chapter is about the construction of embryos, we are here mostly concerned with constructive enzymes, which broker the synthesis of new chemical compounds. One such process is shown in action on colour page 12.

A problem may have occurred to you. It's all very well to talk of jigsaw dents and sockets, highly specific active sites capable of speeding up a particular chemical reaction a trillionfold. But doesn't it sound too good to be true? How do enzyme molecules of exactly the right shape evolve from less perfect beginnings? What is the probability that a socket, shaped at random, will have just the right shape, and just the right chemical properties, to arrange a marriage between two molecules, P and Q, finessing their encounter at exactly the right angle? Not very great if you think 'finished jigsaw' – or, indeed, if you think 'great plug conspiracy'. Instead, you have to think 'smooth gradient of improvement'. As so often when we are faced with the riddle of how complex and improbable things can arise in evolution, it is a fallacy to assume that the final perfection that we see today is the way it always was. Fully fashioned, highly evolved enzyme molecules achieve trillionfold speedups of the reactions they catalyse, and they do so by being beautifully crafted to exactly the right shape. But you don't need a trillionfold speedup in order to be favoured by natural selection. A millionfold will do nicely! So will a thousandfold. And even tenfold or twofold would be enough for natural selection to get an adequate grip. There is a smooth gradient of improvement in an enzyme's performance, all the way from almost no dent at all, through a crudely shaped dent, to a socket of exactly the right shape and chemical signature. 'Gradient' means that each step is a noticeable improvement, however slight, over the one before. And 'noticeable' for natural selection can mean an improvement smaller than the minimum that would be required for us to notice it.

So, you see the way it works. Elegant! A cell is a versatile chemical factory, capable of spewing out massive quantities of a wide variety of different substances, the choice being made by which enzyme is present. And how is *that* choice made? By which gene is *turned on*.

Just as the cell is a vat filled with lots of chemicals, only a minority of which react with each other, so every cell nucleus contains the entire genome, but with only a minority of genes turned on. When a gene is turned on in, say, a cell of the pancreas, its sequence of code letters directly determines the sequence of amino acids in a protein; and the sequence of amino acids determines (remember the image of the magnetic necklace?) the shape into which the protein folds itself; and the shape into which the protein folds itself determines the precisely shaped sockets that marry up substances drifting around in the cell. Every cell, with very few exceptions such as red blood corpuscles, which lack a nucleus, contains the genes for making all the enzymes. But in any one cell, only a few genes will be turned on at any one time. In, say, thyroid cells, the genes that make the right enzymes for catalysing the manufacture of thyroid hormone are turned on. And correspondingly for all the different kinds of cells. Finally, the chemical reactions that go on in a cell determine the way that cell is shaped and the way it behaves, and the way it participates in origami-style interactions with other cells. So the whole course of embryonic development is controlled, via an intricate sequence of events, by genes. It is genes which determine sequences of amino acids, which determine tertiary structures of proteins, which determine the socket-like shapes of active sites, which determine cell chemistry, which determine 'starling-like' cell behaviour in embryonic development. So, differences in genes can, at the originating end of the complex chain of events, cause differences in the way embryos develop, and hence differences in the form and behaviour of adults. The survival and reproductive success of those adults then feeds back

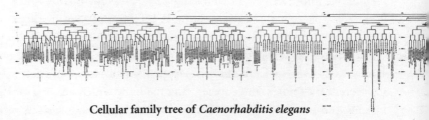

Cellular family tree of *Caenorhabditis elegans*

on the survival in the gene pool of the genes that made the difference between success and failure. And that is natural selection.

Embryology seems complicated – is complicated – but it is easy to grasp the important point, which is that we are dealing with local self-assembly processes all the way. It's a separate question, given that (almost) all the cells contain all the genes, how it is decided which genes are turned on in each different kind of cell. I must briefly deal with that now.

THEN WORMS SHALL TRY

Whether or not a given gene is turned on in a given cell at a given time is determined, often via a cascade of other genes called switch genes or controller genes, by the chemical environment of the cell. Thyroid cells are quite different from muscle cells, and so on, even though their genes are the same. That's all very well, you may say, once the development of the embryo is under way, and the different kinds of tissues such as thyroid and muscle already exist. But every embryo starts out as a single cell. Thyroid cells and muscle cells, liver cells and bone cells, pancreas cells and skin cells, all are descended from a single fertilized egg cell, via a branching family tree. This is a cellular family tree going back no further than the moment of conception, nothing to do with the evolutionary tree going back millions of years, which keeps cropping up in other chapters. Let me show you, for example, the complete family tree of all 558 cells of a newly hatched larva of the nematode worm, *Caenorhabditis elegans* (below: please pay close attention to every detail of this diagram). By the way, I don't know

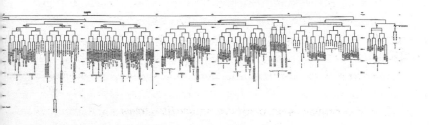

what this tiny worm did to earn its species name of *elegans*, but I can think of a good reason why it might have deserved it retrospectively. I know that not all my readers like my digressions, but the research that has been done on *Caenorhabditis elegans* is such a ringing triumph of science that you aren't going to stop me.

Caenorhabditis elegans was chosen in the 1960s as an ideal experimental animal by the formidably brilliant South African biologist Sydney Brenner. He had recently completed his work, with Francis Crick and others at Cambridge, on cracking the genetic code, and was looking around for a new big problem to solve. His inspired choice, and his own pioneering research on its genetics and neuro-anatomy, has led to a worldwide community of *Caenorhabditis* researchers that has grown into the thousands. It is only a bit of an exaggeration to say that we now know *everything* about *Caenorhabditis elegans*! We know its entire genome. We know exactly where every one of its 558 cells (in the larva; 959 in the adult hermaphroditic form, not counting reproductive cells) is in the body, and we know the exact 'family history' of every one of those cells, through embryonic development. We know of a large number of mutant genes, which produce abnormal worms, and we know exactly where the mutation acts in the body and the exact cellular history of how the abnormality develops. This little animal is known from start to finish, known inside out, known from head to tail and all stations in between, known through and through ('O frabjous day!'). Brenner was belatedly recognized with the Nobel Prize for Physiology in 2002, and a related species was named in his honour, *Caenorhabditis brenneri*. His regular column in the journal *Current Biology*, under the byline 'Uncle Syd', is a model of intelligent and irreverent scientific wit – as elegant as the worldwide research effort on *C. elegans* that he inspired. But I do wish molecular biologists would talk to some zoologists (like Brenner himself) and learn not to refer to *Caenorhabditis* as 'the' nematode, or even 'the' worm, as though there were no others.

Of course you can't read the names of the cell types at the bottom of the diagram (it would take seven pages to print the whole thing out legibly), but they say things like 'pharynx', 'intestinal muscle', 'body muscle', 'sphincter muscle', 'ring ganglion', 'lumbar ganglion'. The cells of all these types are literally cousins of one another: cousins by virtue of their ancestry within the lifetime of the individual worm. For example, I am looking at a particular body muscle cell called MSpappppa, which is a sibling of another body muscle cell, first cousin of two more body muscle cells, first cousin once removed of two more body muscle cells, second cousin of six pharynx cells, third cousin of seventeen pharynx cells . . . and so on. Isn't it amazing that we can actually use words like 'second cousin once removed', with the utmost precision and certainty, to refer to named and repeatably identifiable cells in an animal's body? The number of cell 'generations' that separates the tissues from the original egg is not that great. After all, there are only 558 cells in the body, and you can theoretically make 1,024 (2 to the power 10) in ten generations of cell splitting. The numbers of cell generations for human cells would be much larger. Nevertheless, you could in theory make a similar family tree for every one of your trillion-odd cells (as opposed to the 558 cells of a *C. elegans* female larva), tracing each one's descent back to the one fertilized egg cell. In mammals, however, it is not possible to identify particular, repeatably named cells. In us, it is more a case of statistical populations of cells, whose details are different in different people.

I hope my euphoric digression on the elegance of *Caenorhabditis* research has not distracted us too far from the point I was making about how cell types change in their shape and character as they branch away from one another in the embryonic family tree. At the branching point between a clone that is destined to become pharynx cells, and a 'cousin' clone that is destined to become ring ganglion cells, there has to be something to distinguish them, otherwise how would they know to turn on different genes? The answer is that, when the most recent common ancestor of the two clones divided, the two halves of the cell before division were different. So, when the cell divided, the two daughter cells, though identical in their genes (every daughter cell receives a full complement of genes), were not identical in the surrounding chemicals. And this meant that the same genes were not turned on – which changed the fate of their descendants. The same principle applies right through embryology, including its very start. The key to differentiation, in all animals, is asymmetric cell division.*

Sir John Sulston and his colleagues traced each of the cells in the body of the worm back to one and only one of six founder cells – we might even call them 'matriarch' cells – called AB, MS, E, D, C and P4. †
In naming the cells, they used a neat notation that summarized the history of each one. Every cell's name begins with the name of one of those six founder cells, the one from which it is descended. Thereafter, its name is a string of letters, the initial letters of the direction of cell division that gave rise to it: anterior, posterior, dorsal, ventral, left, right. For example, Ca and Cp are the two daughters of matriarch C, the anterior and posterior daughter respectively. Notice that every cell has no more than two daughters (of which one may die). I am

---

* In *Caenorhabditis* the original cell, called Z, has a front end which is different from its rear end, and this difference will come to represent the eventual fore-and-aft body axis – anterior (front) and posterior (rear). When the cell divides, the anterior daughter cell, which is called AB, has more front-end substance than the posterior daughter cell, which is called P1, and this difference will be bootstrapped to make more differences down the line. AB is destined to give rise to well over half the cells of the body, including most of the nervous system, and I won't discuss it further. P1 has two children, again different from each other, called EMS (defining the ventral or belly side of the eventual worm) and P2 (defining the dorsal side). They are grandchildren of Z (remember, when I use words like 'children' and 'grandchildren', I am talking about cells within a developing embryo, not individual worms). EMS now has two children called E and MS, while P2 has two children called C and P3. E, MS, C and P3 are great-grandchildren of Z (the other great-grandchildren are descended from AB, and I am not writing them down, except to say that two of them, called ABal and ABpl, define the left side, and their cousins, ABar and Abpr, define the right side of the eventual worm). P3 has two children called D and P4, which are great-great-grandchildren of Z. MS and C also have children, but I shan't name them here. P4 is destined to give rise to the so-called germ line. The germ line consists of cells that are not involved in building the body, but instead are going to make the reproductive cells. Obviously there is no need to remember or take note of these cell names. The point is only that, although genetically identical to each other, they differ in their chemical nature, as a cumulatively bootstrapped consequence of their history in the sequence of cell divisions within the embryo.

† Sulston, who stayed at Cambridge after Brenner left for America, was another of the triumvirate who won the Nobel Prize for the *Caenorhabditis* work. Sulston went on to lead the British end of the official Human Genome Project, the American end of which was headed first by James Watson and later by Francis Collins.

now looking at a particular body muscle cell, whose name, Cappppv, succinctly discloses its history: C had an anterior daughter, which had a posterior daughter, which had a posterior daughter, which had a posterior daughter, which had a posterior daughter, which had a ventral daughter, which is the body muscle cell in question. Every cell in the body is denoted by a comparable string of letters headed by one of the six founder cells. ABprpapppap, to take another example, is a nerve cell that sits in the ventral nerve cord running along the length of the worm. Needless to say, it is not necessary to take in the details. The beautiful point is that every single cell in the body has such a name, which totally describes its history during embryology. Every one of the ten cell divisions that gave rise to ABprpapppap, and every other cell, was an asymmetric division with the potential for different genes to be switched on in each of the two daughter cells. And in all animals that is the principle by which tissues differentiate, even though all their cells contain the same genes. Most animals, of course, have far more cells than *Caenorhabditis*' 558, and their embryonic development is in most cases less rigidly determined. In particular, as Sir John Sulston kindly reminds me, and as I have already briefly mentioned, in a mammal the 'family trees' of our cells are different for every individual, whereas in *Caenorhabditis* they are almost identical (except in mutant individuals). Nevertheless, the principle remains the same. In any animal, cells differ from each other in different parts of the body, even though they are genetically identical, because of their history of asymmetric cell division during the short course of embryonic development.

Let us hear the conclusion of the whole matter. There is no overall plan of development, no blueprint, no architect's plan, no architect. The development of the embryo, and ultimately of the adult, is achieved by local rules implemented by cells, interacting with other cells on a local basis. What goes on inside cells, similarly, is governed by local rules that apply to molecules, especially protein molecules, within the cells and in the cell membranes, interacting with other such molecules. Again, the rules are all local, local, local. Nobody, reading the sequence of letters in the DNA of a fertilized egg, could predict the shape of the animal it is going to grow into. The only way to discover that is to grow the egg, in the natural way, and see what it turns into. No electronic computer could work it out, unless it was programmed to simulate the natural biological process itself, in which case you might as well dispense with the electronic version and use the developing embryo as its own computer. This way of generating large and complex structures purely by the execution of local rules is deeply distinct from the blueprint way of doing things. If the DNA were some kind of linearized blueprint, it would be a relatively trivial exercise to program a computer to read the letters and draw the animal. But it would not be at all easy – indeed, it might be impossible – for the animal to have evolved in the first place.

And now, so that this chapter on embryos should not end up as a mere digression in a book on evolution, I must return to the sincere dilemma of Haldane's questioner. Given that genes control processes of embryonic development rather than adult shape; given that natural selection – like God – doesn't build tiny wings, but embryology does; how does natural selection go to work on animals to shape their bodies and their behaviour? How does natural selection go to work on embryos, in other words, to rejig them so they become ever more proficient at building successful bodies, with wings, or fins, leaves or armour plating, stings or tentacles or whatever it takes to survive?

Natural selection is the differential survival of successful genes rather than alternative, less successful genes in gene pools. Natural selection doesn't choose genes directly. Instead it chooses their proxies, individual bodies; and those individuals are chosen – obviously and automatically and without deliberative intervention – by whether they survive to reproduce copies of the very same genes. A gene's survival is intimately bound up with the survival of the bodies that it helps to build, because it rides inside those bodies, and dies with them. Any given gene can expect to find itself, in the form of copies of itself, riding inside a large number of bodies, both simultaneously in a population of contemporaries, and successively as generation gives way to generation. Statistically, therefore, a gene that tends, on average, to have a good effect on the survival prospects of the bodies in which it finds itself will tend to increase in frequency in the gene pool. So, on average, the genes that we encounter in a gene pool will tend to be those genes that are good at building bodies. This chapter has been about the procedures by which genes build bodies.

Haldane's interlocutor found it implausible that natural selection could put together in, say, a billion years, a genetic recipe for building her. I find it plausible, although of course neither I nor anybody else can tell you the details of how it happened. The reason it is plausible is precisely that it is all done by local rules. In any one act of natural selection, the mutation that is selected has had – in lots of cells and in lots of individuals in parallel – a very *simple effect* on the shape into which a protein chain spontaneously coils up. This, in turn, through catalytic action, speeds up, say, a particular chemical reaction in all the cells in which the gene is turned on. This changes, perhaps, the rate of growth of the embryonic primordium of the jaw. And this has consequential effects on the shape of the whole face, perhaps shortening the muzzle and giving a more human and less 'ape-like' profile. Now, the natural selection pressures that favour or disfavour the gene can be as complicated as you like. They might involve sexual selection, perhaps aesthetic choice of a high order by would-be sexual partners. Or the change in jaw shape might have a subtle effect on the animal's ability to crack nuts, or its ability to fight rivals. Some hugely elaborate combination of selection pressures, conflicting and compromising with one another in bewildering complexity, can bear upon the statistical success of this particular gene, as it propagates itself through the gene pool. But the gene knows nothing of this. All it is doing, within different bodies and in successive generations, is rejigging a carefully sculpted dent in a protein molecule. The rest of the story follows automatically, in branching cascades of local consequences, from which, eventually, a whole body emerges.

Even more complicated than the selection pressures in the ecological, sexual and social environments of the animals is the phantasmagoric network of influences that go on within and among the developing cells: influences of genes on proteins, genes on genes, proteins on the expression of genes, proteins on proteins; membranes, chemical gradients, physical and chemical guide rails in embryos, hormones and other mediators of action at a distance, labelled cells seeking others with identical or complementary labels. Nobody understands the whole picture, and nobody needs to understand it in order to accept the exquisite plausibility of natural selection. Natural selection favours the survival in the gene pool of the genetic mutations responsible for making crucial changes in embryos. The whole picture emerges as a consequence of hundreds of thousands of small, local interactions, each one comprehensible in principle (although it may be too hard or too time-consuming to unravel in practice) to anyone with sufficient patience to examine it. The whole may be baffling and mysterious in practice, but there is no mystery in principle, either in embryology itself, or in the evolutionary history by which the controlling genes came to prominence in the gene pool. The complications accumulated gradually over evolutionary time: each step was only a tiny bit different from the one before, and each step was accomplished by a small, subtle change in an existing local rule. When you have a sufficient number of small entities – cells, protein molecules, membranes – each at its own level obeying local rules and influencing others – then the eventual consequence is dramatic. If genes survive or fail to survive as a consequence of their influence on such local entities and their behaviour, natural selection of successful genes – and the emergence of their successful products – will inevitably follow. Haldane's questioner was wrong. It is not in principle difficult to make something like her.

And, as Haldane said, it only takes nine months.

CHAPTER 9

THE ARK
OF THE
CONTINENTS

I MAGINE a world without islands.

Biologists often use the word 'island' to mean something other than just a piece of land surrounded by water. From the point of view of a freshwater fish, a lake is an island: an island of habitable water surrounded by inhospitable land. From the point of view of an Alpine beetle, incapable of flourishing below a certain altitude, each high peak is an island, with almost impassable valleys between. There are tiny nematode worms (related to the elegant *Caenorhabditis*) which live inside leaves (as many as 10,000 of them in a single badly infected leaf), diving into them through the stomata, which are the microscopic holes through which leaves take in carbon dioxide and release oxygen. To a leaf-dwelling nematode worm such as *Apheloncoides*, a single foxglove is an island. To a louse, a single human head or crotch might be an island. There must be lots of animals and plants that regard an oasis in a desert as an island of cool, green habitability surrounded by a hostile sea of sand. And, while we are redefining words from an animal's point of view, since an archipelago is a chain or cluster of islands, I suppose a freshwater fish might define an archipelago as a chain or cluster of lakes, such as the lakes along the Great Rift Valley in Africa. An Alpine marmot might define a chain of mountain peaks separated by valleys as an archipelago. A leaf-mining insect might regard an avenue of trees as an archipelago. A botfly might regard a herd of cattle as a moving archipelago.

Having redefined the word 'island' (the sabbath was made for man, not man for the sabbath) let me return to my opening. Imagine a world without islands.

> He had bought a large map representing the sea
> Without the least vestige of land:
> And the crew were much pleased when they found it to be
> A map they could all understand.

We won't go quite as far as the Bellman, but imagine if all the land were gathered together in one great continent in the middle of a featureless sea. There are no islands offshore, no lakes or mountain ranges on the land: nothing to break the monotonous sweep of smooth uniformity. In this world an animal can easily go from anywhere to anywhere else, limited only by sheer distance, untroubled by inhospitable barriers. This is not a world friendly to evolution. Life on Earth would be extremely boring if there were no islands, and I want to begin this chapter by explaining why.

HOW NEW SPECIES ARE BORN

Every species is a cousin of every other. Any two species are descended from an ancestral species, which split in two. For example, the common ancestor of people and budgerigars lived about 310 million years ago. The ancestral species split in two, and the two strands went their separate ways for the rest of time. I chose human and budgie to make it vivid, but that same ancestral species is shared by all mammals on one side of that early divide, and all reptiles (zoologically speaking, birds are reptiles, as we saw in Chapter 6) on the other side. In the unlikely event that a fossil of this ancestral species was ever found, it would need a name. Let's call it *Protamnio darwinii*. We don't know any details about it, and the details don't matter at all for the argument, but we won't go far wrong if we imagine it as a sprawling lizard-like creature, scurrying about catching insects. Now, here's the point. When *Protamnio darwinii* split into two sub-populations they would have looked just the same as each other, and could have happily interbred with each other; but one lot were destined to give rise to the mammals, and the other lot were destined to give rise to the birds (and dinosaurs and snakes and crocodiles). These two sub-populations of *Protamnio darwinii* were about to diverge from each other, over a very long time and in a very big way. But they couldn't diverge if they kept on interbreeding with each other. The two gene pools would continually flood each other with genes. So any tendency to diverge would be nipped in the bud before it could get going, swamped by gene flow from the other population.

What actually happened at this epic parting of the ways, nobody knows. It happened a very long time ago, and we have no idea where. But modern evolutionary theory would confidently reconstruct something like the following history. The two sub-populations of *Protamnio darwinii* somehow became separated from each other, most likely by a geographical barrier such as a strip of sea separating two islands, or separating an island from a mainland. It could have been a mountain range that separated two valleys, or a river separating two forests: two 'islands' in the general sense I defined. All that matters is that the two populations were isolated from one another for long enough so that, when time and chance eventually reunited them, they found they had diverged so much that they couldn't interbreed any more. How long is long enough? Well, if they were subject to strong and contrasting selection pressures, it could be as little as a few centuries, or even less. For example, an island might lack a voracious predator that prowled the mainland. Or the island population might have shifted from an insectivorous to a vegetarian diet, like the Adriatic lizards of Chapter 5. Once again, we can't know the details of how *Protamnio darwinii* split, and we don't need to. The evidence from modern animals gives us every reason to think that something like the story I have just told is what happened in the past, for every one of the divergences between the ancestry of any animal and any other.

Even if conditions on either side of the barrier are identical, two geographically separated gene pools of the same species will eventually drift apart from one another, to the point where they can no longer interbreed when the geographical isolation eventually comes to an end. Random changes in the two gene pools will gradually build up, to the point where, if a male and a female from the two sides meet, their genomes will be too different to combine to make a fertile offspring. Whether by random drift alone, or with the aid of differential natural selection, once the two gene pools have reached the point where they no longer need the geographical isolation to stay genetically separate, we call them two different species. In our hypothetical case, perhaps the island population changed more than the mainland population, because of the lack of predators and the switch to a more vegetarian diet. So, a zoologist of the time might have recognized that the island population had become a new species and given it a new name, say *Protamnio saurops*, while the old name, *Protamnio darwinii*, might have continued to serve for the mainland population. In our hypothetical scenario, perhaps it was the island population that was destined to give rise to the sauropsid reptiles (that's everything we call reptiles today plus birds), while the mainland population eventually gave rise to the mammals.

Once again, I must stress, the *details* of my little story are pure fiction. It could equally well have been the island population that gave rise to the mammals. The 'island' could have been an oasis surrounded by desert, rather than land surrounded by water. And of course we haven't the faintest idea whereabouts on the Earth's surface this great divide took place – indeed, the world map would have looked so different that the question scarcely means anything. What is not fiction is the major lesson: most, if not all, of the millions of evolutionary divergences that have populated the Earth with such luxuriant diversity began with the chance separation of two sub-populations of a species, often, though not always, on either side of a geographical barrier such as a sea, a river, a mountain range or a desert valley. Biologists use the word 'speciation' for the splitting of a species into two daughter species. Most biologists will tell you that geographical isolation is the normal prelude to speciation, although some, especially entomologists, may chime in with the reservation that 'sympatric speciation' can also be important. Sympatric speciation, too, requires some kind of initial, incidental separation to get the ball rolling, but it is something other than geographic separation. It could be a local change in microclimate. I won't go into the details, but will just say that sympatric speciation seems to be especially important for insects. Nevertheless, for simplicity's sake, I shall in the rest of this chapter assume that the initial separation that precedes speciation is normally geographical. You may remember that, in Chapter 2's treatment of domestic dog breeds, I likened the effect of the rules imposed by pedigree breeders to the creation of 'virtual islands'.

'ONE MIGHT REALLY FANCY . . .'

How, then, do two populations of a species find themselves on opposite sides of a geographical barrier? Sometimes the barrier itself is the novelty. An earthquake opens up an impassable gorge, or changes the course of a river, and a species that had been a single breeding population finds itself severed in two. More usually, the barrier was there all along, and it is the animals themselves that cross it, in a rare freak event. It has to be rare, otherwise it doesn't deserve to be called a barrier at all. Before 4 October 1995 there were no members of the species *Iguana iguana* on the Caribbean island of Anguilla. On that date, a population of these large lizards suddenly appeared on the eastern side of the island. Fortuitously, they were actually seen arriving. They were clinging to a mat of driftwood and uprooted trees, some more than 30 feet long, that had drifted from a neighbouring island, probably Guadeloupe 160 miles away. The previous month two hurricanes, Luis on 4–5 September, and Marilyn two weeks later, had ripped through the area and could easily have uprooted the trees, complete with iguanas, which habitually spend time up trees. The new population on Anguilla was still going strong in 1998, and Dr Ellen Censky, who led the original study, informs me that they are flourishing to this day, seemingly even more so than the other species of iguana that lived on Anguilla before the new invaders arrived.

The point about such freak dispersal events is that they must be common enough to account for speciation, but not too common. If they were too common – if, say, iguanas drifted from Guadeloupe to Anguilla every year – the incipiently speciating population on Anguilla would be continually swamped by incoming gene flow and therefore could not diverge from the Guadeloupe population. By the way, please don't be misled by my use of a phrase like 'must be common enough'. It could be misunderstood to mean that steps of some kind were taken to ensure that the islands were just the right distance apart to facilitate speciation! Of course that puts the cart before the horse. It is rather that, wherever there happen to be islands (islands in the broad sense, as always) spaced out at an appropriate distance to facilitate speciation, there speciation will occur. And the appropriate distance will depend on how easy it is for the animals concerned to travel. The 160 miles that separate Guadeloupe from Anguilla would be child's play to any strong flying bird such as a petrel. But even a sea crossing of a few hundred yards might be difficult enough to midwife a new species of, say, frogs or wingless insects.

The Galapagos archipelago is separated from the mainland of South America by about 600 miles of open water, nearly four times as far as those iguanas sailed on their uprooted raft to Anguilla. The islands are all volcanic, and young by geological standards. None of them has ever been connected to any mainland. The entire fauna and flora of the islands must have travelled there, presumably from mainland South America. Even though small birds can fly, 600 miles is enough to make a crossing by finches a very rare event. Not so rare that it couldn't happen, however, and there are finches on Galapagos, whose ancestors, at some point in history, were presumably blown across, perhaps by a freak storm. These finches are all of a recognizably South American type, although the species themselves are unique to the Galapagos islands. Look at Darwin's map which I have adopted for sentimental reasons and because he used the magnificently

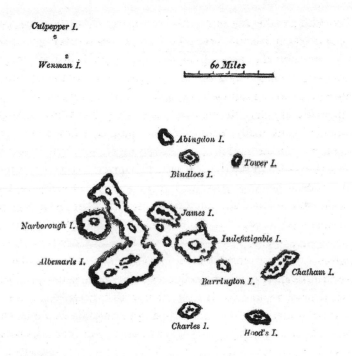

Darwin's map of the Galapagos islands with English names,
now seldom used

naval-sounding English names for the islands, rather than the
modern Spanish names. Notice that the 60-mile scale is about a
tenth of the distance an animal would have had to travel to arrive
on the archipelago from the mainland in the first place. The islands
themselves are only tens of miles from each other, but hundreds of
miles from the mainland. What a wonderful recipe for speciation. It
would be too simple to say that the chance of being accidentally blown
or rafted across a sea barrier to an island is inversely proportional to
the width of the barrier. Nevertheless, there will clearly be some sort
of inverse correlation between distance and probability of crossing.
The difference between the average inter-island distance of a few tens
of miles, and the 600-mile distance to the mainland, is so large that

you would expect the archipelago to be a powerhouse of speciation. And so it is, as Darwin eventually realized, although not until after he had left the islands, never to return.

This disparity, between tens of miles as the distance between islands within the archipelago, and hundreds of miles as the distance of the whole archipelago from the mainland, leads the evolutionist to expect that the different islands might house species that are pretty similar to each other but more different from their counterparts on the mainland. And that is exactly what we do find. Darwin himself put it well, coming tantalizingly close to evolutionary language, even before he had properly formulated his ideas. I have placed the key clause in italics, and shall repeat it throughout this chapter in different contexts.

> Seeing this gradation and diversity of structure in one small, intimately related group of birds, *one might really fancy that from an original paucity of birds in this archipelago, one species had been taken and modified for different ends.* In a like manner it might be fancied that a bird, originally a buzzard, had been induced here to undertake the office of the carrion-feeding Polybori of the American continent.

The last sentence is a reference to the Galapagos hawk, *Buteo galapagoensis*, another species that is found only on Galapagos, but which somewhat resembles species on the mainland, especially *Buteo swainsoni*, which annually migrates between North and South America and could well have been blown off course on one or two freak occasions. Nowadays, we should refer to the Galapagos hawk and the flightless cormorant as 'endemic' to the islands, meaning that this is the only place where they are found. Darwin himself, who had not yet fully embraced evolution, used the then current phrase 'aboriginal creations', which meant that God had created them here and nowhere else. He used the same phrase of the giant tortoises, which then abounded on all the islands, and also of the two species of iguana, the Galapagos land iguana and the Galapagos marine iguana. The marine iguanas are truly remarkable creatures, quite different from anything seen anywhere else in the world. They dive to the sea bottom and graze seaweed, which seems to be their only food. They are graceful swimmers, although not, in Darwin's outspoken view, beautiful to look at:

> It is a hideous looking creature, of a dirty black colour, stupid,* and sluggish in its movements. The usual length of a full-grown one is about a yard, but there are some even four feet long . . . their tails are flattened sideways, and all four feet partially webbed . . . When in the water this lizard swims with perfect ease and quickness, by a serpentine movement of its body and flattened tail – the legs being motionless and closely collapsed on its sides.

Since marine iguanas are so good at swimming, it might be supposed that they, rather than the land iguanas, made the long crossing from the mainland and subsequently speciated, in the archipelago, to give rise to the land iguana. This is almost certainly not the case, however. The Galapagos land iguana is not greatly different from iguanas still living on the mainland, whereas the marine iguanas are unique to the Galapagos archipelago. No lizard with the same marine habits has ever been found elsewhere in the world. We are nowadays confident that it was the land iguana that originally arrived from the South American mainland, perhaps carted on driftwood like the modern ones from Guadeloupe that were blown to Anguilla. On Galapagos, they subsequently speciated to give rise to the marine iguana. And it was almost certainly the geographical isolation permitted by the spaced-out pattern of the islands that made possible the initial

---

* *The Voyage of the Beagle.* Victorian naturalists were given to value judgements of this kind in their books. My grandparents possessed a bird book in which the entry on the cormorant frankly began, 'There is nothing to be said for this deplorable bird.'

separation between the ancestral land iguanas and the newly speciating marine iguanas. Presumably some land iguanas were accidentally rafted across to a hitherto iguana-free island, and there adopted a marine habit, free from contamination by genes flowing in from the land iguanas on the original island. Much later, they spread to other islands, eventually returning to the island from which their land ancestors had originally hailed. By now they could no longer interbreed with them, and their genetically inherited marine habits were safe from contamination by land iguana genes.

In example after example, Darwin noticed the same thing. The animals and plants of each island of Galapagos are largely endemic to the archipelago ('aboriginal creations'), but they are also for the most part unique, in detail, from island to island. He was especially impressed with the plants in this respect:

> Hence we have the truly wonderful fact, that in James Island [Santiago], of the thirty-eight Galapageian plants, or those found in no other part of the world, thirty are exclusively confined to this one island; and in Albemarle Island [Isabela], of the twenty-six aboriginal Galapageian plants, twenty-two are confined to this one island, that is, only four are at present known to grow in the other islands of the archipelago; and so on . . . with the plants from Chatham [San Cristobal] and Charles [Floreana] Islands.

He noticed the same thing with the distribution of mockingbirds over the islands.

> My attention was first thoroughly aroused, by comparing together the numerous specimens, shot by myself and several other parties on board, of the mocking-thrushes, when, to my astonishment, I discovered that all those from Charles Island belonged to one species (*Mimus trifasciatus*); all from Albemarle Island to *M. parvulus*; and all from James

and Chatham Islands (between which two other islands are situated, as connecting links) belonged to *M. melanotis*.

So it is, all over the world. The fauna and flora of a particular region are just what we should expect if, to quote Darwin on the finches that now bear his name, 'one species had been taken and modified for different ends'.

The Vice-Governor of the Galapagos Islands, Mr Lawson, intrigued Darwin by informing him

> that the tortoises differed from the different islands, and that he himself could with certainty tell from which island any one was brought. I did not for some time pay sufficient attention to this statement, and I had already partially mingled together the collections from two of the islands. I never dreamed that islands, about fifty or sixty miles apart, and most of them in sight of each other, formed of precisely the same rocks, placed under a quite similar climate, rising to a nearly equal height, would have been differently tenanted.

All the Galapagos giant tortoises are similar to a particular mainland species of land tortoise, *Geochelone chilensis*, which is smaller than any of them. At some point during the few million years that the islands have existed, one or a few of these mainland tortoises inadvertently fell in the sea and floated across. How could they have survived the long and doubtless arduous crossing? Surely most of them didn't. But it would have only taken one female to do the trick. And tortoises are astonishingly well-equipped to survive the crossing.

The early whalers took thousands of giant tortoises from the Galapagos islands away in their ships for food. To keep the meat fresh, the tortoises were not killed until needed, but they were not fed or watered while waiting to be butchered. They were simply turned on their backs, sometimes stacked several deep, so they couldn't walk away. I tell the story not in order to horrify (although I have to say that such barbaric cruelty does horrify me), but to make a point. Tortoises can survive for weeks without food or fresh water, easily long enough to float in the Humboldt Current from South America to the Galapagos archipelago. And tortoises do float.

Having reached and multiplied upon their first Galapagos island, the tortoises would with comparative ease – again accidentally – have island-hopped the much shorter distances to the rest of the archipelago by the same means. And they did what many animals do when they arrive on an island: they evolved to become larger. This is the long-noticed phenomenon of island gigantism (confusingly, there is an equally well-known phenomenon of island dwarfism).* If the tortoises had followed the pattern of Darwin's famous finches, they would have evolved a different species on each of the islands. Then, after subsequent accidental driftings from island to island, they would have been unable to interbreed (that's the definition, remember, of a separate species) and would have been free to evolve a different way of life uncontaminated by genetic swamping.

You could say that the different species' incompatible mating habits and preferences constitute a kind of genetic substitute for the geographic isolation of separate islands. Though they overlap geographically, they are now isolated on separate 'islands' of mating exclusivity. So they can diverge yet further. The Large, the Medium and the Small Ground Finch originally diverged on different islands; the three species now coexist on most of the Galapagos islands, never interbreeding and each specializing in a different kind of seed diet.

The tortoises did something similar, evolving distinctive shell shapes on the different islands. The species on the larger islands have high domes. Those on the smaller islands have saddle-shaped shells with a high-lipped window at the front for the head. The reason for

---

* The rule seems to be that, on islands, big animals get smaller (for example, there were dwarf elephants the height of a large dog on Mediterranean islands such as Sicily and Crete) while small animals get bigger (as in the Galapagos tortoises). There are several theories for this divergent tendency, but the details would take us too far afield.

this seems to be that the large islands are wet enough to grow grass, and the tortoises there are grazers. The smaller islands are mostly too dry for grass, and the tortoises resort to browsing on cactuses. The high-lipped saddle shell allows the neck to reach up to the cactuses which, for their part, grow higher in an evolutionary arms race against the browsing tortoises.

The tortoise story adds to the finch model the further complication that, for tortoises, volcanoes are islands within islands. Volcanoes provide high, cool, damp, green oases, surrounded at low altitude by dry lava fields which, for a grazing giant tortoise, constitute hostile deserts. Each of the smaller islands has a single large volcano and its own single species (or sub-species) of giant tortoise (except those few islands that have none at all). The big island of Isabela ('Albemarle' to Darwin) consists of a string of five major volcanoes, and each volcano has its own species (or sub-species) of tortoise. Truly, Isabela is an archipelago within an archipelago: a system of islands within an island. And the principle of islands in the literal geographical sense, setting the stage for the evolution of islands in the metaphorical genetic sense of species, has never been more elegantly demonstrated than here in the archipelago of Darwin's blest youth.[*]

Islands don't come much more isolated than St Helena, a single volcano in the South Atlantic some 1,200 miles from the coast of Africa. It has about 100 endemic plants (the young Darwin would have called them 'aboriginal creations' and the older Darwin would have said they evolved there). Among these are (or were, for some of them are now extinct) forest trees belonging to the daisy family.

These trees resemble in habit trees on the African mainland to which they are not closely related. The mainland plants to which they *are* related are herbs or small shrubs. What must have happened is that a few seeds of small herbs or shrubs chanced across the thousand-

---

[*] These paragraphs, on giant tortoises, are extracted from an article that I wrote on a boat called the *Beagle* (not the real one, which is unfortunately long extinct) in the Galapagos archipelago, and published in the *Guardian* on 19 February 2005.

Forest trees on St Helena

mile gap from Africa, settled on St Helena and, because the niche of forest trees was unfilled, evolved larger and more woody trunks until they became proper trees. Similar tree-like daisies have evolved independently on the Galapagos archipelago. It is the same pattern on islands the world over.

Each of the great African lakes has its own unique fish fauna, dominated by the group called cichlids. The cichlid faunas of Lake Victoria, Lake Tanganyika and Lake Malawi, each several hundred species strong, are completely distinct from each other. They have evidently evolved separately in the three lakes, which makes it all the more fascinating that they have converged on the same range of 'trades' in all three. In each lake, it looks as though one or two founder species somehow made their way in, perhaps from rivers, in the first place. And in each lake these founders then speciated and speciated again, to populate the lake with the hundreds of species that we see today. How, within the confines of a lake, did the budding species achieve the initial geographical isolation that enabled them to split apart?

When introducing islands, I explained that, from a fish's point of view, a lake surrounded by land is an island. Slightly less obviously, even an island in the conventional sense of land surrounded by water can be an 'island' for a fish, especially a fish that lives only in shallow water. In the sea, think of a coral-reef fish, which never ventures into deep water. From its point of view, the shallow fringe of a coral island is an 'island', and the Great Barrier Reef is an archipelago. Something similar can happen even in a lake. Within a lake, especially a large one, a rocky outcrop can be an 'island' for a fish whose habits confine it to shallow water. This is almost certainly how at least some of the cichlids in the African great lakes achieved their initial isolation. Most individuals were confined to shallow water around islands, or in bays and inlets. This achieved partial isolation from other such pockets of shallow water, linked by occasional traversings of the deeper water between them to form the watery equivalent of a Galapagos-like 'archipelago'.

There's good evidence (for example from sediment core samples) that the level of Lake Malawi (it was called Lake Nyasa when I spent my first bucket-and-spade holidays on its sandy beaches) rises and falls dramatically over the centuries, and reached a low point in the eighteenth century, more than 100 metres lower than the present level. Many of its islands were not islands at all during that time, but hills on the land around the then smaller lake. When the lake level rose, in the nineteenth and twentieth centuries, the hills became islands, ranges of hills became archipelagoes, and the process of speciation took off among the cichlids that live in shallow water, known locally as Mbuna. 'Almost every rocky outcrop and island has a unique Mbuna fauna, with endless colour forms and species. As many of these islands and outcrops were dry land within the last 200–300 years, the establishment of the faunas has taken place within that time.'

Such rapid speciation is something the cichlid fishes are extremely good at. Lake Malawi and Lake Tanganyika are old, but Lake Victoria is extremely young. The lake basin was formed only about 400,000 years ago, and it has dried up several times since then, most recently about 17,000 years ago. This seems to mean that its endemic fauna of 450 or so species of cichlid fishes have all evolved over a timescale of centuries, not the millions of years that we usually associate with evolutionary divergence on this grand scale. The cichlids of Africa's lakes impress us mightily with what evolution can do in a short space of time. They almost qualified for inclusion in the 'before our very eyes' chapter.

The woods and forests of Australia are dominated by trees of a single genus, *Eucalyptus*, and there are more than 700 species of them, filling a huge range of niches. Once again, Darwin's dictum about finches can be coopted: one could almost imagine that one species of eucalypt had been 'taken and modified for different ends'. And, along parallel lines, an even more famous example is the Australian mammal fauna. In Australia there are, or were until recent extinctions possibly caused by the arrival of aboriginal people, the ecological equivalents of wolves, cats, rabbits, moles, shrews, lions, flying squirrels and many others. Yet they are marsupials, quite different from the wolves, cats, rabbits, moles, shrews, lions and flying squirrels with which we are familiar in the rest of the world, the so-called eutherian mammals. The Australian equivalents are all descended from just a few, or even one, ancestral marsupial species, 'taken and modified for different ends'. This beautiful marsupial fauna has also produced creatures for which it is harder to find a counterpart outside Australia. The many species of kangaroo mostly fill antelope-like niches (or monkey or lemur-like niches in the case of the tree kangaroos) but get about by hopping rather than galloping. They range from the large red kangaroo (and some even larger extinct ones, including a fearsome, bounding carnivore) to the small wallabies and tree kangaroos. There were giant, rhinoceros-sized marsupials, Diprotodonts, related to modern wombats but 3 yards long, 6 feet tall at the shoulder, and weighing 2 tons. I shall return to the marsupials of Australia in the next chapter.

It is almost too ridiculous to mention it, but I'm afraid I have to because of the more than 40 per cent of the American population who, as I lamented in Chapter 1, accept the Bible literally: think what the geographical distribution of animals should look like if they'd all dispersed from Noah's Ark. Shouldn't there be some sort of law of decreasing species diversity as we move away from an epicentre – perhaps Mount Ararat? I don't need to tell you that that is not what we see.

Why would all those marsupials – ranging from tiny pouched mice through koalas and bilbys to giant kangaroos and Diprotodonts – why would all those marsupials, but no eutherians at all, have migrated en masse from Mount Ararat to Australia? Which route did they take?

And why did not a single member of their straggling caravan pause on the way, and settle – in India, perhaps, or China, or some haven along the Great Silk Road? Why did the entire order Edentata (all twenty species of armadillo, including the extinct giant armadillo, all six species of sloth, including extinct giant sloths, and all four species of anteater) troop off unerringly for South America, leaving not a rack behind, leaving no hide nor hair nor armour plate of settlers somewhere along the way? Why were they joined by the entire infraorder of caviomorph rodents, including guinea pigs, agoutis, pacas, maras, capybaras, chinchillas and lots of others, a large group of characteristically South American rodents, found nowhere else? Why did an entire sub-order of monkeys, the platyrrhine monkeys, end up in South America and nowhere else? Shouldn't at least a few of them have joined the rest of the monkeys, the catarrhines, in Asia or Africa? And shouldn't at least one species of catarrhine have found itself in the New World, along with the platyrrhines? Why did all the penguins undertake the long waddle south to the Antarctic, not a single one to the equally hospitable Arctic?

An ancestral lemur, again very possibly just a single species, found itself in Madagascar. Now there are thirty-seven species of lemur (plus some extinct ones). They range in size from the pygmy mouse lemur, smaller than a hamster, to a giant lemur, larger than a gorilla and resembling a bear, which went extinct quite recently. And they are all, every last one of them, in Madagascar. There are no lemurs anywhere else in the world, and there are no monkeys in Madagascar. How on Earth do the 40 per cent history-deniers think this state of affairs came about? Did all thirty-seven and more species of lemur troop in a body down Noah's gangplank and hightail it (literally in the case of the ringtail) for Madagascar, leaving not a single straggler by the wayside, anywhere throughout the length and breadth of Africa?

Once again, I am sorry to take a sledgehammer to so small and fragile a nut, but I have to do so because more than 40 per cent of the American people believe literally in the story of Noah's Ark. We should be able to ignore them and get on with our science, but we can't afford to because they control school boards, they home-school their children to deprive them of access to proper science teachers, and they include many members of the United States Congress, some state governors and even presidential and vice-presidential candidates. They have the money and the power to build institutions, universities, even a museum where children ride life-size mechanical models of dinosaurs, which, they are solemnly told, coexisted with humans. And, as recent polls have shown, Britain is not far behind (or should that read 'ahead'?), along with parts of Europe and most of the Islamic world.

Even if we leave Mount Ararat to one side; even if we refrain from lampooning those who take the Noah's Ark myth literally, similar problems apply to any theory of the separate creation of species. Why would an all-powerful creator decide to plant his carefully crafted species on islands and continents in exactly the appropriate pattern to suggest, irresistibly, that they had evolved and dispersed from the site of their evolution? Why would he put lemurs in Madagascar and nowhere else? Why put platyrrhine monkeys in South America only, and catarrhine monkeys in Africa and Asia only? Why no mammals in New Zealand, except bats who could fly there? Why do the animals in island chains most closely resemble those on neighbouring islands, and why do they nearly always resemble – less strongly but still unmistakably – those on the nearest continent or large island? Why would the creator put only marsupial mammals in Australia, again except bats who could fly there, and those who could arrive in man-made canoes? The fact is that, if we survey every continent and every island, every lake and every river, every mountaintop and every Alpine valley, every forest and every desert, the only way to make sense of the distribution of animals and plants is, yet again, to follow Darwin's insight about the Galapagos finches: 'One might really fancy that from an original paucity . . . one species had been taken and modified for different ends.'

Darwin was fascinated by islands, and he tramped the length and breadth of a good few during the voyage of the *Beagle*. He even worked out the surprising truth about how islands of one major class, those built by the animals called corals, are formed. Darwin later came to recognize the crucial importance of islands and archipelagoes for his theory, and he did several experiments to settle questions about the theory of geographical isolation as a prelude to speciation (he didn't use the word). For example, in a number of experiments he kept seeds in sea water for long periods, and demonstrated that some retained the power to germinate even after immersion for long enough to have drifted from continents to neighbouring islands. Frogspawn, on the other hand, he found to be immediately killed by sea water, and he made good use of this to explain a signal fact about the geographical distribution of frogs:

> With respect to the absence of whole orders on oceanic islands, Bory St. Vincent long ago remarked that Batrachians (frogs, toads, newts) have never been found on any of the many islands with which the great oceans are studded. I have taken pains to verify this assertion, and I have found it strictly true. I have, however, been assured that a frog exists on the mountains of the great island of New Zealand; but I suspect that this exception (if the information be correct) may be explained through glacial agency. This general absence of frogs, toads, and newts on so many oceanic islands cannot be accounted for by their physical conditions; indeed it seems that islands are peculiarly well fitted for these animals; for frogs have been introduced into Madeira, the Azores, and Mauritius, and have multiplied so as to become a nuisance. But as these animals and their spawn are known to be immediately killed by sea-water, on my view we can see that there would be great difficulty in their transportal across the sea, and therefore why they do not exist on any oceanic island. But why, on the theory of creation, they should not have been created there, it would be very difficult to explain.

Darwin was well aware of the significance of the geographical distribution of species for his theory of evolution. He noted that most of the facts could be accounted for if we assume that animals and plants have evolved. From this, we should expect – and we find – that modern animals tend to live on the same continent as fossils that could plausibly be their ancestors, or close to their ancestors. We should expect, and we find, that animals share the same continent with species that resemble them. Here is Darwin on the subject, paying special attention to the animals of South America that he knew so well:

> the naturalist in travelling, for instance, from north to south never fails to be struck by the manner in which successive groups of beings, specifically distinct, yet clearly related, replace each other. He hears from closely allied, yet distinct kinds of birds, notes nearly similar, and sees their nests similarly constructed, but not quite alike, with eggs coloured in nearly the same manner. The plains near the Straits of Magellan are inhabited by one species of Rhea (American ostrich), and northward the plains of La Plata by another species of the same genus; and not by a true ostrich or emeu, like those found in Africa and Australia under the same latitude. On these same plains of La Plata, we see the agouti and bizcacha, animals having nearly the same habits as our hares and rabbits, . . . but they plainly display an American type of structure. We ascend the lofty peaks of the Cordillera and we find an alpine species of bizcacha; we look to the waters, and we do not find the beaver or musk-rat, but the coypu and capybara, rodents of the American type.

This is mostly common sense, and Darwin was able to account for an enormous range of observations by means of it. But there are certain facts about the geographical distribution of animals and plants, and the distribution of rocks, that need a different kind of explanation:

one that is anything but common sense, and which would have staggered and enthralled Darwin, if only he had known about it.

DID THE EARTH MOVE?

Everybody in Darwin's time thought that the map of the world was pretty much a constant. Some of Darwin's contemporaries did countenance the possibility of large land bridges, now submerged, to explain, for example, the similarities between the floras of South America and Africa. Darwin himself was not greatly enamoured of the land bridge idea, but he surely would have exulted in the modern evidence that entire continents move over the face of the Earth. This provides by far the best explanation of certain major facts of animal and plant dispersion, especially of fossils. For example, there are similarities between the fossils of South America, Africa, Antarctica, Madagascar, India and Australia, which nowadays we explain by invoking the once great southern continent of Gondwana, uniting all those modern lands. Once again, our late-coming detective is forced to the conclusion that evolution is a fact.

The theory of 'continental drift', as it used to be called, was first championed by the German climatologist Alfred Wegener (1880–1930). Wegener was not the first to look at a map of the world and notice that the shape of a continent or island often matches the coastline opposite as if the two land masses were pieces of a jigsaw puzzle, even when the opposite coastline is far away. I'm not talking about little local examples, such as the Isle of Wight's neat dovetailing into the Hampshire coast, almost as though the Solent wasn't there. What Wegener and his predecessors noticed was that something along the same lines seemed to be true of the whole facing sides of the giant continents of Africa and America. The Brazilian coast looks tailor-cut to fit under the bulge of West Africa, while the northern part of Africa's bulge is a nice fit to the North American coast from Florida to Canada. Not only do the shapes roughly match: Wegener also pointed to matching geological formations up and down the east side of South America and corresponding parts of the west side of Africa. Slightly less clearly, the west coast of Madagascar forms a pretty good fit to the east coast of Africa (not the section of South African coast that is opposite it today but the coast of Tanzania and Kenya further north), while the long, straight line of Madagascar's east side is comparable to the straight edge of western India. Wegener also pointed out that the ancient fossils to be found in Africa and South America were more similar than you would expect if the map of the world had always been the way it is today. How could this be, given the width of the South Atlantic ocean? Were the two continents once much closer, or even joined? The idea was tantalizing, but ahead of its time. Wegener also noticed matchings between the fossils of Madagascar and India. And there are similarly telling affinities between the fossils of northern North America and of Europe.

Such observations led Wegener to propose his daringly heretical hypothesis of continental drift. All the great continents of the world, he suggested, had once been joined up in a gigantic super-continent, which he called Pangaea. Over an immense span of geological time, he proposed, Pangaea gradually dismembered itself to form the continents we know today, which then slowly drifted to their present positions and have not finished drifting yet.

One can almost hear Wegener's sceptical contemporaries wondering, to use the street-talk of today, what he had been smoking. Yet we now know that he was right. Or almost right. Far-sighted and imaginative as Wegener was, I must make it clear that his hypothesis of continental drift was significantly different from our modern theory of plate tectonics. Wegener thought that the continents ploughed through the oceans like gigantic ships, not quite floating in water like Dr Dolittle's hollow island of Popsipetl, but floating atop the semi-liquid mantle of the planet. Reasonably enough, other scientists erected fortresses of scepticism. What titanic forces could propel an object the size of South America or Africa for thousands of miles? I shall explain how the modern theory of plate tectonics differs from Wegener's theory before coming to its supporting evidence.

Cartoon inspired by Wegener's 'continental drift' theory

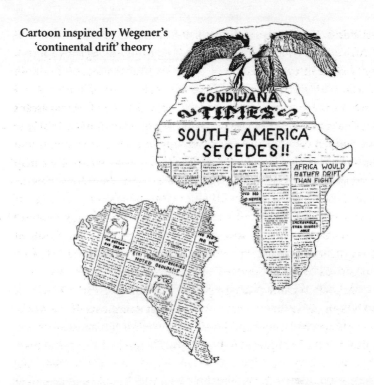

In the theory of plate tectonics the whole of the Earth's surface, including the bottoms of the various oceans, consists of a series of overlapping rocky plates like a suit of armour. The continents that we see are thickenings of the plates that rise above sea level. The greater part of the area of each plate lies under the sea. Unlike Wegener's continents, the plates do not sail through the sea, or plough through the surface of the Earth, they *are* the surface of the Earth. Don't think, like Wegener, of the continents themselves as being jigsawed together or being pulled apart from each other; it isn't like that. Think of a plate, instead, as being continuously manufactured at a growing edge, in a remarkable process called sea-floor spreading, which I shall explain in a moment. At other edges, a plate may be 'subducted' under a neighbouring plate. Or neighbouring plates may slide alongside one another. The picture on colour page 17 shows a portion of the San Andreas Fault in California, which is where the edges of the Pacific plate and the North American plate shear past each other. The combination of sea-floor spreading and subduction means that there are no gaps between plates. The entire surface of the planet is covered in plates, each one typically disappearing by subduction underneath a neighbouring plate on one side, or sliding past another plate, while it grows out from a sea-floor spreading zone elsewhere.

It is inspiring to think of the huge rift valley that must once have snaked its way down the continent of Gondwana between the future Africa and future South America. No doubt it was at first dotted with lakes like the present rift valley of East Africa. Later, it filled with sea water as South America sheared away with rending tectonic agony. Imagine the view that greeted some stout dinosaurian Cortez as he gazed across the long, narrow straits at the slowly departing 'West Gondwana'. Wegener was right that the jigsaw complementarity of their shapes is no accident. But he was wrong to think of the continents as gigantic rafts, ploughing their way through the sea-filled gaps between them. South America and Africa, and their continental shelves, are but the thickened regions of two plates, much of whose rocky surfaces lies under the sea. The plates constitute the hard lithosphere – literally, 'sphere of rock' – which floats atop the hot, semi-molten asthenosphere – 'sphere of weakness'. The asthenosphere is weak in the sense that it is not rigid and brittle like the rocky plates of the lithosphere but behaves somewhat like a liquid: yielding, like putty or toffee, if not necessarily molten. A little confusingly, this distinction between two concentric spheres doesn't wholly correspond to the more familiar distinction (based on chemical composition rather than physical strength) between the 'crust' and the 'mantle'.

Most plates consist of two distinct kinds of lithospheric rock. The deep ocean bottoms are covered in a rather uniform layer of very dense igneous rock, about 10 kilometres thick. This igneous layer is overlain by a superficial layer of sedimentary rock and mud. A continent, to repeat, is the area of a plate visible above sea level, rising to this height where the plate is thickened by additional layers of less dense rock. The undersea parts of the plates are continuously being created at their margins – the eastern margin in the case of the South American plate, the western margin in the case of the African plate. These two margins comprise the mid-Atlantic ridge, which snakes its way down the middle of the Atlantic from Iceland (which is, indeed, the only substantial part of the ridge that reaches the surface) to the far south.

Similar undersea ridges are rolling out other plates in other parts of the world (see colour pages 18–19). These undersea ridges work like elongated fountains (on the slow timescale of geology), welling up molten rock in the process I have already mentioned called sea-floor spreading. The spreading sea-floor ridge in mid-Atlantic seems to push the African plate eastwards and the South American plate westwards. The image of a pair of rolltop desks spreading in divergent directions has been suggested, and it conveys the idea, provided we remember that it is all happening on a timescale too slow for humans to see. Indeed, the speed at which South America and Africa pull apart has been memorably likened – so memorably that it has become almost a cliché – to the speed at which fingernails grow. The fact that they are now thousands of miles apart is further testimony to the vast and unbiblical age of the Earth, comparable to the evidence from radioactivity which we met in Chapter 4.

I used the phrase 'seems to push' just now, and I must hasten to backtrack. It is tempting to think of those upwelling 'rolltop desks' as pushing their respective continental plates along from behind. This is unrealistic; the scale is all wrong. Tectonic plates are much too massive to be pushed from behind by upwelling volcanic forces along a mid-ocean ridge. A swimming tadpole might as well try to push a supertanker. But now, here's the key point. The asthenosphere, in its capacity as a quasi-liquid, has convection currents that extend throughout its whole surface, under the entire area of the plates. In any one region, the asthenosphere is slowly moving in a consistent direction, and then circling back in the opposite direction down in its deeper layers. The upper layer of asthenosphere under the South American plate, for example, is moving inexorably westward. And, while it is inconceivable that an upwelling 'rolltop desk' could have the strength to push the whole South American plate before it, it is not at all inconceivable that a convection current inching its way steadily in a consistent direction *under the entire lower surface* of a plate could carry its 'floating' continental burden along with it. We aren't talking tadpoles now. A supertanker in the Humboldt Current, with its engines switched off, will indeed go with the flow.

That is, in summary, the modern theory of plate tectonics. I must now turn to the evidence that it is true. Actually, as is normal with established scientific facts,* there are lots of different kinds of evidence, but I am only going to talk about the most strikingly elegant kind. This is the evidence from the ages of the rocks, and especially from the magnetic stripes in them. It's almost too good to be true, a perfect illustration of my 'detective coming late to the scene of the crime' and being driven inexorably to only one conclusion. We even have something that looks very like fingerprints: giant magnetic fingerprints in the rocks.

We shall accompany our metaphorical detective on a voyage across the South Atlantic, in a custom-built submarine capable of withstanding the daunting pressures of the deep sea. The submarine is equipped to drill down for rock samples, through the superficial sediments of the sea bottom down to the volcanic rocks of the lithosphere itself, and it also has an onboard laboratory for dating rock samples radiometrically (see Chapter 4). The detective sets a course due east from the Brazilian port of Maceio, 10 degrees of latitude south of the equator. Having travelled 50 kilometres or so through the shallow waters of the continental shelf (which for present purposes counts as part of South America) we batten down the high-

---

* Like the modern 'theory' of evolution, it is an established fact in the normal sense of the word: a theory in the first of the *OED*'s definitions that I quoted in Chapter 1, and renamed 'theorum'.

pressure hatches and dive (what understatement!), dive down into the depths where the only light normally seen is the occasional spark of greenish luminescence from the grotesques that inhabit this alien world.

When we hit bottom at nearly 20,000 feet (full fathom 3,000), we drill down to the volcanic lithosphere and take a core sample of the rock. The onboard radioactive dating lab goes to work, and reports a Lower Cretaceous age, about 140 million years. The submarine grinds on eastwards along the tenth parallel, taking rock samples at frequent intervals. The age of each sample is carefully measured and the detective pores over the datings, looking for a pattern. He doesn't have to look far. Even Dr Watson couldn't miss it. As we travel east along the great plains of the sea bottom, the rocks are plainly getting younger and younger, steadily younger. About 730 kilometres into our journey, the rock samples are of late Cretaceous age, about 65 million years old, which happens to be when the last of the dinosaurs went extinct. The trend towards younger and younger rocks continues as we approach the middle of the Atlantic and the submarine's searchlights start to pick out the foothills of a gigantic underwater mountain range. This is the mid-Atlantic ridge (see colour pages 18 19), which our submarine must now start to climb. Up and up we crawl, still taking rock samples, and still noticing that the rocks are getting younger and younger. By the time we reach the peaks of the ridge, the rocks are so young they might as well have only just welled up as fresh lava from volcanoes. Indeed, that is pretty much what has happened. Ascension Island is a part of the mid-Atlantic ridge which protruded above sea level as a result of a recent series of eruptions – well, recent: maybe 6 million years ago; that's recent by the standards of the rocks we have been sampling along our submarine way.

We now push on towards Africa, over the other side of the ridge, down to the deep plains at the bottom of the eastern Atlantic. We continue to take rock samples and – you've guessed it – the rocks now become steadily older as we move towards Africa. It is the mirror image of the pattern we noticed before we reached the mid-Atlantic ridge. The detective is in no doubt of the explanation. The two plates are moving apart as the sea floor spreads away from the ridge. All the new rock that is being added to the two diverging plates comes from the volcanic activity of the ridge itself, and it is then carried away, in opposite directions, on one or other of the gigantic rolltop desks that we call the African plate and the South American plate. The false colours in the pictures on colour pages 18–19 illustrating this process denote the age of the rocks, red being the youngest. You can see how beautifully the age profiles on the two sides of the mid-Atlantic ridge mirror each other.

What an elegant story! But it gets better. The detective notices a more subtle pattern in the rock samples as they are processed in the onboard laboratory. The rock cores pulled up from the deep lithosphere are slightly magnetic, like compass needles. The phenomenon is well understood. When molten rock solidifies, the Earth's magnetic field becomes imprinted into it, in the form of a polarization of the fine crystals of which igneous rock is made. The crystals behave like tiny frozen compass needles, locked into the direction they were pointing in at the moment when the molten lava solidified. Now, it has long been known that Earth's magnetic pole is not fixed but wanders, probably because of slowly oozing currents in the mixture of molten iron and nickel in the planet's core. At present, magnetic north lies near Ellesmere Island in northern Canada, but it won't stay there. To determine true north using a magnetic compass, sailors need to look up a correction factor, and the correction factor changes from year to year as the planet's magnetic field fluctuates.

So long as our detective meticulously records the exact angle at which his rocky cores sat when he drilled them out, the frozen magnetic field in each core tells him the position of the Earth's magnetic field on the day that the rock solidified out of lava. And now for the clincher. It happens that, at irregular intervals of tens or hundreds of thousands of years, the Earth's magnetic field completely reverses, presumably because of major shifts in the molten nickel/iron core. What was magnetic north flips over to a position near the true South Pole, and what was magnetic south flips to the north. And of course the rocks pick up the current position of magnetic north on the day that they solidify from lava welling up from the deep sea bottom. Because the polarization reverses every few tens of thousands of years, a magnetometer can detect stripes running along the bedrock: stripes in which the rock samples' magnetic fields all point in one direction, alternating with stripes in which the magnetic fields all point in the opposite direction. Our detective colours them black and white on the map. And when he looks at the stripes on the map like fingerprints, he notices an unmistakable pattern. As with the false colour stripes denoting the absolute age of the rocks, the magnetic fingerprint stripes on the west side of the mid-Atlantic ridge are an elegant mirror image of the stripes on the east side. Exactly what you'd expect if the magnetic polarity of the rock was laid down when the lava first solidified in the ridge and then slowly moved outwards from the ridge, in opposite directions, at a fixed and very slow rate. Elementary, my dear Watson.*

To revert to the terminology of Chapter 1, the morphing of Wegener's hypothesis of continental drift into the modern theory of plate tectonics is a textbook example of the solidification of a tentative hypothesis into a universally accepted theorum or fact. Plate tectonic movements are important in this chapter, because without them we cannot fully understand the distribution of animals and plants over the continents and islands of the world. When I spoke of the initial geographical barrier that separated two incipient species, I proposed an earthquake diverting the course of a river. I could also have mentioned plate tectonic forces, splitting a continent in two and ferrying the two gigantic fragments in opposite directions, complete with animal and plant passengers – the ark of the continents.

Madagascar and Africa were once part of the great southern continent of Gondwana, together with South America, Antarctica,

---

* Alas, Holmes never said it (just as Burns never wrote 'for the sake of' Auld Lang Syne), but the allusion works because everybody thinks he did.

India and Australia. Gondwana began to break up – painfully slowly by the standards of our perception – about 165 million years ago. At this point Madagascar, while still joined to India, Australia and Antarctica as East Gondwana, pulled away from the eastern side of Africa. At about the same time, South America pulled away from West Africa in the other direction. East Gondwana itself broke up rather later, and Madagascar finally became separated from India about 90 million years ago. Each of the fragments of old Gondwana carried with it its cargo of animals and plants. Madagascar was a real 'ark', and India was another. It is probable, for example, that the ancestors of ostriches and elephant birds originated in Madagascar/India when they were still united. Later they split. Those that remained on the giant raft called Madagascar evolved to become the elephant birds, while the ancestors of ostriches sailed off on the good ship India and subsequently – when India collided with Asia and raised the Himalayas – were liberated on to the mainland of Asia, whence they eventually found their way to Africa, their main stamping ground today (yes, the males really do stamp their feet, to impress females). Elephant birds, alas, we no longer see (or hear, more's the pity, for if they stamped the very ground must have shaken). Far more massive than the largest ostriches, these Madagascar giants are the probable origin of the legendary 'roc', which features in the Second Voyage of Sinbad the Sailor. Although large enough for a man to have ridden, they had no wings, so could never have carried Sinbad aloft as advertised.*

Not only does the now solidly established theory of plate tectonics account for numerous facts about the distribution of fossils and living creatures, it also provides yet more evidence of the extreme antiquity of the Earth. It ought, therefore, to be a major thorn in the side of creationists, at least creationists of the 'young Earth' persuasion. How

---

* Indeed, physical laws of scaling ensure that birds as big as an elephant bird couldn't indulge in powered, flapping flight at all, no matter how big their wing span. This is because the muscles needed to power such massive wings would need to be so big they couldn't lift their own weight.

do they cope with it? Very weirdly indeed. They don't deny the shifting of the continents, but they think it all happened at high speed very recently, at the time of Noah's flood.* You'd think that, since they are conspicuously happy to discount evidence that doesn't suit them in the case of the massive quantity and range of evidence for the fact of evolution, they'd pull the same trick with the evidence for plate tectonics too. But no: oddly, they accept as a fact that South America once fitted snugly into Africa. They seem to regard the evidence for this as conclusive, even though the evidence for the fact of evolution is, if anything, even stronger, and they gaily deny that. Since evidence means so little to them, one wonders why they don't go the whole hog and simply deny the whole of plate tectonics too.

Jerry Coyne's *Why Evolution is True* offers a masterly treatment of the evidence from geographical distribution (as you'd expect from the senior author of the most authoritative recent book on speciation). He also hits the nail on the head with respect to the creationists' penchant for ignoring evidence when it doesn't support the position that they *know*, from Scripture, has got to be true: 'The biogeographic evidence for evolution is now so powerful that I have never seen a creationist book, article, or lecture that has tried to refute it. Creationists simply pretend that the evidence doesn't exist.' Creationists act as though fossils provide the only evidence for evolution. The fossil evidence is indeed very strong. Truckloads of fossils have been uncovered since Darwin's time, and all this evidence either actively supports, or is compatible with, evolution. More tellingly, as I have already emphasized, not a single fossil contradicts evolution. Nevertheless, strong as the fossil evidence is, I again want to emphasize that it is not the strongest we have. Even if not a single fossil had ever been found, the evidence from surviving animals would still overwhelmingly force the conclusion that Darwin was right. The detective coming on the scene of the crime after the event

---

* It is an arresting image: South America and Africa speeding away from each other faster than a man can swim, for forty days continuously.

can amass surviving clues that are even more incontrovertible than fossils. In this chapter we have seen that the distribution of animals on islands and continents is exactly what we should expect if they are all cousins that have evolved from shared ancestors over very long periods. In the next chapter we shall compare modern animals with each other, looking at the distribution of characteristics in the animal kingdom, especially comparing their sequences of genetic code, and shall come to the same conclusion.

CHAPTER 10

THE TREE OF COUSINSHIP

BONE TO HIS BONE

What a piece of work is the mammalian skeleton. I don't mean it is beautiful in itself, although I think it is. I mean the fact that we can talk about 'the' mammalian skeleton at all: the fact that such a complicatedly interlocking thing is so gloriously different across the mammals, in all its parts, while simultaneously being so obviously the *same* thing throughout the mammals. Our own skeleton is familiar enough to need no picture, but look at this skeleton of a bat. Isn't it fascinating how every bone has its own identifiable

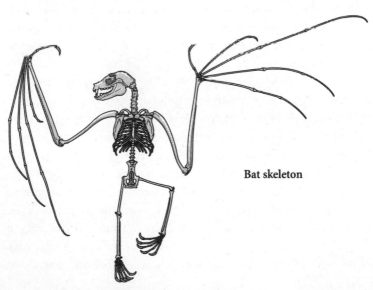

Bat skeleton

counterpart in the human skeleton? Identifiable, because of the order in which they join up to each other. Only the proportions are different. The bat's hands are hugely enlarged (relative to its total size, of course) but nobody could possibly miss the correspondence between our fingers and those long bones in the wings. The human hand and the bat hand are obviously – no sane person could deny it – two versions of the same thing. The technical term for this kind of sameness is 'homology'. The bat's flying wing and our grasping hand are 'homologous'. The hands of the shared ancestor – and the rest of the skeleton – were taken and pulled, or compressed, part by part, in different directions and by different amounts, along different descendant lineages.

The same applies – although with different proportions again – in the wing of a pterodactyl (not a mammal, but the principle still holds, which makes it all the more impressive). This pterodactyl's wing membrane is largely borne by a single finger, what we would call the 'little' finger or 'pinky'. I confess to a homology-inspired neurosis about so much weight being borne by the fifth finger, because in humans it seems so fragile. Silly, of course, because to a pterodactyl the fifth finger, far from being 'little', stretched most of the length of the body, and it presumably would have felt stout and strong, as our arm feels to us. Yet again, it goes to illustrate the point I am making. The fifth finger is *modified* to bear the wing membrane. All the details have become different, but it is still recognizably the fifth finger because of its spatial relationship to the other bones of the skeleton. This long, stout, wing-supporting strut is 'homologous' to our little finger. The word for 'little finger' in pterodactylese means 'ruddy great strut'.

In addition to the true fliers – birds, bats, pterosaurs and insects – lots of other animals glide: a habit that might tell us something about the origins of true flight. They have gliding membranes, which need skeletal support; but it doesn't have to come from the finger bones as it does in the wings of bats and pterosaurs. Flying squirrels (two independent groups of rodents), and flying

Pterodactyl skeleton

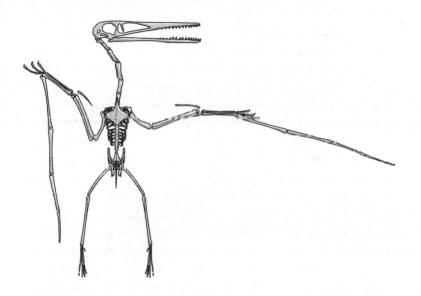

phalangers (Australian marsupials, looking almost exactly like flying squirrels but not closely related) stretch a membrane of skin between the arms and the legs. Individual fingers are not required to bear much load, and they are not enlarged. I, with my little-finger neurosis, would be happier as a flying squirrel than as a pterodactyl, because it feels 'right' to use whole arms and whole legs to do load-bearing work.

Overleaf is the skeleton of a so-called flying lizard, another elegant forest glider. You can immediately see that it is the ribs, rather than the fingers, or the arms and legs, that have become modified to bear the 'wings' – the flight membranes. Once again, the resemblance of the skeleton as a whole to other vertebrate skeletons is completely clear. You could go through every bone, one by one, identifying, in each case, the precise bone to which it corresponds in the human or bat or pterosaur skeleton.

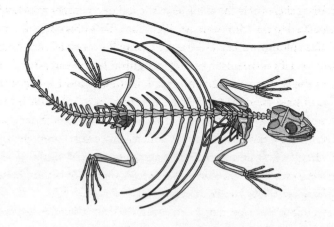

'Flying lizard' skeleton

The colugo, or so-called 'flying lemur', of the south-east Asian forests resembles the flying squirrels and flying phalangers, except that the tail, as well as the arms and legs, is included in the support structure of the flight membrane. That doesn't feel right to me, because I can't imagine what it is like to have a tail at all, although we humans, along with all the other 'tail-less' apes, have a vestigial tail, the coccyx, buried beneath the skin. Almost tail-less as we apes are, it is hard for us to imagine what it must be like to be a spider monkey, whose tail dominates the entire spinal column. You can see from the picture on colour page 26 how much longer it is even than the already long arms and legs. As in many New World monkeys (indeed, many New World mammals generally, which is a curious fact, hard to interpret), the spider monkey's tail is 'prehensile', meaning that it is modified for grasping, and it almost seems to end in an extra hand, although it is not homologous to a real hand, and has no fingers. Indeed, the spider monkey's tail looks very much like an extra leg or arm.

I probably don't need to spell out the message again. The underlying skeleton is the same as in the tail of any other mammal, but modified to do a different job. Well, the tail itself is not quite the same: the spider monkey tail has an extra allowance of vertebrae, but the vertebrae themselves are recognizably the same kind of thing as the vertebrae in any other tail, including our own coccyx. Can you imagine what it would be like to be a monkey with five grasping 'hands' – one at the end of each leg as well as at the end of each arm, and a tail – from any of which you could happily hang? I can't. But I know that the tail of a spider monkey is homologous to my coccyx, just as the enormously long and strong wing bone of a pterodactyl is homologous to my little finger.

Here's another surprising fact. A horse's hoof is homologous to the fingernail of your middle finger (or the toenail of your middle toe). Horses walk on tiptoe, literally, unlike us when we walk on what we *call* tiptoe. They have almost entirely lost their other toes and fingers. In a horse, the homologues of our index finger and our ring finger, and their hind-leg equivalents, survive as tiny 'splint' bones, joined to the 'cannon' bone, and not visible outside the skin. The cannon bone is homologous to our middle metacarpal which is buried in our hand (or metatarsal, buried in our foot). The entire weight of the horse – very substantial in the case of a Shire or a Clydesdale – is borne on the middle fingers and middle toes. The homologies, for example to our middle fingers or those of a bat, are completely clear. Nobody could doubt them; and, as if to ram the point home, freak horses are sometimes born with three toes on each leg, the middle one serving as a normal 'foot', the two side ones having miniature hooves (see picture overleaf).

Can you see how beautiful it is, this idea of near-indefinite modification over immensities of time, each modified form retaining unmistakable traces of the original? I glory in the litopterns, extinct South American herbivores, not closely related to any modern animals, and very different from horses – except that

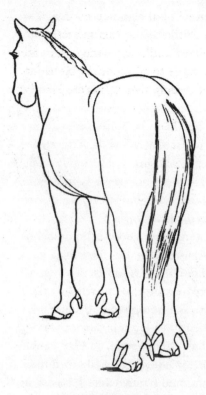

Polydactylic horse

they had almost identical legs and hooves. Horses (in North America*) and litopterns (in South America, which was in those days a gigantic island, the Panama isthmus being way in the future) independently evolved exactly the same reduction of all the fingers and toes except the middle ones, and sprouted identical hooves on the ends of those. Presumably there aren't all that many ways for a herbivorous mammal to become a fast runner. Horses and litopterns hit upon the same way – reducing all digits except the middle one – and they carried it to the same conclusion. Cows and antelopes hit upon another solution: reducing all but two digits.

The following statement sounds paradoxical but you can see how it makes sense, and also how important it is as an observation. The skeletons of all mammals are identical, but their individual bones are different. The resolution of the paradox lies in my calculated usage of 'skeleton' for the *assemblage* of bones, in ordered attachment one

---

* You may be surprised to hear that horses evolved in North America, because it is widely known that when the European invaders first came to the Americas, the sight of them on horseback amazed the natives. The bulk of horse evolution did indeed take place in America. Then horses spread to the rest of the world, shortly (by geological standards) before going extinct in America. They are American animals that have been re-introduced to America by man.

to the other. The shapes of individual bones are not, on this view, properties of the 'skeleton' at all. 'Skeleton', in this special sense, ignores the shapes of individual bones, and is concerned only with the order in which they join up: 'bone to his bone' in the words of Ezekiel, and, more vividly, in the song that is based upon the passage:

> Your toe bone connected to your foot bone,
> Your foot bone connected to your ankle bone,
> Your ankle bone connected to your leg bone,
> Your leg bone connected to your knee bone,
> Your knee bone connected to your thigh bone,
> Your thigh bone connected to your hip bone,
> Your hip bone connected to your back bone,
> Your back bone connected to your shoulder bone,
> Your shoulder bone connected to your neck bone,
> Your neck bone connected to your head bone,
> I hear the word of the Lord!

The point is that this song could apply to literally any mammal, indeed any land vertebrate, and in far more detail than these words suggest. For example your 'head bone', or skull, contains twenty-eight bones, mostly joined together in rigid 'sutures', but with one major moving bone (the lower jaw*). And the wonderful thing is that, give or take the odd bone here and there, the same set of twenty-eight bones, which can clearly be labelled with the same names, is found across all the mammals.

---

* A single bone in mammals. The reptilian lower jaw is more complicated – and thereby hangs a fascinating tale that I reluctantly omitted from this book (you can't have everything). In an amazing feat of evolutionary legerdemain, the smaller bones of the reptilian lower jaw were coopted into the mammalian ear, where they constitute an exquisitely delicate bridge, to transport sound from the eardrum to the inner ear.

Human skull

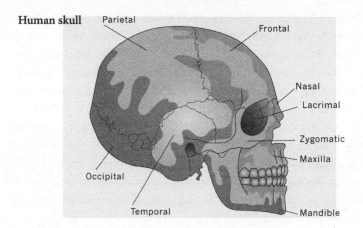

Horse skull

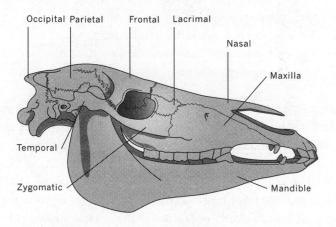

Your neck bone connected to your occipital bone
Your occiput connected to your parietal bone
Your parietal connected to your frontal bone
Your frontal bone connected to your nasal bone
. . .
Your 27th bone connected to your 28th bone . . .

All this is the same, regardless of the fact that the shapes of the particular bones are radically different across the mammals.

What do we conclude from all this? We have here confined ourselves to modern animals, so we are not seeing evolution in action. We are the detectives, come late to the scene. And the pattern of resemblances among the skeletons of modern animals is exactly the pattern we should expect if they are all descended from a common ancestor, some of them more recently than others. The ancestral skeleton has been gradually modified down the ages. Some pairs of animals, for example giraffes and okapis, share a recent ancestor. It is not strictly correct to describe a giraffe as a vertically stretched

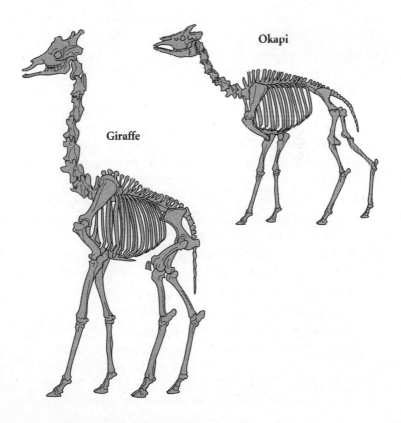

Okapi

Giraffe

okapi, for both are modern animals. But it would be a good guess (supported by fossil evidence, as it happens, but we aren't talking about fossils in this chapter) that the shared ancestor probably looked more like the okapi than the giraffe. Similarly, impalas and gnus[*] are close cousins of each other, and slightly more distant cousins of giraffes and okapis. All four of them are more distant cousins still of other cloven-hoofed animals, such as pigs and warthogs (which are cousins of each other and of peccaries). All the cloven-hoofed animals are more distant cousins of horses and zebras (which don't have cloven hooves and are close cousins of each other). We can go on as long as we like, bracketing pairs of cousins into groups, and groups of groups of cousins, and (groups of (groups of (groups of cousins))). I have slipped into using brackets automatically, and you know just what they signify. The meaning of the brackets in the following is immediately clear to you, because you already know all about cousins sharing grandparents, and second cousins sharing great-grandparents, and so on:

{(wolf fox)(lion leopard)}{(giraffe okapi) (impala gnu)}

Everything points to a simple branching tree of ancestry – a family tree.

I have implied that the tree of resemblances is really a family tree, but are we forced to this conclusion? Are there any alternative interpretations? Well, just barely! The hierarchical pattern of resemblances was spotted by creationists in pre-Darwinian times, and they did have a non-evolutionary explanation – an embarrassingly far-fetched one. Patterns of resemblance, in their view, reflected *themes* in the mind of the designer. He had various ideas for how to make animals. His thoughts ran along a mammal theme, and,

---

[*] The Dutch 'wildebeest' is increasingly used in preference to 'gnu'. I am trying to save 'gnu' because, if it dies out altogether, the witty song by Flanders and Swann won't make sense any more. ('Gnor am I in the least / Like that dreadful hartebeest / Oh gno gno gno, I'm a gnu!')

independently, they ran along an insect theme. Within the mammal theme, the designer's ideas were neatly and hierarchically bisected into sub-themes (say, the cloven-hoofed theme) and sub-sub-themes (say, the pig theme). There is a strong element of special pleading and wishful thinking about this, and nowadays creationists seldom resort to it. Indeed, as with the evidence from geographical distribution, which we discussed in the last chapter, they rarely discuss comparative evidence at all, preferring to stick to fossils, where they have been taught (wrongly) to think they are on promising ground.

NO BORROWING

To emphasize how odd the idea of a creator sticking rigidly to 'themes' is, reflect that any sensible human designer is quite happy to borrow an idea from one of his inventions, if it would benefit another. Maybe there is a 'theme' of aircraft design, which is separate from the 'theme' of train design. But a component of a plane, say an improved design for the reading lights above the seats, might as well be borrowed for use in trains. Why should it not, if it serves the same purpose in both? When motor cars were first invented, the name 'horseless carriage' tells us where some of the inspiration came from. But horse-drawn vehicles don't need steering wheels – you use reins to steer horses – so the steering wheel had to have another source. I don't know where it came from, but I suspect that it was borrowed from a completely different technology, that of the boat. Before being superseded by the steering wheel, which was introduced around the end of the nineteenth century, the original steering device of the car was the tiller, also borrowed from boats, but moved from the rear to the front of the vehicle.

If feathers are a good idea within the bird 'theme', such that every single bird, without exception, has them whether it flies or not, why do literally no mammals have them? Why would the designer not borrow that ingenious invention, the feather, for at least one bat? The evolutionist's answer is clear. All birds have inherited their feathers from their shared ancestor, which had feathers. No mammal is descended from that ancestor. It's as simple as that.* The tree of resemblances is a family tree. It is the same kind of story for every branch and every sub-branch and every sub-sub-branch of the tree of life.

Now we come to an interesting point. There are plenty of beautiful examples where it looks, superficially, as though ideas might have been 'borrowed' from one part of the tree and grafted on to another, like an apple variety grafted on to a stock. A dolphin, which is a small whale, looks superficially like various kinds of large fish. One of these fish, the dorado (*Coryphæna hippuris*) is even sometimes called a 'dolphin'. Dorados and true dolphins have the same streamlined shape, suited to their similar ways of life as fast hunters near the surface of the sea. But their swimming technique, though superficially similar, was not borrowed from one by the other, as you can quickly see if you look at the details. Although both derive their speed mostly from the tail, the dorado, like all fish, moves its tail from side to side. But the true dolphin betrays its mammal history by beating its tail up and down. The side-to-side wave travelling down the ancestral fish backbone has been inherited by lizards and snakes, which could almost be said to 'swim' on land. Contrast that with a galloping horse or cheetah. The speed comes from bending of the spine, as it does with fish and snakes; but in mammals the spine bends up and down, not side to side. It is an interesting question how the transition was made in the ancestry of mammals. Maybe there was an intermediate stage, which hardly bent its spine at all, in either direction, like a frog. On the other hand, crocodiles are capable of galloping (frighteningly fast) as well as using the lizard-like gait more conventional among reptiles. The ancestors of mammals were nothing like crocodiles, but

---

* I presume my readers know better than the author(s) of Leviticus, who thought that bats were birds. In chapter 11, verses 13–19, is a long list of birds that are an abomination, beginning with the eagle and ending with 'the stork, the heron after her kind, and the lapwing, and the bat'. It is a separate question why it was necessary to condemn any animals as abominations. It was a common practice in many religions.

maybe crocodiles show us how an intermediate ancestor might have combined the two gaits.

Anyway, the ancestors of whales and dolphins were fully paid-up land mammals, who surely galloped across the prairies, deserts or tundras with an up-and-down flexion of the spine. And when they returned to the sea, they retained their ancestral up-and-down spinal motion. If snakes 'swim' on land, dolphins 'gallop' through the sea! Accordingly, the fluke of a dolphin may look superficially like the forked tail of a dorado, but it is set horizontally, whereas the dorado's tail fins are aligned in the vertical plane. There are numerous other respects in which the dolphin's history is written all over it, and I shall come to them in the chapter of that title.

There are other examples where the superficial resemblance is so great that it seems quite hard to reject the 'borrowing' hypothesis, but a closer inspection shows that we must. Animals can look so alike that you feel they must be related. But it then turns out that the similarities, though impressive, are outnumbered by the differences when you look at the whole body. 'Pill bugs' (see over) are familiar little creatures, with lots of legs, who habitually roll up into a protective ball, like armadillos. Indeed, this may be the origin of the Latin name *Armadillidium*. That is the name of one kind of 'pill bug', which is a crustacean, a woodlouse, related to shrimps but living on land – where it betrays its recent aquatic ancestry by breathing with gills, which have to be kept moist. But the point of the story is that there is a completely different kind of 'pill bug' which is not a crustacean at all but a millipede. When you see them rolled up, you'd think they were almost identical. Yet one is a modified woodlouse, while the other is a modified (modified in the same direction) millipede. If you unroll them and look carefully, you will immediately see at least one important difference. The pill millipede has two pairs of legs on most segments, the pill woodlouse only one. Isn't it beautiful, all this endless modification? A more detailed examination will show that, in hundreds of respects, the pill millipede really does resemble a more conventional millipede.

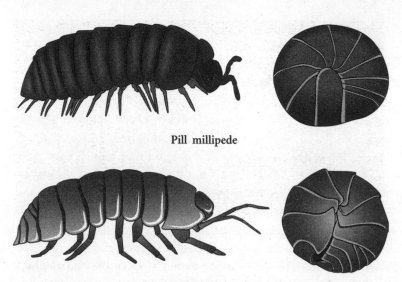

Pill millipede

Pill woodlouse

The resemblance to a woodlouse is superficial – convergent.

Almost any zoologist who was not a specialist would say that the skull on the opposite page belongs to a dog. The specialist would discover that it isn't actually a dog skull by noting the two prominent holes in the roof of the mouth. These are tell-tale signs of marsupials, the large group of mammals nowadays found mostly in Australia. It is in fact the skull of *Thylacinus*, the 'Tasmanian wolf'. Thylacines and true dogs (for example dingos, with which they competed in Australia and Tasmania) have converged on a very similar skull because they have (had, alas, in the case of the unfortunate thylacine) a similar lifestyle.

I have already mentioned the magnificent marsupial mammal fauna of Australia, in the chapter on the geographical distribution of animals. The relevant point for this chapter is the repeated convergences between these marsupials and a great variety of opposite numbers among the 'eutherian' (i.e. non-marsupial) mammals, which dominate the rest of the world. Though far from identical, even in superficial characteristics, each marsupial in the

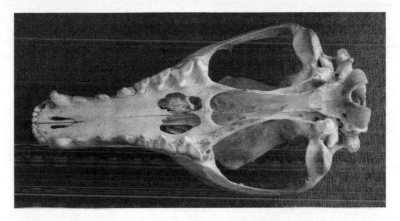

Thylacine 'marsupial wolf' or 'Tasmanian tiger' skull

illustration overleaf is sufficiently similar to its eutherian equivalent
– that is, the eutherian that most closely practises the same 'trade'
– to impress us, but certainly not sufficiently similar to suggest
'borrowing' by a creator.

The sexual shuffling of the genes in a gene pool could be
regarded as a kind of borrowing or sharing of genetic 'ideas',
but sexual recombination is confined within one species and is
therefore irrelevant to this chapter, which is about comparisons
between species: for example, comparisons between marsupial and
eutherian mammals. Interestingly, high-level borrowing of DNA
is rife among bacteria. In a process that is sometimes regarded
as a kind of precursor to sexual reproduction, bacteria – even
quite distantly related strains of bacteria – swap DNA 'ideas' with
promiscuous abandon. 'Borrowing ideas' is indeed one of the main
ways by which bacteria pick up useful 'tricks' such as resistance to
particular antibiotics.

The phenomenon is often called by the rather unhelpful name
of 'transformation'. That's because, when it was discovered in
1928 by Frederick Griffith, nobody understood about DNA. What
Griffith found was that a non-virulent strain of *Streptococcus* could
pick up virulence from a completely different strain, even though

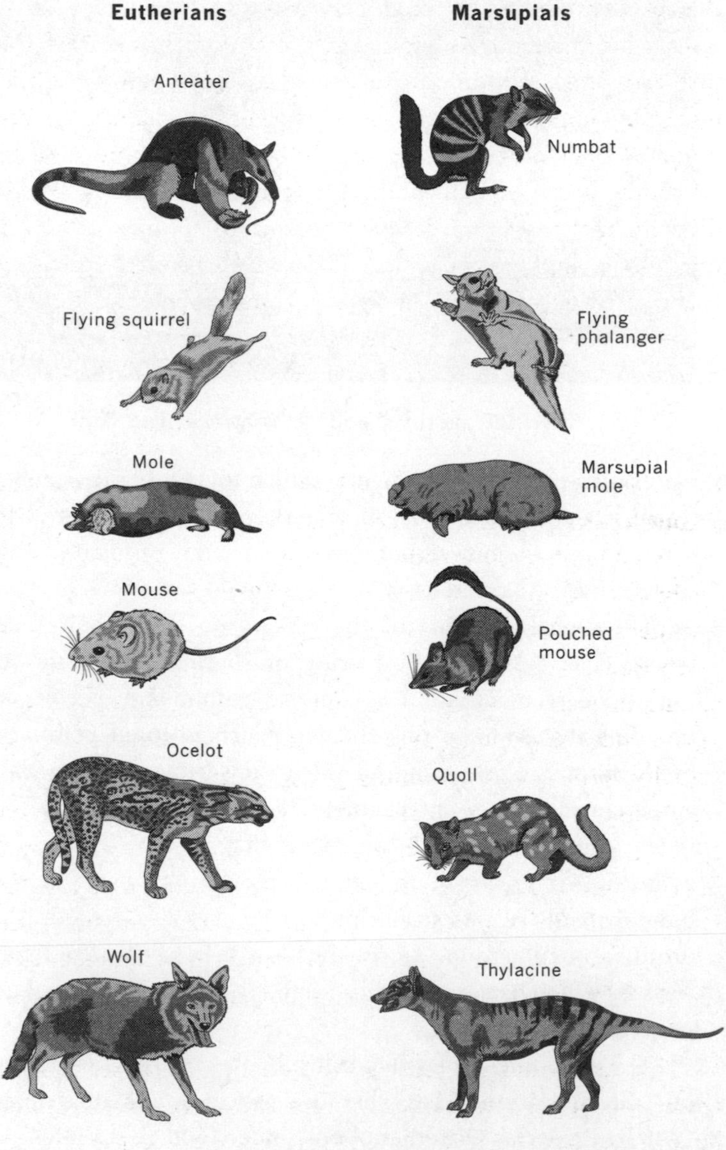

Eutherians

Anteater

Flying squirrel

Mole

Mouse

Ocelot

Wolf

Marsupials

Numbat

Flying phalanger

Marsupial mole

Pouched mouse

Quoll

Thylacine

Eutherian and marsupial
opposite numbers

that virulent strain was dead. Nowadays we would say that the non-virulent strain incorporated into its genome some DNA from the dead virulent strain (DNA doesn't care about being 'dead', it is just coded information). In the language of this chapter, the non-virulent strain 'borrowed' a genetic 'idea' from the virulent strain. Of course, bacteria borrowing genes from other bacteria is a very different matter from a designer borrowing his own ideas from one 'theme' and re-using them in another theme. Nevertheless, it is interesting because, if it were as common in animals as it is in bacteria, it would make it harder to disprove the 'designer borrowing' hypothesis. What if bats and birds behaved like bacteria in this respect? What if chunks of bird genome could be ferried across, perhaps by bacterial or viral infection, and implanted in a bat's genome? Maybe a single species of bats might suddenly sprout feathers, the feather-coding DNA information having been borrowed in a genetic version of a computer's 'Copy and Paste'.

In animals, unlike bacteria, gene transfer seems almost entirely confined to sexual congress within species. Indeed, a species can pretty well be defined as a set of animals that engage in gene transfer among themselves. Once two populations of a species have been separated for long enough that they can no longer exchange genes sexually (usually after an initial period of enforced geographical separation, as we saw in Chapter 9), we now define them as separate species, and they will never again exchange genes, other than by the intervention of human genetic engineers. My colleague Jonathan Hodgkin, Oxford's Professor of Genetics, knows of only three tentative exceptions to the rule that gene transfer is confined within species: in nematode worms, in fruit flies, and (in a bigger way) in bdelloid rotifers.

This last group is especially interesting because, uniquely among major groupings of eucaryotes, they have no sex. Could it be that they have been able to dispense with sex because they have reverted to the ancient bacterial way of exchanging genes? Cross-species gene transfer seems to be commoner in plants. The parasitic

Bdelloid rotifer

plant dodder (*Cuscuta*) donates genes to the host plants around which it is entwined.*

I am undecided about the politics of GM foods, torn between the potential benefits to agriculture on the one hand and precautionary instincts on the other. But one argument I haven't heard before is worth a brief mention. Today we curse the way our predecessors introduced species of animals into alien lands just for the fun of it. The American grey squirrel was introduced to Britain by a former Duke of Bedford: a frivolous whim that we now see as disastrously irresponsible. It is interesting to wonder whether taxonomists of the future may regret the way our generation messed around with genomes: transporting, for example, 'anti-freeze' genes from Arctic fish into tomatoes to protect them from frost. A gene that gives jellyfish a fluorescent glow has been borrowed from them by scientists and inserted into the genome of potatoes, in the hope of making them light up when they need watering. I have even read of an 'artist' who plans an 'installation'

---

* Biologists used to cite plant haemoglobin as a possible example of DNA borrowing by plants from the animal kingdom. Plants of the pea family (Leguminosae) have on their roots 'nodules', in which dwell bacteria that capture nitrogen from the atmosphere and make it available to the plants. This is why farmers often include a leguminous crop, such as clover or a vetch, in their rotation. It puts valuable nitrogen in the soil, especially if the clover crop is ploughed under. The nodules are a reddish colour because they contain a form of haemoglobin, similar to the oxygen-transporting molecule that makes our blood red. The genes for making haemoglobin are in the plant genome, not the bacterial genome. Haemoglobin is important to the bacteria, which need oxygen, and it can be regarded as part of the deal between bacteria and plants: the bacteria give the plants usable nitrogen, while the plants give the bacteria a house, and usable oxygen delivered via haemoglobin. Since we are accustomed to associating haemoglobin with blood, it was natural to wonder whether a gene for making it had somehow been 'borrowed' from an animal genome, perhaps ferried by a bacterium. It would, indeed, have been a very valuable idea to 'borrow'. Unfortunately for this appealing idea – the ultimate blood transfusion – molecular biological evidence shows that haemoglobins are ancient denizens of plant genomes. They are not borrowed. They have been there from ancient times.

consisting of luminous dogs, glowing with the aid of jellyfish genes. Such debauchery of science in the name of pretentious 'art' offends all my sensibilities. But could the damage go further? Could these frivolous caprices undermine the validity of future studies of evolutionary relationships? Actually I doubt it, but perhaps the point is at least worth raising, in a precautionary spirit. The whole point of the precautionary principle, after all, is to avoid future repercussions of choices and actions that may not be obviously dangerous now.

CRUSTACEANS

I began the chapter with the vertebrate skeleton, which is a lovely example of an invariant pattern linking variable detail. Almost any other major group of animals would show the same kind of thing. I'll take just one other favourite example: the decapod crustaceans, the group that includes lobsters, prawns, crabs and hermit crabs (which are not crabs, by the way). The body plan of all crustaceans is the same. Whereas our vertebrate skeleton consists of hard bones inside an otherwise soft body, crustaceans have an *exoskeleton* consisting of hard tubes, inside which the animal keeps and protects its soft bits. The hard tubes are jointed and hinged, in something like the same way as our bones are. Think, for example, of the delicate hinges in the legs of a crab or lobster, and the more robust hinge of the claw. The muscles that power the pinch of a large lobster are inside the tubes that make up the claw. The equivalent muscles when a human hand pinches something attach to the bones that run through the middle of the finger and thumb.

Like vertebrates, but unlike sea urchins or jellyfish, crustaceans are left/right symmetrical, with a train of segments running the length of the body from head to tail. The segments are the same as each other in their underlying plan, but often differ in detail. Each segment consists of a short tube joined, either rigidly or by a hinge, to the two neighbouring segments. As with vertebrates, the organs and organ systems of a crustacean show a repeat pattern as you move from front to rear. For example, the main nerve trunk, which runs the length of the body on the ventral side (not the dorsal side, as the vertebrate spinal cord does), has a pair of ganglia (sort of mini-brains*) in each segment, from which sprout nerves supplying the segment. Most of the segments have a limb on each side, each limb again consisting of a series of tubes joined by hinges. Crustacean limbs usually terminate in a two-way branch, which in many cases you could call a claw. The head is segmented too although, as with the vertebrate head, the segmental pattern is more disguised here than in the rest of the body. There are five pairs of limbs lurking in the head, although it might sound a bit strange to call them limbs since they are modified to become antennae or components of the jaw apparatus.

---

* It is a little-known fact that some dinosaurs had a ganglion in the pelvis, which was so large (at least relative to the brain in the head) as almost to deserve the title of second brain. This prompted the following delightfully witty verse by the American comic writer Bert Leston Taylor (1866–1921):

Behold the mighty dinosaur,
Famous in prehistoric lore,
Not only for his power and strength
But for his intellectual length.
You will observe by these remains
The creature had two sets of brains –
One in his head (the usual place),
The other at his spinal base,
Thus he could reason A priori
As well as A posteriori.
No problem bothered him a bit
He made both head and tail of it.
So wise was he, so wise and solemn,
Each thought filled just a spinal column.
If one brain found the pressure strong
It passed a few ideas along.
If something slipped his forward mind
'Twas rescued by the one behind.
And if in error he was caught
He had a saving afterthought.
As he thought twice before he spoke
He had no judgment to revoke.
Thus he could think without congestion
Upon both sides of every question.
Oh, gaze upon this model beast,
Defunct ten million years at least.

They are therefore usually called appendages rather than limbs. More or less invariably, the five segmental appendages of the head, reading from the front, consist of first antennae (or antennules), second antennae (often just called antennae), mandibles, first maxillae (or maxillules) and second maxillae. The antennules and antennae are mostly engaged in sensing things. The mandibles and maxillae are concerned with chewing, milling or otherwise processing food. As we proceed back along the body, the segmental appendages or limbs are pretty variable, the middle ones often consisting of walking legs, while those sprouting from the rearmost segments are often pressed into service doing other things such as swimming.

In a lobster or a prawn, after the usual five head segment appendages, the first body segment appendages are the claws. The next four pairs are walking legs. The segments bearing claws and walking legs are bunched together as the thorax. The rest of the body is called the abdomen. Its segments, at least until you reach the tip of the tail, are the 'swimmerets', feathery appendages that help with swimming, quite importantly so in some delicately graceful prawns. In crabs, the head and thorax have merged into a single large unit, to which all the first ten pairs of limbs are attached. The abdomen is doubled back under the head/thorax so that you can't see it at all from above. But if you turn a crab over, you can clearly see the abdomen's segmental pattern. The picture below shows the typical narrow abdomen of a male crab. The female abdomen is wider and resembles an apron, which it is indeed called. Hermit crabs are unusual in that the abdomen is asymmetrical (to fit into the empty

Male crab showing narrow,
folded-back abdomen

mollusc shell which is its house), and soft and unarmoured (because the mollusc shell provides protection).

To get an idea of some of the wonderful ways in which the crustacean body is modified in detail, while the body plan itself is not modified at all, look at the set of drawings opposite by the famous nineteenth-century zoologist Ernst Haeckel, perhaps Darwin's most devoted disciple in Germany (the devotion was not reciprocated, but even Darwin would surely have admired Haeckel's draughtsmanship). Just as we did with the vertebrate skeleton, look at each body part of these crabs and crayfish, and see how, without fail, you can find its exact opposite number in all the rest. Every bit of the exoskeleton is joined to the 'same' bits, but the shapes of the bits themselves are very different. Once again, the 'skeleton' is invariant, while its parts are anything but. And once again the obvious – I would say the only sensible – interpretation is that all these crustaceans have inherited the plan of their skeleton from a common ancestor. They have moulded the individual components into a rich variety of shapes. But the plan itself remains, exactly as inherited from the ancestor.

WHAT WOULD D'ARCY THOMPSON HAVE DONE WITH A COMPUTER?

In 1917 the great Scottish zoologist D'Arcy Thompson wrote a book called *On Growth and Form*, in the last chapter of which he introduced his famous 'method of transformations'.* He would draw an animal on graph paper, and then he would distort the graph paper in a mathematically specifiable way and show that the form of the original animal had turned into another, related animal. You could

---

* D'Arcy Thompson was surely one of the most erudite scientists ever. Not only did he write famously beautiful English of a patrician character, not only was he a published mathematician and classical scholar as well as Professor of Natural History at Scotland's oldest university, but his book is laced with quotations, which he assumed he had no need to translate (how times have changed), in Latin, Greek, Italian, German, French and even Provençal (the last he *did* deign to translate – into French!).

Haeckel's crustaceans. Ernst Haeckel was a distinguished German zoologist and an excellent zoological artist.

think of the original graph paper as a piece of rubber, on which you draw your first animal. Then the transformed graph paper would be equivalent to the same piece of rubber, stretched or pulled out of shape in some mathematically defined way. For example, he took six species of crab and drew one of them, *Geryon*, on ordinary graph paper (the undistorted sheet of rubber). He then distorted his mathematical 'rubber sheet' in five separate ways, to achieve an approximate representation of the other five species of crab. The details of the mathematics don't matter, although they are fascinating. What you can clearly see is that it doesn't take much to transform one crab into another. D'Arcy Thompson himself wasn't very interested in evolution, but it is easy for us to imagine what the genetic mutations would have to do in order to bring about changes like this. That doesn't mean we should think of *Geryon*, or any other one of these six crabs, as being ancestral to the others. None of them was, and in any case that is not the point. The point is that whatever the ancestral crab looked like, transformations of this *kind* could change any one of these six species (or a putative ancestor) into any other.

Evolution never happened by taking one adult form and coaxing it into the shape of another. Remember that every adult grows as an embryo. The mutations selected would have worked in the developing embryo by changing the rate of growth of parts of the body relative to other parts. In Chapter 7 we interpreted the evolution of the human skull as a series of changes in the rates of growth of some parts relative to other parts, controlled by genes in the developing embryo. We should expect, therefore, that if we draw a human skull on a sheet of 'mathematical rubber', it should be possible to distort the rubber in some mathematically tidy way and achieve an approximate likeness to the skull of a close cousin, such as a chimpanzee, or – perhaps with a bigger distortion – a more distant cousin such as a baboon. And this is just what D'Arcy Thompson showed. Note, once again, that it was an arbitrary decision to draw the human skull first, and then transform it into the chimpanzee and the baboon. He could equally well have drawn, say, the chimpanzee first and then

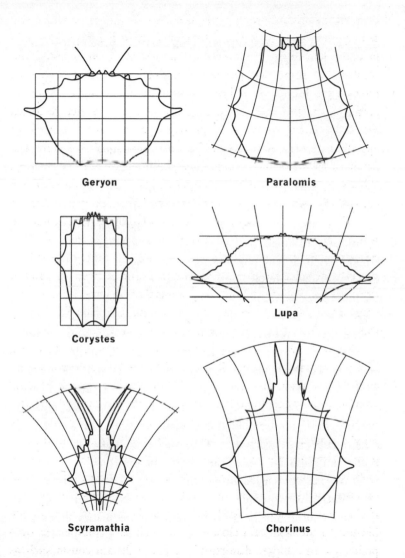

Geryon

Paralomis

Corystes

Lupa

Scyramathia

Chorinus

D'Arcy Thompson's crab 'transformations'

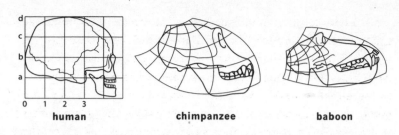

human   **chimpanzee**   **baboon**

D'Arcy Thompson's skull 'transformation'

worked out the necessary distortions to make the human and the baboon. Or, more interestingly for a book on evolution, which his was not, he might have drawn, say, an *Australopithecus* skull first on the undistorted rubber, and worked out how to transform it to make a modern human skull. This would surely have worked just as well as the pictures above, and it would have been evolutionarily meaningful in a more direct way.

At the beginning of this chapter I introduced the idea of 'homology', using the arms of bats and humans as an example. Indulging an idiosyncratic use of language, I said that the skeletons were identical while the bones were different. D'Arcy Thompson's transformations furnish us with a way to make this idea more precise. In this formulation, two organs – for example, bat hand and human hand – are homologous if it is possible to draw one on a sheet of rubber and then distort the rubber to make the other one. Mathematicians have a word for this: 'homeomorphic'.*

Zoologists recognized homology in pre-Darwinian times, and pre-evolutionists would describe, say, bat wings and human hands as homologous. If they had known enough mathematics, they would have been happy to use the word 'homeomorphic'. In post-Darwinian times, when it became generally accepted that bats and humans share a common ancestor, zoologists started to define homology in

---

* Strictly speaking, two shapes are homeomorphic if you can distort one to become the other without breaking it, and without any new touchings.

evolutionary terms. Homologous resemblances are those inherited from the shared ancestor. The word 'analogous' came to be used for resemblances due to shared function, not ancestry. For example, a bat wing and an insect wing would be described as analogous, as opposed to the homologous bat wing and human arm. If we want to use homology as evidence for the fact of evolution, we can't use evolution to define it. For this purpose, therefore, it is convenient to revert to the pre-evolutionary definition of homology. The bat wing and human arm are homeomorphic: you can transform one into the other by distorting the rubber on which it is drawn. You cannot transform a bat wing into an insect wing in this way, because there are no corresponding parts. The widespread existence of homeomorphisms, which are not defined in terms of evolution, can be used as evidence for evolution. It is easy to see how evolution could go to work on any vertebrate arm and transform it into any other vertebrate arm, simply by changing relative rates of growth in the embryo.

Ever since becoming acquainted with computers as a graduate student in the 1960s, I have wondered what D'Arcy Thompson might have done with a computer. The question became pressing in the 1980s, when affordable computers with screens (as opposed to just paper printers) became common. Drawing on stretched rubber and then distorting the drawing surface in a mathematical way – it was just *begging* for the computer treatment! I suggested that Oxford University should bid for a grant to employ a programmer to put D'Arcy Thompson's transformations on a computer screen, and make them available in a user-friendly manner. We got the money, and employed Will Atkinson, a first-class programmer and biologist, who became a friend and an adviser to me on my own programming projects. Once he had solved the difficult problem of programming a rich repertoire of mathematical distortions of the 'rubber', it was then a relatively simple task for him to incorporate this mathematical wizardry into a biomorph-style artificial selection program, similar to my own 'biomorph' programs, here described in Chapter 2. As with my programs, the 'player' was confronted with a screen full of animal forms, and invited to choose one of them for 'breeding', generation after generation. Once again there were 'genes' that persisted through the generations, and once again the genes influenced the form of the 'animals'. But in this case, the way the genes influenced animal form was by controlling the distortion of the 'rubber' on which an animal's form had been drawn. Theoretically, therefore, it should have been possible to start with, say, an *Australopithecus* skull drawn on the undistorted 'rubber', and breed your way through creatures with progressively larger braincases and progressively shorter muzzles – progressively more human-like, in other words. In practice it proved very difficult to do anything like that, and I think the fact is, in itself, interesting.

I think one reason it was difficult is, yet again, that D'Arcy Thompson's transformations change one *adult* form into another adult form. As I emphasized in Chapter 8, that is not how genes in evolution work. Every individual animal has a developmental history. It starts as an embryo and grows, by disproportionate growth of different parts of the body, into an adult. Evolution is not a genetically controlled distortion of one adult form into another; it is a genetically controlled alteration in a developmental program. Julian Huxley (grandson of T.H. and brother of Aldous) recognized this when, soon after publication of the first edition of D'Arcy Thompson's book, he modified the 'method of transformations' to study the way early embryos turn into later embryos or adults. That's all I want to say about D'Arcy Thompson's method of transformations here. I'll return to the topic in the final chapter, to make a related point.

Comparative evidence has always, as I suggested at the beginning of this chapter, told even more compellingly than fossil evidence in favour of the fact of evolution. Darwin himself took a similar view, at the end of his chapter in *On the Origin of Species* on the 'Mutual Affinities of Organic Beings':

> Finally, the several classes of facts which have been considered
> in this chapter, seem to me to proclaim so plainly, that the

innumerable species, genera, and families of organic beings, with which this world is peopled, have all descended, each within its own class or group, from common parents, and have all been modified in the course of descent, that I should without hesitation adopt this view even if it were unsupported by other facts or arguments.

MOLECULAR COMPARISONS

What Darwin didn't – couldn't – know is that the comparative evidence becomes even more convincing when we include molecular genetics, in addition to the anatomical comparisons that were available to him.

Just as the vertebrate skeleton is invariant across all vertebrates while the individual bones differ, and just as the crustacean exoskeleton is invariant across all crustaceans while the individual 'tubes' vary, so the DNA code is invariant across all living creatures, while the individual genes themselves vary. This is a truly astounding fact, which shows more clearly than anything else that all living creatures are descended from a single ancestor. Not just the genetic code itself, but the whole gene/protein system for running life, which we dealt with in Chapter 8, is the same in all animals, plants, fungi, bacteria, archaea and viruses. What varies is what is written in the code, not the code itself. And when we look comparatively at what is written in the code – the actual genetic sequences in all these different creatures – we find the same kind of hierarchical tree of resemblance. We find the same *family tree* – albeit much more thoroughly and convincingly laid out – as we did with the vertebrate skeleton, the crustacean skeleton, and indeed the whole pattern of anatomical resemblances through all the living kingdoms.

If we want to work out how closely related any pair of species is – say, how close a hedgehog is to a monkey – the ideal would be to look at the complete molecular texts of every gene of both species, and compare every jot and tittle, as a biblical scholar might compare two scrolls or fragments of Isaiah. But it is time-consuming and expensive. The Human Genome Project took about ten years, representing many person-centuries. Although it would now be possible to achieve the same result in a fraction of the time, it would still be a large and expensive undertaking, as would the hedgehog genome project. Like the Apollo moon landings, and like the Large Hadron Collider (which has just been switched on in Geneva as I write – the gigantic scale of this international endeavour moved me to tears when I visited), the complete deciphering of the human genome is one of those achievements that makes me proud to be human. I am delighted that the chimpanzee genome project has now been successfully accomplished, and the equivalent for various other species. If the present rate of progress continues (see 'Hodgkin's Law' below), it will soon be economically feasible to sequence the genome of every pair of species whose closeness of cousinship we might want to measure. Meanwhile, for the most part we have to resort to sampling particular parts of their genomes, and it works pretty well.

We can sample by picking out a few choice genes (or proteins, whose sequences are directly translated from genes) and comparing them across species. I'll come to that in a moment. But there are other ways of doing a kind of crude, automatic sampling, and the technologies to do that have been around for longer. An early method, which works surprisingly well, exploits the immune system of rabbits (you could actually use any animal you like, but rabbits do the job nicely). As part of the body's natural defence against pathogens, the rabbit's immune system manufactures antibodies against any foreign protein that enters the bloodstream. Just as you could tell that I have had whooping cough by looking at the antibodies in my blood, so you can tell what a rabbit has been exposed to in the past by looking at its immune response in the present. The antibodies present in the rabbit constitute a history of the natural shocks to which its flesh has been heir – including artificially injected proteins. If you inject, say, a chimpanzee protein into a rabbit, the antibodies that it makes will subsequently attack the same protein if it is injected again. But suppose your second injection is of the equivalent protein, not from a chimpanzee but from a gorilla? The rabbit's prior exposure to the chimpanzee protein will have *partially* forearmed it against the gorilla version, but the reaction will be weaker. And it will also have forearmed it against the kangaroo version of the protein, but the reaction will be weaker still, given that the kangaroo is much less closely related to the chimpanzee that did the priming than the gorilla is. The strength of the rabbit's immune response to a subsequent injection of a protein is a measure of the resemblance of that protein to the original to which the rabbit was first exposed. It was by this method, using rabbits, that Vincent Sarich and Allan Wilson, at the University of California at Berkeley, demonstrated in the 1960s that humans and chimpanzees are much more closely related to each other than anybody had previously realized.

There are also methods that use the genes themselves, comparing them across species directly rather than comparing the proteins they encode. One of the oldest and most effective of these methods is called DNA hybridization. DNA hybridization is usually what lies behind those statements one often sees along the lines of: 'Humans and chimpanzees share 98 per cent of their genes.' There is some confusion, by the way, about exactly what is meant by percentage figures such as these. Ninety-eight per cent of *what* is identical? The exact figure depends on how large the units are that we are counting. A simple analogy makes this clear, and it does so in an interesting way, because the differences between the analogy and the real thing are as revealing as the similarities. Suppose we have two versions of the same book and we want to compare them. Perhaps it is the book of Daniel, and we want to compare the canonical version with an ancient scroll that has just been discovered in a cave overlooking the Dead Sea. What percentage of the chapters of the two books are identical? Probably zero, for it takes only one discrepancy, anywhere in the whole chapter, for us to say the two are not identical. What percentage of their *sentences* are identical? The percentage will now be much higher. Even higher will be the percentage of words that are identical, because words have fewer letters than sentences – fewer opportunities to bust the identity. But a word resemblance is still broken if any one letter in the word differs. Therefore, if you line the two texts up side by side and compare them letter by letter, the percentage of identical letters will be even higher than the percentage of identical words. So an estimate like '98 per cent in common' doesn't mean anything unless we specify the size of the unit we are comparing. Are we counting chapters, words, letters or what? And the same is true when we compare DNA from two species. If you are comparing whole chromosomes, the percentage shared is zero, because it only takes one tiny difference, somewhere along the chromosomes, to define the chromosomes as different.

The often-quoted figure of about 98 per cent for the shared genetic material of humans and chimps actually refers neither to numbers of chromosomes nor to numbers of whole genes, but to numbers of DNA 'letters' (technically, base pairs) that match each other within the respective human and chimp genes. But there is a pitfall. If you do the lining up naïvely, a *missing* letter (or an added letter), as opposed to a mistaken letter, will cause all subsequent letters to mismatch, because they will then all be staggered, one step out (until there is a mistake in the other direction to bring them back into step again). It is clearly unfair to let the estimate of discrepancies be inflated in this way. A scholar's eye, scanning two scrolls of Daniel, automatically copes with this, in a way that is hard to quantify. How can we do it with DNA? This is where we leave our analogy with books and scrolls and go straight to the real thing because, as it happens, the real thing – DNA – is easier to understand than the analogy!

If you gradually heat DNA, there comes a point – somewhere around 85°C – when the bonding between the two strands of the double helix breaks, and the two helices separate. You can think of 85°C, or whatever the temperature turns out to be, as a 'melting point'. If you let it cool again, each single helix spontaneously joins up again with another single helix, or fragment of single helix, wherever it finds one with which it can pair, using the ordinary base-pairing rules of the double helix. You might think that this would always be the partner from which it lately separated, and with which, of course, it is perfectly matched. Indeed it could be, but it usually isn't as tidy as that. Fragments of DNA will find other fragments with which they can pair, and they will usually not be exactly their original partners. And indeed, if you add separated fragments of DNA from another species, fragments of the single strands are quite capable of joining up with fragments of single strands from the wrong species, in just the same way as they will join up with single strands from the right species. Why should they not? It is the remarkable conclusion of the Watson–Crick molecular biology revolution that DNA is just DNA. It doesn't 'care' whether it is human DNA, chimp DNA or apple DNA. Fragments will happily pair off with complementary fragments wherever they find them. Nevertheless, the strength of bonding is not always equal. Single-stranded lengths of DNA bond more tightly with matching single strands than they do with less similar single strands. This is because more of the 'letters' of the DNA (Watson and Crick's 'bases') find themselves opposite partners with which they cannot pair. The bonding of the strands is therefore weakened – like a zip fastener with some teeth missing.

How shall we measure this strength of bonding, after fragments from different species have found each other and united? By an almost ludicrously simple method. We measure the 'melting point' of the bonds. You remember I said that the melting point of double-stranded DNA is about 85°C. This is true of normal, properly matched double-stranded DNA, as when a strand of human DNA is 'melted' away from a complementary strand of human DNA. But when the bonding is weaker – as when a human strand has bonded with a chimpanzee strand – a slightly lower temperature is sufficient to break the bond. And when human DNA has bonded with DNA from a more distant cousin like a fish or a toad, an even lower temperature suffices to separate them. The difference between the melting point when a strand is bonded to one of its own kind, and the melting point when it is bonded to a strand from another species, is our measure of the genetic distance between the two species. As a rule of thumb, each decrease by 1° Celsius in the 'melting point' is approximately equivalent to a drop of 1 per cent in the number of DNA letters matched (or an increase of 1 per cent in the number of missing teeth in the zip fastener).

There are complications in the method, which I haven't gone into, and tricky problems, which have ingenious solutions. For instance, if you mix human with chimp DNA, much of the fragmented human DNA will bond with other human DNA fragments, and much of the chimp DNA will bond with its own kind. How do you separate off the hybrid DNA, whose 'melting point' is what you really want to measure, from the 'same-kind' DNA? The answer is by a clever trick involving previous radioactive labelling. But the details would take us too far off our path. The main point here is that DNA hybridization is the technique that leads scientists to figures like 98 per cent for the genetic similarity between humans and chimpanzees, and it yields predictably lower percentages as you move to more distantly related pairs of animals.

The newest method of measuring the similarity between a pair of matching genes from different species is the most direct, and the most expensive: actually read the sequence of letters in the genes themselves, using the same methods as were used for the Human Genome Project. Although it is still expensive to compare the entire genome, you can get a good approximation by comparing a sample of genes, and this is now increasingly done.

Whichever technique we use for measuring similarity between two species, whether it is rabbit antibodies, or melting points, or direct sequencing, the next step is pretty much the same. Having obtained a single number representing the similarity between each pair of species, we then place the figures in a table. Take a set of species and write their names, in the same order, as both the column headings and the row headings. Then place the percentage similarities in the appropriate cells. The table will be triangular (half of a square) because, for example, the percentage similarity between human and dog will be the same as the similarity between dog and human. So if you filled in all of a square table each of the two halves either side of the diagonal would mirror the other.

Now, what sort of results should we expect? On the evolution model we should predict that you'll find yourself putting a high score in the cell connecting human and chimpanzee; a lower score in the cell connecting human and dog. The human/dog cell should theoretically have an identical resemblance score to the chimpanzee/dog cell because humans and chimpanzees have exactly the same degree of relation to dogs. It should be identical, too, to the monkey/dog cell and the lemur/dog cell. This is because humans, chimpanzees, monkeys and lemurs are all connected to the dog via their common ancestor, an early primate (which probably looked a bit like a lemur). The same score should show up in the human/cat, chimpanzee/cat, monkey/cat and lemur/cat cells, because cats and dogs are related to all primates via the shared ancestor of all carnivores. There should be a much lower score – ideally equally low – in all the cells uniting, say, a squid with any mammal. And it shouldn't matter which mammal you choose, since all are equally distant from a squid.

These are strong theoretical expectations, but there is no reason why, in practice, they should not be violated. If they were violated, it would be evidence against evolution. What actually happens turns out to be – within statistical margins of error – just what we should expect on the assumption that evolution has happened. This is another way of saying that, if you put the genetic distances between pairs of species on the limbs of a tree, everything adds up in a satisfying way. Of course the adding up is not quite perfect. Numerical expectations in biology are seldom realized with better than approximate accuracy.

Comparative DNA (or protein) evidence can be used to decide – on the evolutionary assumption – which pairs of animals are closer cousins than which others. What turns this into extremely powerful evidence for evolution is that you can construct a tree of genetic resemblances separately for each gene in turn. And the important result is that almost every gene delivers the same tree of life. Once again, this is exactly what you would expect if you were dealing with a true family tree. It is not what you would expect if a designer had surveyed the whole animal kingdom and picked and chosen – or 'borrowed' – the best proteins for the job, wherever in the animal kingdom they might be found.

The earliest large-scale study along these lines was done by a group of geneticists in New Zealand led by Professor David Penny. Penny's group took five genes which, although not identical across all mammals, were similar enough to have earned the same name in all. The details don't matter, but for the record the five genes were those for haemoglobin A, haemoglobin B (haemoglobins give blood its red colour), fibrinopeptide A, fibrinopeptide B (fibrinopeptides are used in clotting blood) and cytochrome C (which plays an important role in cellular biochemistry). They chose eleven mammals to compare: rhesus monkey, sheep, horse, kangaroo, rat, rabbit, dog, pig, human, cow and chimpanzee.

Penny and his colleagues thought statistically. They wanted to calculate the probability that, purely by chance, two molecules would yield the same family tree, if evolution wasn't true. So they tried to imagine all possible trees that could terminate in eleven descendants. It's a surprisingly large number. Even if you limit yourself to 'binary trees' (that is, trees with branches that only bifurcate – no tri-furcating or higher-furcating), the total number of possible trees is more than 34 million. The scientists patiently looked at every one of the 34 million trees and compared each one with the other 33,999,999 trees. No, of course they didn't! It would take too much computer time. They did, however, devise a clever statistical approximation, a short-cut equivalent to that mammoth calculation.

This is how the method of approximation worked. They took the first of the five genes, say haemoglobin-A (in all cases I use the name of the protein to stand for the gene that codes for that protein). Of all those millions of trees, they wanted to find which was the most 'parsimonious' where haemoglobin-A was concerned. Parsimonious here means 'needing to postulate the minimum amount of evolutionary change'. For example, all those thousands of trees that assumed that the closest cousin to a human was a kangaroo while humans and chimpanzees are more distantly related, proved to be very unparsimonious trees: they needed to assume a lot of evolutionary change, in order to yield the result that kangaroos and humans had a recent common ancestor. Haemoglobin-A's verdict would be along these lines:

> This is a terribly unparsimonious tree. Not only do I have to put in lots of mutational work in order to end up so different in humans and kangaroos, despite our close cousinship according to this tree, I also have to put in lots of mutational work in the other direction, in order to ensure that, despite their great separation on this particular tree, humans and chimps somehow ended up with such similar haemoglobin-A. I vote against this tree.

Haemoglobin-A delivers a verdict of this kind, some verdicts more favourable than others, on each of the 34 million trees, and finally ends up choosing a few dozen top-ranking trees. Of each of these top-ranking trees, haemoglobin-A would say something like this:

> This tree puts humans and chimpanzees as close cousins, and it puts sheep and cows as close cousins, and it puts kangaroos out on a limb. This turns out to be a very good tree, because it makes me do hardly any mutational work at all to explain the evolutionary changes. This is an excellently parsimonious tree. It gets the haemoglobin-A vote!

Of course, it would have been nice if haemoglobin-A, and every other gene, could have come up with a single most parsimonious tree, but that is too much to ask. Among the 34 million trees, it is only to be expected that several slightly different trees should tie for haemoglobin-A's top-ranking slot.

Now, how about haemoglobin-B? How about cytochrome-C? Each one of the five proteins is entitled to its own separate vote, to find its own preferred (that is, most parsimonious) trees from among the 34 million trees. It would be perfectly possible for cytochrome-C to come up with a completely different vote on which is the most parsimonious tree. It could turn out that the cytochrome-C of humans really is very similar to that of kangaroos, and very different from that of chimpanzees. Far from saluting the close pairing of sheep and cow discerned by haemoglobin-A, cytochrome-C might find that it hardly needs to mutate at all in order to place sheep very close to, say, monkeys, and in order to place cows very close to rabbits. On the creation hypothesis there is no reason why that shouldn't happen. But what Penny and his colleagues actually found was that there was astonishingly high agreement among all five proteins (and they used yet more clever statistics to show how unlikely such concordance would be by chance). All five proteins 'voted' for pretty much the same subset of trees from among the 34 million possible trees. This is, of course, exactly what we would expect on the assumption that there really is only one true tree relating all eleven animals, and it is the *family* tree: the tree of evolutionary relationships. What is more, the consensus tree that the five molecules all voted for turned out to be the same as zoologists had already worked out on anatomic and palaeontological, not molecular, grounds.

The Penny study was published in 1982, quite a while ago now. The intervening years have seen a prolific multiplication of detailed evidence on the exact sequences of genes of lots and lots of species of animals and plants. Agreement on the most parsimonious trees now extends far beyond the eleven species and five molecules that Penny and his colleagues studied. Theirs was just a nice example, overwhelming as their statistical evidence proved. The sum total of genetic sequence data now available puts the matter beyond all conceivable doubt. Far more convincingly even than the (also highly

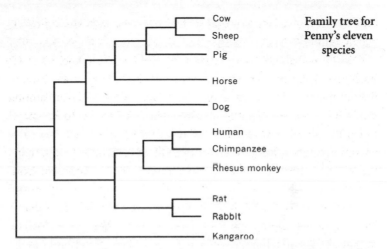

**Family tree for
Penny's eleven
species**

Cow
Sheep
Pig
Horse
Dog
Human
Chimpanzee
Rhesus monkey
Rat
Rabbit
Kangaroo

convincing) fossil evidence, the evidence from comparisons among
genes is converging, rapidly and decisively, on a single great tree of
life. Above is a tree for the eleven species of the Penny study, which
represents a modern consensus vote from many different parts of
the mammalian genome. It is the consistency of agreement* among
different genes in the genome that gives us confidence, not only in
the historical accuracy of the consensus tree itself, but also in the fact
that evolution has occurred.

If molecular genetic technology continues to expand at its present
exponential rate, by the year 2050 deriving the complete sequence
of an animal's genome will be cheap and quick, scarcely any more
trouble than taking its temperature or its blood pressure. Why do
I say that genetic technology is expanding exponentially? Could we
even measure it? There is a parallel in computer technology called
Moore's Law. Named after Gordon Moore, one of the founders of
the Intel computer chip company, it can be expressed in various
ways because several measures of computer power are linked to each
other. One version of the law states that the number of units that

---

* A rare case where 'the exception' really does 'prove the rule' was published just before this
paperback edition went to press. See page 437 for details.

can be packed into an integrated circuit of a given size doubles every eighteen months to two years or so. It is an empirical law, meaning that, rather than deriving from some piece of theory, it just turns out to be true when you measure the data. It has held good over a period of about fifty years so far, and many experts think it will do so for at least a few more decades. Other exponential trends, with a similar doubling time, which can be regarded as versions of Moore's Law, include the increase in speed of computation, and size of memory, per unit cost. Exponential trends always lead to startling results, as Darwin demonstrated when, with the aid of his mathematician son George, he took the elephant as an example of a slow-breeding animal and showed that, in just a few centuries of unrestricted exponential growth, the descendants of just one pair of elephants would carpet the earth. Needless to say, population growth of elephants is not, in practice, exponential. It is limited by competition for food and space, by disease, and by many other things. That, indeed, was Darwin's whole point, for that is where natural selection steps in.

But Moore's Law really has remained in force, at least approximately, for fifty years. Although nobody has a very clear idea why, various measures of computer power actually have increased exponentially in practice, where Darwin's elephant trend is exponential only in theory. It occurred to me that there might be a similar law in force for genetic technology and the sequencing of DNA. I suggested it to Jonathan Hodgkin, Oxford's Professor of Genetics (who had once been an undergraduate pupil of mine). To my delight, it turned out that he had already thought of it – and measured it, in preparation for a lecture at his old school. He estimated the cost of sequencing a standard length of DNA at four dates in history, 1965, 1975, 1995 and 2000. I inverted his figures to 'bangs for the buck', or 'How much DNA could you sequence for £1,000?' I plotted the figures on a logarithmic scale, chosen because an exponential trend will always show up as a straight line when plotted logarithmically. Sure enough, Hodgkin's four points fall pretty well on a straight line. I fitted a line to the points (for the technique of linear regression, see

The San Andreas Fault, a great gash up the length of California. One day the western part of the state, with Baja California, will be an island in the Pacific.

(**a**) Colour coding the age of the rocks under the sea. Chapter 9's hypothetical submarine sets a course due east from the bulge of Brazil, reaching the young rocks of the mid-Atlantic ridge halfway across. (**b**) Sea-floor spreading and (**c**) the deep and slow convection currents that drive the movements of the plates.

a

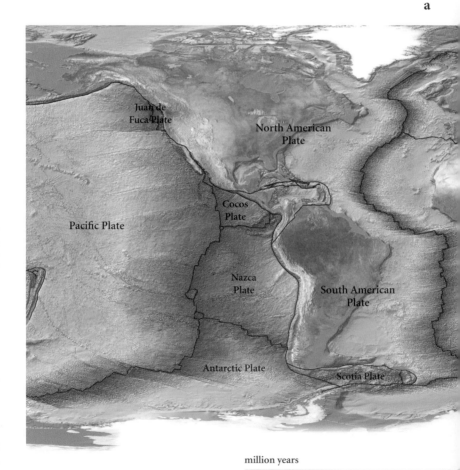

million years

0   20   40   60   80   100   120

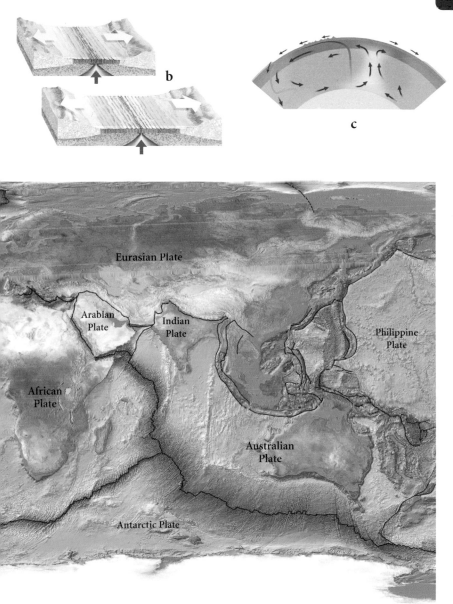

b

c

Eurasian Plate

Arabian
Plate

Indian
Plate

Philippine
Plate

African
Plate

Australian
Plate

Antarctic Plate

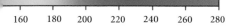

160   180   200   220   240   260   280

a

b

c

d

Galapagos: young showcase of evolution?

(**a**) Caldera on Fernandina, youngest and most volcanically active of the Galapagos islands. (**b**) Aerial view of Galapagos, showing the green of the highlands (volcanoes) and the dark colour of the lava plains. (**c**) Galapagos pelican plunge-diving for fish. This Galapagos sub-species of the brown pelican bears, for some reason, has the sub-specific name *urinator*. (**d**) Galapagos marine iguana swimming. This habit is unique among lizards. The giant tortoises of Galapagos vary from island to island. The saddleback shape (**e**), unlike the dome shape of grazing tortoises (**f**), is characteristic of islands where the tortoises browse on cactuses and therefore have to stretch their necks high. (**g**) Typical Galapagos scene. Galapagos brown pelican, Galapagos penguin (the only penguin to reach – just – the Northern Hemisphere) and 'Sally Lightfoot' crabs on black lava rocks.

Australia and Madagascar: two 'islands' of evolution.

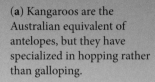

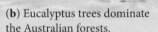

(**a**) Kangaroos are the Australian equivalent of antelopes, but they have specialized in hopping rather than galloping.

a

(**b**) Eucalyptus trees dominate the Australian forests.

b

(**c**) Koalas are the sloths of Australian forests, with a similarly slow metabolic rate. They have specialized in eating eucalpyt leaves, perhaps because few other animals can deal with their toxins. Note the baby in the pouch, which points backwards, probably for reasons of historical accident.

c

(**d**) The platypus is a survivor from ancient times when the mammals of Gondwana still laid eggs.

d

(e) Ringtailed lemur. If the *Beagle* had
visited Madagascar instead of Galapagos,
would we now speak of 'Darwin's lemurs'?
(f) Would Martian trees be any more
strange than this Madagascan baobab?
(g) Possibly my favourite species in all the
world: the dancing sifaka.

Don't laugh at the foot-raising, sky-pointing displays of the blue-footed booby. They impress other boobies, and that is all that matters.

note on p. 112) and then took the liberty of projecting it on into the future. More recently, just as this book was going to press, I showed this section to Professor Hodgkin, and he told me the most recent data of which he was aware: the duckbilled platypus genome, which was sequenced in 2008 (the platypus was a good choice, because of its strategic position in the tree of life: the ancestor that it shares with us lived 180 million years ago, which is nearly three times as long ago as the extinction of the dinosaurs). I've drawn the platypus's point as a star on the graph, and you can see that it fits pretty well near the projected line that was calculated from the earlier data.

The slope of the line for what I am now calling (without permission) Hodgkin's Law is only slightly shallower than that for Moore's Law. The doubling time is a bit more than two years, where the Moore's Law doubling time is a bit less than two years. DNA technology is intensely dependent on computers, so it's a good guess that Hodgkin's Law is at least partly dependent on Moore's Law. The arrows on the right indicate the genome sizes of various creatures. If you follow the arrow towards the left until it hits the sloping line of Hodgkin's Law, you can read off an estimate of when it will be possible to sequence a genome the same size as the creature concerned for only £1,000 (of today's money). For a genome the size of yeast's, we

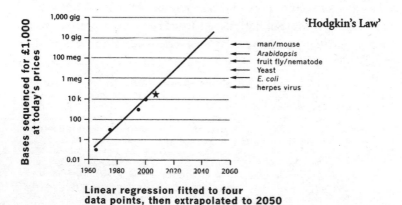

need wait only till about 2020. For a new mammal genome (as far as this kind of back-of-envelope calculation is concerned, all mammals are equally expensive), the estimated date is just this side of 2040. It's an exhilarating prospect: a massive database of DNA sequences, cheaply and easily obtained from all corners of the animal and plant kingdoms. Detailed DNA comparisons will fill in all the gaps in our knowledge about the actual evolutionary relatedness of every species to every other: we shall know, with complete certainty, the entire family tree of all living creatures.* Goodness knows how we'll plot it; it won't fit on any practical-sized sheet of paper.

The largest-scale attempt in that direction so far has been made by a group associated with David Hillis, brother of Danny Hillis who pioneered one of the first supercomputers. The Hillis plot makes the tree diagram more compact by wrapping it around in a circle. You can't see the gap, where the two ends almost meet, but it lies between the 'bacteria' and the 'archaea'. To see how the circular plot works, look at the greatly stripped-down version tattooed on the back of Dr Clare D'Alberto of the University of Melbourne, whose enthusiasm for zoology is more than skin deep. Clare has graciously allowed me to reproduce the photograph in this book (see colour page 25). Her tattoo includes a small sample of eighty-six species (the number of terminal twigs). You can see the gap in the circular plot, and imagine the circle opened out. The smaller number of illustrations around the edge are strategically chosen from bacteria, protozoa, plants, fungi, and four animal phyla. The vertebrates are represented by

---

* Perhaps 'all living creatures' calls for a note of caution. In an earlier section of this chapter we saw that while the principle of 'No borrowing' is almost completely appropriate for animals and plants, bacteria are different. Among bacteria (and archaea, which are superficially like bacteria but rather distantly related) there is plenty of sharing of genes. Whereas animals use sexual coupling to exchange DNA within species, bacteria use their own form of 'Copy and Paste' to pass DNA around, even between distantly related species. Although I was right to extol the 'one true tree of life' for animals and plants, the whole business gets messier when we turn to micro-organisms. As my colleague the philosopher Dan Dennett has put it, where the tree of life for animals is a majestically spreading oak, that for bacteria is more like a banyan. Where bacteria are concerned, there is something to be said for compiling a 'one true tree' for each gene separately, regardless of which particular kinds of bacteria it happens to be travelling around in. What a fascinating prospect. How Darwin would have loved it.

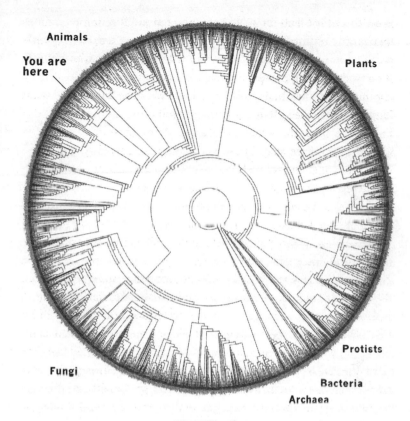

Animals

You are here

Plants

Fungi

Protists

Bacteria

Archaea

The Hillis plot

the weedy sea dragon on the right, a surprising fish, protected by its resemblance to seaweed. The Hillis circular plot is the same, except that it has three thousand species. Their names appear around the outside edge of the circle above, far too small to read – though *Homo sapiens* is helpfully marked 'You are here'. You can get an idea of how sparse a sampling of the tree even this huge plot is when I tell you that the closest relatives of humans that it can fit in the circle are rats and mice. The mammals had to be stripped down drastically, in order to fit in all the other branches of the tree to the same depth. Just imagine trying to plot a similar tree with ten million species instead of the three thousand included here. And ten million is not the most extravagant estimate of the number of surviving species. It's well worth downloading the Hillis tree from his website (see endnotes), and then printing it as a wall hanging, on a piece of paper which, they recommend, should be at least 54 inches wide (even bigger would be an advantage).

THE MOLECULAR CLOCK

Now, while we are talking molecules, we have some unfinished business left over from the chapter on evolutionary clocks. There, we looked at tree rings, and at various kinds of radioactive clocks, but we deferred consideration of the so-called molecular clock until we had learned about some other aspects of molecular genetics. The time has now come. Think of this section as an appendix to the chapter on clocks.

The molecular clock assumes that evolution is true, and that it proceeds at a sufficiently constant rate through geological time to be used as a clock in its own right, provided that it can be calibrated using fossils, which are in turn calibrated with radioactive clocks. Just as a candle clock assumes that candles burn at a fixed and known rate, and a water clock assumes that water drains from a bucket at a rate that can be calibrated, and a grandfather clock assumes that a pendulum swings at a fixed rate, so the molecular clock assumes that there are certain aspects of evolution *itself* that proceed at a fixed rate. That fixed rate can be calibrated against those parts of the evolutionary record that are well documented with (radioactively datable) fossils. Once calibrated, the molecular clock can then be used for other parts of evolution that are not well documented by fossils. For example, it can be used for animals that don't have hard skeletons and seldom fossilize.

Nice idea, but what gives us the right to hope that we can find evolutionary processes that go at a fixed rate? Indeed, much evidence suggests that evolutionary rates are highly variable. Long before the modern era of molecular biology, J. B. S. Haldane proposed the *darwin* as a measure of evolutionary rates. Suppose that, over evolutionary time, some measured characteristic of an animal is changing in a consistent direction. For example, suppose the mean leg length is increasing. If, over a period of a million years, leg length increases by a factor of e (2.718 . . ., a number chosen for reasons of mathematical convenience, which we needn't go into),* the rate of evolutionary change is said to be one darwin. Haldane himself assessed the rate of evolution of the horse as approximately 40 millidarwins, while it has been suggested that the evolution of domestic animals under artificial selection should be measured in kilodarwins. The rate of evolution of guppies transplanted to a predator-free stream, as described in Chapter 5, has been estimated as 45 kilodarwins. The evolution of 'living fossils' such as *Lingula* (page 140) is probably to be measured in microdarwins. You get the point: rates of evolution of things that you can see and measure, like legs and beaks, are hugely variable.

If rates of evolution are so variable, how can we hope to use them as a clock? This is where molecular genetics comes to the rescue. At first sight, it will not be clear how this can be so. When measurable characteristics like leg length evolve, what we are seeing is the outward and visible manifestation of an underlying genetic change. How, then, can it be the case that rates of change at the molecular level provide a good clock while rates of leg or wing evolution don't? If legs and beaks undergo change at rates ranging from microdarwins to kilodarwins, why should molecules be any more reliable as clocks? The answer is that the genetic changes that manifest themselves in outward and visible evolution – of things like legs and arms – are a very small tip of the iceberg, and they are the tip that is heavily influenced by varying natural selection. The majority of genetic change at the molecular level is *neutral*, and can therefore

---

* When I first read *Calculus Made Easy* by Silvanus P. Thompson, on the recommendation of my engineer grandfather, it gave me goosebumps when Thompson introduced e in italics as '*a number never to be forgotten*'. One consequence of using e rather than, say, 2, as the factor of choice, is that you can calculate the darwins directly by subtracting natural logarithms from one another. Other scientists have proposed, as a unit of evolutionary rate, the haldane.

be expected to proceed at a rate that is independent of usefulness and might even be approximately constant within any one gene. A neutral genetic change has no effect on the survival of the animal, and this is a helpful credential for a clock. This is because genes that affect survival, positively or negatively, would be expected to evolve at a changed rate, reflecting this.

When the neutral theory of molecular evolution was first proposed by, among others, the great Japanese geneticist Motoo Kimura, it was controversial. Some version of it is now widely accepted and, without going into the detailed evidence here, I am going to accept it in this book. Since I have a reputation as an arch-'adaptationist' (allegedly obsessed with natural selection as the major or even only driving force of evolution) you can have some confidence that if even I support the neutral theory it is unlikely that many other biologists will oppose it!*

A neutral mutation is one that, although easily measurable by molecular genetic techniques, is not subject to natural selection, either positive or negative. 'Pseudogenes' are neutral for one kind of reason. They are genes that once did something useful but have now been sidelined and are never transcribed or translated. They might as well not exist, as far as the animal's welfare is concerned. But as far as the scientist is concerned they very much exist, and they are exactly what we need for an evolutionary clock. Pseudogenes are only one class of those genes that are never translated in embryology. There are other classes which are preferred by scientists for molecular clocks, but I won't go into detail. What pseudogenes are useful for is embarrassing creationists. It stretches even their creative ingenuity to make up a convincing reason why an intelligent designer should have created a pseudogene – a gene that does absolutely nothing and gives every appearance of being a superannuated version of a gene that used to do something – unless he was deliberately setting out to fool us.

_____

* I have even been called an 'ultra-Darwinist', a gibe that I find less insulting than its coiners perhaps intended.

Leaving pseudogenes aside, it is a remarkable fact that the greater part (95 per cent in the case of humans) of the genome might as well not be there, for all the difference it makes. The neutral theory applies even to many of the genes in the remaining 5 per cent – the genes that are read and used. It applies even to genes that are totally vital for survival. I must be clear here. We are not saying that a gene to which the neutral theory applies has no effect on the body. What we are saying is that a mutant version of the gene has exactly the same effect as the unmutated version. However important or unimportant the gene itself may be, the mutated version has the same effect as the unmutated version. Unlike pseudogenes, where the gene itself can properly be described as neutral, we are now talking about cases where it is only *mutations* (i.e. changes in genes) that can strictly be described as neutral, not genes themselves.

Mutations can be neutral for various reasons. The DNA code is a 'degenerate code'. This is a technical term meaning that some code 'words' are exact synonyms of each other.* When a gene mutates into one of its synonyms, you might as well not bother to call it a mutation at all. Indeed, it isn't a mutation, as far as consequences on the body are concerned. And for the same reason it isn't a mutation at all as far as natural selection is concerned. But it is a mutation as far as molecular geneticists are concerned, for they can see it using their methods. It is as though I were to change the font in which I write a word, say kangaroo to kangaroo. You can still read the word, and it still means the same Australian hopping animal. The change of typeface from Minion to Helvetica is detectable but irrelevant to the meaning.

---

* 'Degenerate' is not the same (though the two terms are often confused) as 'redundant', another technical term of Information Theory. A redundant code is one in which the same message is conveyed more than once (e.g. 'She is a female woman' conveys the message of her sex three times). Redundancy is used by engineers to guard against transmission errors. A degenerate code is one in which more than one 'word' is used to mean the same thing. In the genetic code, for example, CUC and CUG both spell 'Leucine': a mutation from CUC to CUG therefore makes no difference. That's 'degenerate'.

Not all neutral mutations are quite so neutral as that. Sometimes the new gene translates into a different protein, but the 'active site' (remember the carefully shaped 'dents' that we met in Chapter 8) of the new protein remains the same as the old one. Consequently, there is literally no effect on the embryonic development of the body. The unmutated and the mutated form of the gene are still synonyms as far as their effects on bodies are concerned. It is also possible (although 'ultra-Darwinists' like me incline against the idea) that some mutations really do change the body, but in such a way as to have no effect on survival, one way or the other.

So, to sum up on the neutral theory, to say that a gene, or a mutation, is 'neutral' doesn't necessarily mean that the gene itself is useless. It could be vitally important to the animal's survival. What it means is that the mutated form of a gene – which might or might not be important for survival – is no *different* from the unmutated form with respect to its effects (which might be very important) on survival. As it happens, it is probably true to say that most mutations are neutral. They are undetectable by natural selection, but detectable by molecular geneticists; and that is an ideal combination for an evolutionary clock.

None of this is to downgrade the all-important tip of the iceberg – the minority of mutations that are not neutral. It is they that are selected, positively or negatively, in the evolution of improvements. They are the ones whose effects we actually see – and natural selection 'sees' too. They are the ones whose selection gives living things their breathtaking illusion of design. But it is the rest of the iceberg – the neutral mutations, which are in the majority – that concern us when we are talking about the molecular clock.

As geological time goes by, the genome is subjected to a rain of attrition in the form of mutations. In that small portion of the genome where the mutations really matter for survival, natural selection soon gets rid of the bad ones and favours the good ones. The neutral mutations, on the other hand, simply pile up, unpunished and unnoticed – except by molecular geneticists. And now we need a new technical term: *fixation*. A new mutation, if it is genuinely new, will have a low frequency in the gene pool. If you revisit the gene pool a million years later, it is possible that the mutation will have increased in frequency to 100 per cent or something close to it. If that happens, the mutation is said to have 'gone to fixation'. We shall no longer think of it as a mutation. It has become the norm. The obvious way for a mutation to go to fixation is for natural selection to favour it. But there is another way. It can go to fixation by chance. Just as a once proud surname can die out for lack of male heirs, so the alternatives to the mutation we are talking about can just happen to disappear from the gene pool. The mutation itself can become frequent in the gene pool, by the same luck as has led 'Smith' to emerge as the commonest surname in England. Of course it is much more interesting if the gene goes to fixation for a good reason – that's natural selection – but it can also happen by chance, given a large enough number of generations. And geological time is vast enough for neutral mutations to go to fixation at a predictable rate. The rate at which they do so varies, but it is characteristic of particular genes, and, given that most mutations are neutral, this is precisely what makes the molecular clock possible.

It's fixation that matters for the molecular clock, because 'fixed' genes are the ones that we look at when we compare two modern animals to try to estimate how long ago their ancestors split apart. Fixed genes are the genes that characterize a species. They are the ones that are all but universal in the gene pool. And we can compare the genes that have become fixed in one species with the genes that have become fixed in another, in order to estimate how recently the two species split apart. There are complications, which I won't go into because Yan Wong and I discussed them fully in 'The Epilogue to the Velvet Worm's Tale'. With reservations, and with various important correction factors, the molecular clock works.

Just as radioactive clocks tick at hugely variable speeds, with half-lives ranging from fractions of a second through to tens of billions of years, so different genes provide a marvellous spread of molecular clocks, suitable for timing evolutionary change on scales ranging from a million to a billion years, and all stages in between. Just as each radioactive isotope has its characteristic half-life, so each gene has a characteristic turnover rate – the rate at which new mutations typically go to fixation by random chance. Histone genes characteristically turn over at a rate of one mutation per billion years. Fibrinopeptide genes are a thousand times faster, with a turnover of one new mutation fixed per million years. Cytochrome-C and the suite of haemoglobin genes have intermediate turnovers, with times to fixation measured in millions to tens of millions of years.

Neither radioactive clocks nor molecular clocks tick in a regular fashion like a pendulum clock or a watch. If you could hear them ticking, they'd sound like a Geiger counter, the radioactive clocks literally so since a Geiger counter is precisely what you would use to listen to them. A Geiger counter doesn't tick regularly, like a watch; it ticks at random, the ticks coming in strange, stuttering bursts. That's how mutations, and fixations, would sound, if we could hear them on the immensely long timescale of geology. But, whether stuttering like a Geiger counter or ticking metronomically like a watch, the important thing about a timekeeper is that it should tick at a known *average* rate. That's what radioactive clocks do, and that's what molecular clocks do.

I introduced the molecular clock by saying that it assumes the fact of evolution and therefore can't be used in evidence of it. But now, having understood how the clock works, we can see that I was too pessimistic. The very existence of pseudogenes – useless, untranscribed genes that bear a marked resemblance to useful genes – is a perfect example of the way animals and plants have their history written all over them. But that is a topic that must wait for the next chapter.

CHAPTER 11

HISTORY WRITTEN ALL OVER US

I BEGAN this book by imagining a teacher of Latin forced to waste time and energy defending the proposition that the Romans and their language ever existed. Let's return to that thought, and ask what actually is the evidence for the Roman Empire and the Latin language. I live in Britain where, as in the rest of Europe, Rome wrote her signature all over the map, carved her ways into the landscape, wove her language through ours and her history through our literature. Walk the length of Hadrian's Wall, whose preferred local name is still 'The Roman Wall'. Walk, as I walked Sunday after Sunday in crocodile formation from my boarding school in (relatively) new Salisbury, to the Roman flint fort of Old Sarum, and commune with the imagined ghosts of dead legions. Unfold an Ordnance Survey map of England. Wherever you see a long, dead straight country road, especially when there are green field gaps between stretches of road or cart track that you can exactly line up with a ruler, you'll almost always find a Roman label beside it. Vestiges of the Roman Empire are all around us.

Living bodies, too, have their history written all over them. They bristle with the biological equivalent of Roman roads, walls, monuments, potsherds, even ancient inscriptions carved into the living DNA, ready to be deciphered by scholars.

Bristle? Yes, literally. When you are cold, or badly frightened, or haunted by the peerless craftsmanship of a Shakespeare sonnet, you get goosebumps. Why? Because your ancestors were normal mammals with hairs all over, and these were raised or lowered at the behest of sensitive bodily thermostats. Too cold, and the hairs were erected to plump up the layer of insulating trapped air. Too warm, and the coat was flattened to allow body heat to escape more easily. In later evolution, the hair-erection system was hijacked for social communication purposes, and became involved in *The Expression of the Emotions*, as Darwin was one of the first to appreciate in his book of that name. I can't resist sharing with you some lines – vintage Darwin – from that book:

> Mr Sutton, the intelligent keeper in the Zoological Gardens, carefully observed for me the Chimpanzee and Orang; and he states that when they are suddenly frightened, as by a thunderstorm, or when they are made angry, as by being teased, their hair becomes erect. I saw a chimpanzee who was alarmed at the sight of a black coalheaver, and the hair rose all over his body . . . I took a stuffed snake into the monkey-house, and the hair on several of the species instantly became erect . . . When I showed a stuffed snake to a Peccary, the hair rose in a wonderful manner along its back; and so it does with a wild boar when enraged.

The hackles are raised in anger. In fear also, hairs stand on end to increase the body's apparent size and scare off dangerous rivals or predators. Even we naked apes still have the machinery to raise non-existent (or barely-existent) hairs, and we call it goosebumps. The hair-erection machinery is a *vestige*, a non-functional relic of something that did a useful job in our long-dead ancestors. Vestigial hairs are among the many instances of history written all over us. They constitute persuasive evidence that evolution has occurred, and again it comes not from fossils but from modern animals.

As we saw in the previous chapter, where I compared it to a comparably sized fish such as a dorado, you don't have to dig very deep inside a dolphin to uncover its history of life on dry land. Despite its streamlined, fish-like exterior, and despite the fact that it now makes its entire living in the sea and would soon die if beached, a dolphin, but not a dorado, has 'land mammal' woven through its very warp and woof. It has lungs not gills, and will drown like any land animal if prevented from coming up for air, although it can hold its breath for much longer than a land mammal. In all sorts of ways, its air-breathing apparatus is changed to fit its watery world. Instead of breathing through two little nostrils at the end of its nose like any normal land mammal, it has a single nostril in the top of its head, which enables it to breathe while only just breaking the surface. This 'blowhole' has a tight-sealing valve to keep water out, and a wide bore to minimize the time needed for breathing. In an 1845 communication to the Royal Society, which Darwin, as a Fellow, would quite likely have read, Francis Sibson Esq.* wrote: 'The muscles that open and close the blow-hole, and that act upon the various sacs, form one of the most complicated yet most exquisitely adjusted pieces of machinery that either nature or art presents.' The dolphin's blowhole goes to great lengths to correct a problem that would never have arisen at all if only it breathed with gills, like a fish. And many of the details of the blowhole can be seen as corrections to secondary problems that arose when the air intake migrated from the nostrils to the top of the head. A real designer would have planned it in the top of the head in the first place – that's if he hadn't decided to abolish lungs and go for gills anyway. Throughout this chapter, we shall continually find examples of evolution correcting an initial 'mistake' or historical relic by post hoc compensation or tweaking, rather than by going back to the drawing board as a real designer would. In any case, the elaborate and complex gateway to the blowhole is eloquent testimony to the dolphin's remote ancestry on dry land.

---

* In Britain, 'Esq.' meant (still means, although the usage is rapidly becoming extinct) 'gentleman', not 'lawyer' as (I recently discovered) it means in America. I have even encountered female American lawyers referring to themselves as 'Esq.'. This seems to English people about as odd as Americans must find the designation of the first female Law Lord (British equivalent of Supreme Court Justice) as 'Lord Justice Elizabeth Butler-Sloss'. The English use of 'Esq.' seems even odder to many in the rest of the world. I am told that the 'E' pigeonhole in hotels the world over is replete with undelivered letters looking for a Mr 'Esq'.

In countless other ways, dolphins and whales could be said to have their ancient history written all over and through them, like vestiges of Roman roads drawn out in dead straight cart tracks and bridleways across the map of England. Whales have no hind legs, but there are tiny bones, buried deep inside them, which are the remnants of the pelvic girdle and hind legs of their long-gone walking ancestors. The same is true of the sirenians or sea cows (I've already mentioned them several times: the manatees, dugongs and the 8-yard-long Steller's sea cow, hunted to extinction by humans).* Sirenians are very different from whales and dolphins, but they are the only other group of wholly marine mammals that never come ashore. Where dolphins are fast, actively intelligent carnivores, manatees and dugongs are slow, dreamy herbivores. At the manatee aquarium that I visited in western Florida, for once I didn't rage against the loudspeakers playing music. It was sleepy lagoon music and it seemed so languidly appropriate that all was forgiven. Manatees and dugongs float effortlessly in hydrostatic equilibrium, not by means of a swim bladder as fish do (see below), but through being equipped with heavy bones as a counterweight to the natural buoyancy of their blubber. Their specific gravity is therefore very close to that of water, and they can make fine adjustments to it by pulling in or expanding the rib cage. The precision of their buoyancy control is enhanced by the possession of a separate cavity for each lung: they have two independent diaphragms.

Dolphins and whales, dugongs and manatees give birth to live babies, like all mammals. That habit is not actually peculiar to mammals. Many fish are livebearers, but they do it in a very different way (actually a fascinating variety of very different ways, doubtless independently evolved). The dolphin's placenta is unmistakably

---

* The connection with the legendary Sirens may be the habit, which they share with their land relatives the elephants, of suckling their young from pectoral breasts. Perhaps sexually frustrated sailors who had been at sea for a very long time witnessed this from a distance and mistook them for women. Sirenians have sometimes been blamed for the mermaid legend.

mammalian, and so is its habit of suckling the young with milk. Its brain is also beyond question the brain of a mammal, and a very advanced mammal at that. The cerebral cortex of a mammal is a sheet of grey matter, wrapped around the outside of the brain. Getting brainier partly consists in increasing the area of the sheet. This could be done by increasing the total size of the brain, and of the skull that houses it. But there are downsides to having a big skull. It makes it harder to be born, for one thing. As a result, brainy mammals contrive to increase the area of the sheet while staying within limits set by the skull, and they do it by throwing the whole sheet into deep folds and fissures. This is why the human brain looks like a wrinkled walnut; and the brains of dolphins and whales are the only ones to rival those of us apes for wrinkliness. Fish brains don't have

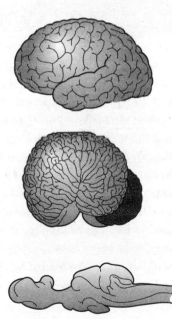

wrinkles at all. Indeed, they don't have a cerebral cortex, and the whole brain is tiny compared to a dolphin's or human's. The dolphin's mammalian history is deeply etched into the wrinkled surface of its brain. It's a part of its mammalness, along with the placenta, milk, a four-chambered heart, a lower jaw having only a single bone, warm-bloodedness, and many other specifically mammalian features.

Warm-blooded is what we call mammals and birds, but really what they have is the ability to keep their temperature constant, regardless of the outside temperature. This is a good idea, because the chemical reactions in a cell can then all be optimized to

Brains of human (top), dolphin (middle), brown trout (bottom)
(not to scale)

work best at a particular temperature. 'Cold-blooded' animals are not necessarily cold. A lizard has warmer blood than a mammal if both happen to be out in the midday sun in the Sahara desert. A lizard has colder blood than a mammal if they are out in the snow. The mammal has the same temperature all the time, and it has to work hard to keep it constant, using internal mechanisms. Lizards use external means to regulate their temperature, moving into the sun when they need to warm themselves up, and into the shade when they need to cool down. Mammals regulate their body temperature more accurately, and dolphins are no exception. Once again, their mammal history is written all over them, even though they have reverted to life in the sea, where most creatures don't maintain a constant temperature.

ONCE PROUD WINGS

The bodies of whales and sirenians abound in historical relics that we notice because they live in a very different environment from their land-dwelling ancestors. A similar principle applies to birds that have lost the habit and equipment of flight. Not all birds fly, but all birds carry at least relics of the apparatus of flight. Ostriches and emus are fast runners that never fly, but they have stubs of wings as a legacy from remote flying ancestors. Ostrich wing stubs, moreover, have not completely lost their usefulness. Although much too small to fly with, they seem to have some sort of balancing and steering role in running, and they enter into social and sexual displays. Kiwi wings are too small to be seen outside the bird's fine coat of feathers, but vestiges of wing bones are there. Moas lost their wings entirely. Their home country of New Zealand, by the way, has more than its fair share of flightless birds, probably because the absence of mammals left wide open niches to be filled by any creature that could get there by flying. But those flying pioneers, having arrived on wings, later lost them as they filled the vacant mammal roles on the ground. This probably doesn't apply to the moas themselves, whose ancestors, as it happened, were already flightless before the great southern continent of Gondwana broke up into fragments, New Zealand among them, each bearing its own cargo of Gondwanan animals. It surely does apply to kakapos, New Zealand's flightless parrots, whose flying ancestors apparently lived so recently that kakapos still try to fly although they lack the equipment to succeed. In the words of the immortal Douglas Adams, in *Last Chance to See*,

> It is an extremely fat bird. A good-sized adult will weigh about six or seven pounds, and its wings are just about good for wiggling about a bit if it thinks it's about to trip over something – but flying is completely out of the question. Sadly, however, it seems that not only has the kakapo forgotten how to fly, but it has also forgotten that it has forgotten how to fly. Apparently a seriously worried kakapo will sometimes run up a tree and jump out of it, whereupon it flies like a brick and lands in a graceless heap on the ground.

While ostriches, emus and rheas are great runners, penguins and Galapagos flightless cormorants are great swimmers. I was privileged to swim with a flightless cormorant in a large rock pool on the island of Isabela, and I was enchanted to witness the speed and agility with which it sought out one undersea crevice after another, staying under for a breathtakingly long time (I had the advantage of a snorkel). Unlike penguins, who use their short wings to 'fly underwater', Galapagos cormorants propel themselves with their powerful legs and huge webbed feet, using their wings only as stabilizers. But all flightless birds, including ostriches and their kind, which lost their wings a very long time ago, are clearly descended from ancestors that used them to fly. No reasonable observer could seriously doubt the truth of that, which means that anybody who thinks about it should find it very hard – why not impossible? – to doubt the fact of evolution.

Numerous different groups of insects, too, have lost their wings, or greatly reduced them. Unlike primitively wingless insects such as silverfish, fleas and lice have lost the wings their ancestors once had. Female gypsy moths have underdeveloped wing muscles and don't fly. They don't need to, for the males fly to them, attracted by a chemical lure which they can detect at astounding dilutions. If the females were to move as well as the males, the system probably wouldn't work, for by the time the male had flown up the slowly drifting chemical gradient, its source would have moved on!

Unlike most insects, which have four wings, the flies, as their Latin name Diptera suggests, have only two. The second pair of wings has become reduced to a pair of 'halteres'. These swing about like very high-speed Indian clubs, which they resemble, functioning as tiny gyroscopes. How do we know that halteres are descended from ancestral wings? Several reasons. They occupy exactly the same place in the third segment of the thorax as the flying wing occupies in the second thoracic segment (and the third too in other insects). They move in the same figure-of-eight pattern as the wings of flies. They have the same embryology as wings and, although they are tiny, if you look at them carefully, especially during development, you can see that they are stunted wings, clearly modified – unless you are an evolution-denier–from ancestral wings. Testifying to the same story, there are mutant fruit flies, so-called homeotic mutants, whose embryology is abnormal and who grow not halteres but a second pair of wings, like a bee or any other kind of insect.

What would the intermediate stages between wings and halteres have looked like, and why would natural selection have favoured the intermediates? What is the use of half a haltere? J. W. S. Pringle, my old Oxford professor whose forbidding

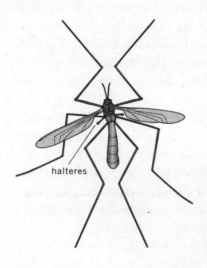

Halteres on a cranefly

mien and stiff bearing earned him the nickname 'Laughing John', was mainly responsible for working out how halteres work. He pointed out that all insect wings have tiny sense organs in the base, which detect twisting and other forces. The sense organs at the base of halteres are very similar – another piece of evidence that halteres are modified wings. Long before halteres evolved, the information streaming into the nervous system from the sense organs at their base would enable fast buzzing wings, while flying, to act as rudimentary gyroscopes. To the extent that any flying machine is naturally unstable, it needs to compensate with sophisticated instrumentation, for example gyroscopes.

The whole question of the evolution of stable and unstable fliers is very interesting. Look at these two pterosaurs, extinct flying reptiles, contemporaries of the dinosaurs. Any aero-engineer could tell you that *Rhamphorhynchus*, the early pterosaur at the top of the picture, must have been a stable flier, because of its long tail with the ping-pong bat

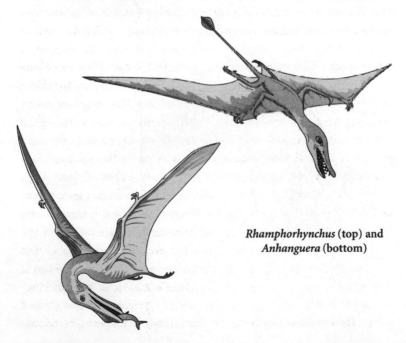

Rhamphorhynchus (top) and
Anhanguera (bottom)

on the end. *Rhamphorhynchus* would not have needed sophisticated gyro-control, such as flies have with their halteres, because its tail made it inherently stable. On the other hand, as the same engineer could tell you, it would not have been very manœuvrable. In any flying machine, there is a trade-off between stability and manœuvrability. The great John Maynard Smith, who worked as an aircraft designer before returning to university to read zoology (on the grounds that aeroplanes were noisy and old-fashioned), pointed out that flying animals can move in evolutionary time, back and forth along the spectrum of this trade-off, sometimes losing inherent stability in the interests of increased manœuvrability, but paying for it in the form of increased instrumentation and computation capability – brain power. At the bottom of the picture on the previous page is *Anhanguera*, a late pterodactyl from the Cretaceous era, some 60 million years after the Jurassic *Rhamphorhynchus*. *Anhanguera* had almost no tail at all, like a modern bat. Like a bat, it would surely have been an unstable aircraft, reliant on instrumentation and computation to exercise subtle, moment-to-moment control over its flight surfaces.

Anhanguera didn't have halteres, of course. It would have used other sense organs to provide the equivalent information, probably the semicircular canals of the inner ear. These were indeed very large in those pterosaurs that have been looked at – although, a touch disappointingly for the Maynard Smith hypothesis, they were large in *Rhamphorhynchus* as well as *Anhanguera*. But, to return to the flies, Pringle suggests that the four-winged ancestors of flies probably had long abdomens, which would have made them stable. All four wings would have acted as rudimentary gyroscopes. Then, he suggests, the ancestors of flies started to move along the stability continuum, becoming more manœuvrable and less stable as the abdomen got shorter. The hind wings started to shift more towards the gyroscopic function (which they had always performed, in a small way, as wings), becoming smaller, and heavier for their size, while the forewings enlarged to take over more of the flying. There would have been a gradual continuum of change, as the forewings assumed ever more of the burden of aviation, while the hind wings shrank to take over the avionics.

Worker ants have lost their wings, but not the capacity to grow wings. Their winged history still lurks within them. We know this because queen ants (and males) have wings, and workers are females who could have been queens but who, for environmental, not genetic, reasons failed to become queens.* Presumably worker ants lost their wings in evolution because they are a nuisance and get in the way underground. Poignant testimony to this is provided by queen ants, who use their wings once only, to fly out of the natal nest, find a mate, and then settle down to dig a hole for a new nest. As they begin their new life underground the first thing they do is lose their wings, in some cases by literally biting them off: painful (perhaps; who knows?) evidence that wings are a nuisance underground. No wonder worker ants never grow wings in the first place.

Probably for similar reasons, ants' nests, and termites' nests, are home to a horde of wingless hangers-on of many different types, feeding on the rich pickings swept in by the ever-rustling streams of returning foragers. And wings are just as much of a hindrance to them as they are to the ants themselves. Who would ever believe that the monstrosity on the right is a fly? Yet we know from a careful and detailed study of its anatomy that not only is it a fly, this parasite of termite nests belongs to a particular family of flies, the Phoridae. On the next page is a

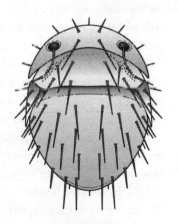

**Parasitic fly from the
Phoridae family**

---

* Larvae destined to become queens are fed special elixirs secreted by glands in the nurse workers' heads. It is very important that the difference between queens and workers is environmentally, not genetically, determined. I have explained why at length in *The Selfish Gene*.

Another fly from the
Phoridae family

more normal member of the same family, which presumably somewhat resembles the winged ancestors of the weirdly wingless creature above, although it too is a parasite of social insects – bees in this case. You can see the similarity to the sickle-shaped head of the weird monster on the previous page. And the monster's stunted wings are just visible as the tiny triangles on either side.

There is an additional reason for winglessness in this riff-raff of lurkers and squatters in ants' and termites' nests. Many of them (not the Phorid flies) have over evolutionary time assumed a protective resemblance to ants, either (or both) to fool the ants or to fool would-be predators who might otherwise pick them out from among the less palatable and better-protected ants. Who, on taking only a casual glance, would notice that the insect below, which lives in ants' nests, is not an ant at all but a beetle? Once again, how do we know? From deep and detailed resemblances to beetles, which hugely outnumber the superficial features in which the insect resembles an ant: exactly the same way as we know that a dolphin is a mammal and not a fish. This creature has its beetle ancestry written all through it, except (again as with dolphins) in those features that define its superficial appearance, such as its winglessness and its ant-like profile.

Beetle disguised as an ant

LOST EYES

Just as ants and their subterranean fellow-travellers lose their wings underground, so numerous different kinds of animals that live in the depths of dark caves where there is no light have reduced or lost their eyes, and are, as Darwin himself noted, more or less completely blind. The word 'troglobite'* has been coined for an animal that lives only in the darkest part of caves and is so specialized that it can live nowhere else. Troglobites include salamanders, fish, shrimps, crayfish, millipedes, spiders, crickets and many other animals. They are very often white, having lost all pigment, and blind. They usually, however, retain vestiges of eyes, and that is the point of mentioning them here. Vestigial eyes are evidence of evolution. Given that a cave salamander lives in perpetual darkness so has no use for eyes, why would a divine creator nevertheless furnish it with dummy eyes, clearly related to eyes but non-functional?

Evolutionists, on their side, need to come up with an explanation for the loss of eyes where they are no longer needed. Why not, it might be said, simply hang on to your eyes, even if you never use them? Might they not come in handy at some point in the future? Why 'bother' to get rid of them? Notice, by the way, how hard it is to resist the language of intention, purpose and personification. Strictly speaking, I should not have used the word 'bother', should I? I should have said something like, 'How does losing its eyes benefit an individual cave salamander so that it is more likely to survive and reproduce than a rival salamander that keeps a perfect pair of eyes, even though it never uses them?'

Well, eyes are almost certainly not cost-free. Setting aside the arguably modest economic costs of making an eye, a moist eye socket, which has to be open to the world to accommodate the swivelling eyeball with its transparent surface, might be vulnerable to infection.

---

* Yes, troglobite, not troglodyte, which means something less extreme.

So a cave salamander that sealed up its eyes behind tough body skin might survive better than a rival individual that kept its eyes.

But there is another way to answer the question and, instructively, it doesn't invoke the language of advantage at all, let alone purpose or personification. When we are talking about natural selection, we think in terms of rare beneficial mutations turning up and being positively favoured by selection. But most mutations are disadvantageous, if only because they are random and there are many more ways of getting worse than there are ways of getting better.* Natural selection promptly penalizes the bad mutations. Individuals possessing them are more likely to die and less likely to reproduce, and this automatically removes the mutations from the gene pool. Every animal and plant genome is subject to a constant bombardment of deleterious mutations: a hailstorm of attrition. It is a bit like the moon's surface, which becomes increasingly pitted with craters due to the steady bombardment of meteorites. With rare exceptions, every time a gene concerned with an eye, for example, is hit by a marauding mutation, the eye becomes a little less functional, a little less capable of seeing, a little less worthy of the name of eye. In an animal that lives in the light and uses the sense of sight, such deleterious mutations (the majority) are swiftly removed from the gene pool by natural selection.

But in total darkness the deleterious mutations that bombard the genes for making eyes are not penalized. Vision is impossible anyway. The eye of a cave salamander is like the moon, pitted with mutational craters that are never removed. The eye of a daylight-dwelling salamander is like the Earth, hit by mutations at the same rate as

---

* This is especially true of mutations of large effect. Think of a delicate machine, like a radio or a computer. A large mutation is equivalent to kicking it with a hobnailed boot, or cutting a wire at random and reconnecting it in a different place. It just *might* improve its performance, but it is not very likely. A *small* mutation, on the other hand, is equivalent to making a tiny adjustment to, say, one resistor, or to the tuning knob of a radio. The smaller the mutation, the more closely the probability of improvement approaches 50 per cent.

cave-dwellers' eyes, but with each deleterious mutation (crater) being cleaned off by natural selection (erosion). Of course, the story of the cave-dweller's eye isn't only a negative one: positive selection comes in too, to favour the growth of protective skin over the vulnerable sockets of the optically deteriorating eyes.

Among the most interesting of historical relics are those features that are used for something (and so are not vestiges in the sense of having outlived their purpose), but seem badly designed for that purpose. The vertebrate eye at its best – say, the eye of a hawk or a human – is a superb precision instrument, capable of feats of fine resolution to rival the best that Zeiss or Nikon can deliver. If it were not so, Zeiss and Nikon would be wasting their time producing high-resolution images for our eyes to look at. On the other hand, Hermann von Helmholtz, the great nineteenth-century German scientist (you could call him a physicist, but his contributions to biology and psychology were greater), said, of the eye: 'If an optician wanted to sell me an instrument which had all these defects, I should think myself quite justified in blaming his carelessness in the strongest terms, and giving him back his instrument.' One reason why the eye seems better than Helmholtz, the physicist, judged it to be is that the brain does an amazing job of cleaning the images up afterwards, like a sort of ultra-sophisticated, automatic Photoshop. As far as optics are concerned, the human eye achieves its Zeiss/Nikon quality only in the fovea, the central part of the retina that we use for reading. When we scan a scene, we move the fovea over different parts, seeing each one in the utmost detail and precision, and the brain's 'Photoshop' fools us into thinking we are seeing the whole scene with the same precision. A top-quality Zeiss or Nikon really *does* show the whole scene with almost equal clarity.

So, what the eye lacks in optics the brain makes up for with its sophisticated image-simulating software. But I haven't yet mentioned the most glaring example of imperfection in the optics. The retina is back to front.

Imagine a latter-day Helmholtz presented by an engineer with a digital camera, with its screen of tiny photocells, set up to capture images projected directly on to the surface of the screen. That makes good sense, and obviously each photocell has a wire connecting it to a computing device of some kind where images are collated. Makes sense again. Helmholtz wouldn't send it back.

But now, suppose I tell you that the eye's 'photocells' are pointing backwards, away from the scene being looked at. The 'wires' connecting

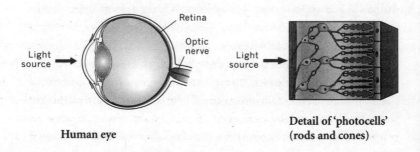

Human eye

Detail of 'photocells' (rods and cones)

the photocells to the brain run all over the surface of the retina, so the light rays have to pass through a carpet of massed wires before they hit the photocells. That doesn't make sense – and it gets even worse. One consequence of the photocells pointing backwards is that the wires that carry their data somehow have to pass through the retina and back to the brain. What they do, in the vertebrate eye, is all converge on a particular hole in the retina, where they dive through it. The hole filled with nerves is called the blind spot, because it is blind, but 'spot' is too flattering, for it is quite large, more like a blind *patch*, which again doesn't actually inconvenience us much because of the 'automatic Photoshop' software in the brain. Once again, send it back, it's not just bad design, it's the design of a complete idiot.

Or is it? If it were, the eye would be terrible at seeing, and it is not. It is actually very good. It is good because natural selection, working as a sweeper-up of countless little details, came along after the big original error of installing the retina backwards, and restored it to a high-quality precision instrument. It reminds me of the saga of the Hubble Space Telescope. You'll remember that, when it was launched in 1990, the Hubble was discovered to possess a major flaw. Owing to an undetected fault in the calibration apparatus when it was being ground and polished, the main mirror was slightly, but seriously, out of shape. The telescope was launched into orbit, and then discovered to be defective. In a daring and resourceful move, astronauts were dispatched to the telescope, and they succeeded in fitting it with what amounted to spectacles. The telescope thereafter worked very well, and further improvements were effected by three more servicing missions. The point I am making is that a major design flaw – catastrophic blunder, even – can be corrected by subsequent tinkering, whose ingenuity and intricacy can, under the right circumstances, perfectly compensate for the initial error. In evolution generally, major mutations, even if they cause improvements in generally the right direction, almost always require a lot of subsequent tinkering – a sweeping-up operation by lots of small mutations that come along later and are favoured by selection because they smooth out the rough edges left by the initial large mutation. This is why humans and hawks see so well, despite the blundering flaw in the initial design. Helmholtz again:

> For the eye has every possible defect that can be found in an optical instrument, and even some which are peculiar to itself; but they are all so counteracted, that the inexactness of the image which results from their presence very little exceeds, under ordinary conditions of illumination, the limits which are set to the delicacy of sensation by the dimensions of the retinal cones. But as soon as we make our observations under somewhat changed conditions, we become aware of the chromatic aberration, the astigmatism, the blind spots, the venous shadows, the imperfect transparency of the media, and all the other defects of which I have spoken.

UNINTELLIGENT DESIGN

This pattern of major design flaws, compensated for by subsequent tinkering, is exactly what we should *not* expect if there really were a designer at work. We might expect unfortunate mistakes, as in the spherical aberration of the Hubble mirror, but we do not expect obvious stupidity, as in the retina being installed back to front. Blunders of this kind come not from poor design but from *history*.

A favourite example, ever since it was pointed out to me by Professor J. D. Currey when he tutored me as an undergraduate, is the recurrent laryngeal nerve.* It is a branch of one of the cranial nerves, those nerves that lead directly from the brain rather than from the spinal cord. One of the cranial nerves, the vagus (the name means 'wandering' and it is apt), has various branches, two of which go to the heart, and two on each side to the larynx (voice box in mammals). On each side of the neck, one of the branches of the laryngeal nerve goes straight to the larynx, following a direct route such as a designer might have chosen. The other one goes to the larynx via an astonishing detour. It dives right down into the chest, loops around one of the main arteries leaving the heart (a different artery on the left and right sides, but the principle is the same), and then heads back up the neck to its destination.

If you think of it as the product of design, the recurrent laryngeal nerve is a disgrace. Helmholtz would have had even more cause to send it back than the eye. But, like the eye, it makes perfect sense the moment you forget design and think history instead. To understand it, we need to go back in time to when our ancestors were fish. Fish have a two-chambered heart, unlike our four-chambered one. It pumps blood forward through a big central artery called the ventral aorta. The ventral aorta usually gives off six pairs of

---

* It is also a favourite of my colleague Jerry Coyne. *Why Evolution is True* gives a wonderfully clear discussion of this example, which I recommend, along with the rest of his excellent book.

branches, leading off to the six gills on either side. The blood then passes up through the gills where it becomes richly laced with oxygen. Above the gills, it is collected by six more pairs of blood vessels into another big vessel running down the middle, called the dorsal aorta, which feeds the rest of the body. The six pairs of gill arteries are evidence of the *segmented* body plan of the vertebrates, which is clearer and more obvious in fish than it is in us. Fascinatingly, it is very obvious in human *embryos*, whose 'pharyngeal arches' are clearly derived from ancestral gills, as one can tell by looking at their detailed anatomy. Of course they don't function as gills, but five-week human embryos can be regarded as little pink fishes, with gills. I can't help wondering, once again, why whales and dolphins, dugongs and manatees have not re-evolved functional gills. The fact that, like all mammals, they have, in the pharyngeal arches, the embryonic scaffolding to grow gills suggests that it should not be too difficult to do so. I don't know why they haven't, but I'm pretty sure there's a good reason, and somebody either knows it or knows how to research it.

All vertebrates have a segmented body plan, but in adult mammals as opposed to embryos this is readily apparent only in the spinal

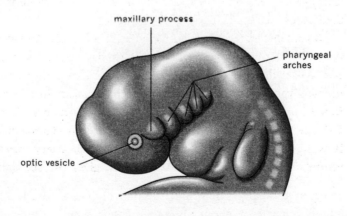

maxillary process

pharyngeal arches

optic vesicle

Pharyngeal arches in human embryo

region, where the vertebrae and the ribs, the blood vessels, muscle blocks (myotomes) and nerves all follow a pattern of modular repetition from front to back. Every segment of the vertebral column has two big nerves sprouting from the spinal cord on either side, called the dorsal root and the ventral root. These nerves mostly do their business, whatever it is, in the vicinity of the vertebrae from which they spring, but some shoot off down the legs and some down the arms.

The head and neck, too, follow the same segmented plan, but it is harder to discern, even in fish, because the segments, instead of being neatly laid out in a fore-and-aft array as they are in the spinal column, have become all jumbled up over evolutionary time. It was one of the triumphs of nineteenth- and early twentieth-century comparative anatomy and embryology to discern the ghostly footprints of segments in the head. For example, the first gill arch in jawless fishes like lampreys (and in embryos of jawed vertebrates) corresponds to the jaws in those vertebrates that have them (that is, all modern vertebrates except lampreys and hagfishes).

Insects, too, and other arthropods such as crustaceans, as we saw in Chapter 10, have a segmented body plan. And it was a similar triumph to show that the insect head contains – again, all jumbled up – the first six segments of what, in their remote ancestors, would have been a train of modules just like the rest of the body. It was a triumph of late twentieth-century embryology and genetics to show that insect segmentation and vertebrate segmentation, far from being independent of each other as I was taught, are actually mediated by parallel sets of genes, the so-called hox genes, which are recognizably similar in insects and vertebrates and many other animals, and that the genes are even laid out in the correct serial order in the chromosomes! That is something none of my teachers would have dreamed of when I was an undergraduate learning, entirely separately, about insect and vertebrate segmentation. Animals of different phyla (for example, insects and vertebrates) are much more united than we ever used to think. And that, too, is because of shared ancestry. The hox plan was already sketched out in the grand ancestor of all bilaterally symmetrical animals. All animals are much closer cousins to each other than we used to think.

To return to the vertebrate head: the cranial nerves are believed to be the much-disguised descendants of segmental nerves, which, in our primitive ancestors, constituted the front end of a train of dorsal roots and ventral roots, just like those we still have sprouting from our spinal column. And the major blood vessels of our chest are the messed-about relics and remnants of the once clearly segmental blood vessels serving the gills. You could say that the mammalian chest has messed up the segmental pattern of the ancestral fish gills, in the same kind of way as, earlier, the fish head messed up the segmented pattern of even earlier ancestors.

Human embryos also have blood vessels supplying their 'gills', which are very similar to those of fish. There are two ventral aortas, one on each side, with segmental aortic arches, one for each 'gill' on each side, connecting to paired dorsal aortas. Most of these segmental blood vessels have disappeared by the end of embryonic development, but it is clearly apparent how the adult pattern is derived from the embryonic – and also from the ancestral – plan. If you were to look at a human embryo about twenty-six days after conception you would see that the blood supply to the 'gills' strongly resembles the segmental blood supply to the gills of a fish. Over the following weeks of gestation the pattern of blood vessels becomes simplified by stages and loses its original symmetry, and by the time the infant is born its circulatory system has become strongly left-biased – quite unlike the neat symmetry of the fish-like early embryo.

I won't go into the messy details of which of our big chest arteries are the survivors of which of the six numbered gill arteries. All that we need to know, to understand the history of our recurrent laryngeal nerves, is that in fish the vagus nerve has branches that supply the last three of the six gills, and it is natural for them, therefore, to pass behind the appropriate gill arteries. There is nothing 'recurrent' about these branches: they seek out their end organs, the gills, by the most direct and logical route.

During the evolution of the mammals, however, the neck stretched (fish don't have necks) and the gills disappeared, some of them turning into useful things such as the thyroid and parathyroid glands, and the various other bits and pieces that combine to form the larynx. Those other useful things, including the parts of the larynx, received their blood supply and their nerve connections from the evolutionary descendants of the blood vessels and nerves that, once upon a time, served the gills in orderly sequence. As the ancestors of mammals evolved further and further away from their fish ancestors, nerves and blood vessels found themselves pulled and stretched in puzzling directions, which distorted their spatial relations one to another. The vertebrate chest and neck became a mess, unlike the tidily symmetrical, serial repetitiveness of fish gills. And the recurrent laryngeal nerves became more than ordinarily exaggerated casualties of this distortion.

The picture opposite, from a 1986 textbook by Berry and Hallam, shows how the laryngeal nerve lacks a detour in a shark. To illustrate the detour in a mammal, Berry and Hallam chose – what more striking example could there be? – a giraffe.

In a person, the route taken by the recurrent laryngeal nerve represents a detour of perhaps several inches. But in a giraffe, it is beyond a joke – many feet beyond – taking a detour of perhaps 15 feet in a large adult! The day after Darwin Day 2009 (his 200th birthday) I was privileged to spend the whole day with a team of comparative anatomists and veterinary pathologists at the Royal Veterinary College near London, dissecting a young giraffe that had unfortunately died at a zoo. It was a memorable day, almost a surreal experience for me. The operating theatre was literally a theatre, with a huge plate-glass wall separating the 'stage' from the raked seats where veterinary students were watching for hours at

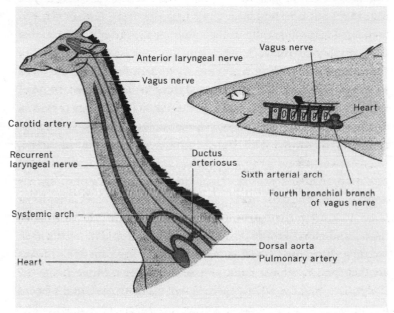

Anterior laryngeal nerve

Vagus nerve

Vagus nerve

Heart

Carotid artery

Recurrent
laryngeal nerve

Ductus
arteriosus

Sixth arterial arch

Fourth branchial branch
of vagus nerve

Systemic arch

Dorsal aorta

Pulmonary artery

Heart

Laryngeal nerve in giraffe and shark

a time. All day – it must have been right out of the normal run of their experience as students – they sat in the darkened theatre and stared through the glass at the brilliantly lit scene, listening to the words spoken by the dissecting team, who all wore throat microphones, as did I and the television production crew filming for a future documentary on Channel Four. The giraffe was laid out on the large, angled dissecting table, with one leg held high in the air by a hook and pulley, its enormous and affectingly vulnerable neck prominently exposed under bright lights. All of us on the giraffe side of the glass wall were under strict orders to wear orange overalls and white boots, which somehow enhanced the dream-like quality of the day.

It is testimony to the length of the detour taken by the recurrent laryngeal that different members of the team of anatomists worked simultaneously on different stretches of the nerve – the larynx near the head, the recurrence itself near the heart, and all stations between – without getting in each other's way, and scarcely needing to communicate with each other. Patiently they teased out the entire course of the recurrent laryngeal nerve: a difficult task that had not, as far as we know, been achieved since Richard Owen, the great Victorian anatomist, did it in 1837. It was difficult, because the nerve is very narrow, even thread-like in its recurrent portion (I suppose I should have known that, but it came as a surprise, nevertheless, when I actually saw it) and it is easily missed in the intricate web of membranes and muscles that surround the windpipe. On its downward journey, the nerve (at this point it is bundled in with the larger vagus nerve) passes within inches of the larynx, which is its final destination. Yet it proceeds down the whole length of the neck before turning round and going all the way back up again. I was very impressed with the skill of Professors Graham Mitchell and Joy Reidenberg, and the other experts doing the dissection, and I found my respect for Richard Owen (a bitter foe of Darwin) going up. The creationist Owen, however, failed to draw the obvious conclusion. Any intelligent designer would have hived off the laryngeal nerve on its way down, replacing a journey of many metres by one of a few centimetres.

Quite apart from the waste of resources involved in making such a long nerve, I can't help wondering whether giraffe vocalizations are subject to a delay, like a foreign correspondent talking over a satellite link. One authority has said, 'Despite possession of a well developed larynx and a gregarious nature, the Giraffe is able to utter only low moans or bleats.' A giraffe with a stutter is an endearing thought, but I won't pursue it. The important point is that this whole story of the detour is a splendid example of how very far living creatures are from having been well designed. And, for an evolutionist, the important question is why natural selection does not do as an engineer would: go back to the drawing board and rejig things in a sensible manner. It is the same question we are meeting over and over in this chapter, and I have attempted to answer it in various ways. The recurrent laryngeal lends itself to an answer in terms of what economists call 'marginal cost'.

As the giraffe's neck slowly lengthened over evolutionary time, the cost of the detour – whether economic cost or cost in terms of 'stuttering' – gradually increased, with the emphasis on 'gradually'. The *marginal* cost of each millimetre of increase was *slight*. As the giraffe's neck began to approach its present impressive length, the *total* cost of the detour might have begun to approach the point where – hypothetically – a mutant individual would survive better if its descending laryngeal nerve fibres hived themselves off from the vagus bundle and hopped across the tiny gap to the larynx. But the mutation needed to achieve this 'hop across' would have to have constituted a major change – upheaval even – in embryonic development. Very probably, the necessary mutation would never

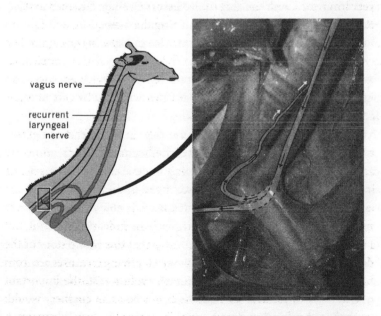

vagus nerve

recurrent
laryngeal
nerve

**Detour made by laryngeal
nerve in giraffe**

happen to arise anyway. Even if it did, it might well have disadvantages – inevitable in any major upheaval during the course of a sensitive and delicate process. And even if these disadvantages might eventually have been outweighed by the advantages of bypassing the detour, the *marginal* cost of each millimetre of *increased* detour *compared with the existing detour* is slight. Even if a 'back to the drawing board' solution would be a better idea if it could be achieved, the competing alternative was just a tiny increase over the existing detour, and the *marginal* cost of this tiny increase would have been small. Smaller, I am conjecturing, than the cost of the 'major upheaval' required to bring about the more elegant solution.

All that is beside the main point, which is that the recurrent laryngeal nerve in any mammal is good evidence against a designer. And in the giraffe it stretches from good to spectacular! That bizarrely long detour down the giraffe's neck and back up again is exactly the kind of thing we expect from evolution by natural selection, and exactly the kind of thing we do *not* expect from any kind of intelligent designer.

George C. Williams is one of the most respected of American evolutionary biologists (his quiet wisdom and craggy features recall one of the most respected of American presidents – who happens to have been born on the same day as Charles Darwin and was also renowned for quiet wisdom). Williams called attention to another detour, similar to that taken by the recurrent laryngeal nerve, but at the other end of the body. The vas deferens is the pipe that carries sperm from the testis to the penis. The most direct route is the fictitious one shown on the left-hand side of the diagram opposite. The actual route taken by the vas deferens is shown on the right of the diagram. It takes a ridiculous detour around the ureter, the pipe that carries urine from the kidney to the bladder. If this were designed, nobody could seriously deny that the designer had made a bad error. But, just as with the recurrent laryngeal nerve, all becomes clear when we look at evolutionary history. The likely original position of the testes is shown in dotted lines. When, in the evolution of mammals, the testes descended to their present position in the scrotum (for reasons that are unclear, but are often thought to be associated with temperature), the vas deferens unfortunately got hooked the wrong way over the ureter. Rather than reroute the pipe, as any sensible engineer would have done, evolution simply kept on lengthening it – once again, the marginal cost of each slight increase in length of detour would have been small. Yet again, it is a beautiful example of an initial mistake compensated for in a post hoc fashion, rather than being properly corrected back on the drawing board. Examples like this must surely undermine the position of those who hanker after 'intelligent design'.

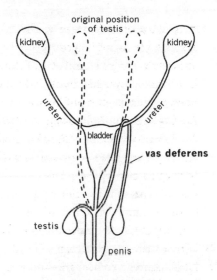

Route of vas deferens from testis to penis

The human body abounds with what, in one sense, we could call imperfections but, in another sense, should be seen as inescapable compromises resulting from our long ancestral history of descent from other kinds of animal. Imperfections are inevitable when 'back to the drawing board' is not an option – when improvements can be achieved only by making ad hoc modifications to what is already there. Imagine what a mess the jet engine would be if Sir Frank Whittle and Dr Hans von Ohain, its two independent inventors, had been forced to abide by a rule that said: 'You are not allowed to start with a clean sheet on your drawing board. You have to start with a propeller engine and change it, one piece at a time, screw by screw, rivet by rivet, from the "ancestral" propeller engine into a "descendant" jet engine.' Even worse, all the intermediates have got to fly, and each one in the chain has got to be at least a slight improvement on its predecessor. You can see that the resulting jet engine would be burdened with all kinds of historical relics and anomalies and imperfections. And each imperfection would be attended by a cumbersome accretion of compensatory bodges and fixes and kludges, each one making the best of the unfortunate prohibition against going right back to the drawing board.

The point is made, but a closer look at biological innovation might draw a different analogy from the propeller engine / jet engine case. An important innovation (the jet engine in our analogy) is quite likely to evolve not from the old organ that did the same job (the propeller engine in this case) but from something completely different, which performed a completely different function. As a nice example, when our fish ancestors took to breathing air, they didn't modify their gills to make a lung (as do some modern air-breathing fish, such as the climbing perch *Anabas*). Instead, they modified a pouch of the gut. And later, by the way, the teleosts – which means just about any fish you are likely to meet, except sharks and their kind – modified the lung (which had previously evolved in ancestors that occasionally breathed air) to become yet another vital organ, which has nothing to do with breathing: the swim bladder.

The swim bladder is perhaps the major key to the teleosts' success, and it is well worth a digression to explain it. It is an internal bladder filled with gas, which can be sensitively adjusted to keep the fish in hydrostatic equilibrium at any desired depth. If you ever played with a Cartesian Diver as a child you'll recognize the principle, but a teleost fish uses an interesting variant of it. A Cartesian Diver is a little toy whose business part is a tiny upended cup, containing a bubble of air, floating at equilibrium in a bottle of water. The number of molecules of air in the bubble is fixed, but you can decrease the volume (and increase the pressure, following Boyle's Law*) by pressing down on the cork in

---

* Boyle's Law states that, for a fixed quantity of gas at a given temperature, the pressure is inversely proportional to the volume. I have never forgotten Boyle's Law since my class at school, Form 4B1, was taught a single lesson by the school's senior science master, whose name was Bunjy. He was standing in for Bufty, our usual physics teacher, and we wrongly thought that, because of Bunjy's extreme age (as we thought) and extreme short sight (as was obvious from his habit of reading a book in contact with his nose), we could ignore his discipline and tease him. How wrong we were. He kept the whole lot of us in for an extra detention lesson that afternoon, which he began by making us write in our notebooks: 'Object of the lesson: To teach 4B1 good manners and Boyle's Law.'

the bottle. Or you can increase the volume of air (and decrease the pressure of the bubble) by raising the cork. The effect is best achieved with one of those stout screw stoppers they put on cider bottles. When you lower or raise the stopper, the diver moves down or up until it reaches its new point of hydrostatic equilibrium. You can coax the diver up and down the bottle by sensitive adjustments to the stopper, and hence to the pressure.

A fish is a Cartesian Diver with a subtle difference. The swim bladder is its 'bubble' and it works in the same way, except that the number of molecules of gas in the bladder is not fixed. When the fish wants to rise to a higher level in the water it releases molecules of gas from the blood into the bladder, thereby increasing the volume. When it wants to sink deeper, it absorbs molecules of gas from the bladder into the blood, thereby decreasing the volume of the bladder. The swim bladder means that a fish doesn't have to do muscular work, as a shark does, in order to stay at a desired depth. It is at hydrostatic equilibrium at whatever depth it chooses. The swim bladder does that job, thereby freeing up the muscles for active propulsion. Sharks, by contrast, have to keep swimming all the time, otherwise they would sink to the bottom, admittedly slowly because they have special low-density substances in their tissues that keep them moderately buoyant. The swim bladder, then, is a coopted lung, which is itself a coopted gut pouch (not, as you might have expected, a coopted gill chamber). And in some fish, the swim bladder itself is yet further coopted into a hearing organ, a kind of eardrum. History is written all over the body, not just once but repeatedly, in exuberant palimpsest.

We've been land animals for about 400 million years, and we've walked on our hind legs for only about the last 1 per cent of that time. For 99 per cent of our time on land, we've had a more-or-less horizontal backbone and walked on four legs. We don't know for certain what selective advantages accrued to the individuals who first rose up and walked on their hind legs, and I am going to leave that matter aside. Jonathan Kingdon has written a whole book on the question (*Lowly Origin*) and I have considered it in some detail in

The Ancestor's Tale. It may not have seemed like a major change when it happened, because other primates such as chimpanzees, some monkeys and the enchanting lemur Verreaux's sifaka do it from time to time. Habitually walking only on two legs as we do, however, had far-reaching ramifications all over the body, which entailed lots of compensatory adjustments. It could be argued that not a single bone or muscle, anywhere in the body, was spared the necessity to change, in order to reconcile some detail, however obscure, however out-of-the-way, and however indirectly or tenuously connected, with the major shift in gait. A similar across-the-board rejigging must attend each and every major change in way of life, from water to land, from land to water, into the air, underground. You cannot separate out the obvious changes in the body and treat them in isolation. To say that there are ramifications of every change is an understatement. There are hundreds, thousands of ramifications, and ramifications of ramifications. Natural selection is forever tweaking, adjusting the trim, 'tinkering' as the great French molecular biologist François Jacob put it.

Here's another good way to look at it. When there's a major shift in the climate, say an ice age, you naturally expect natural selection to adjust the animals to it – grow a thicker coat of hair, for example. But the *external* climate is not the only sort of 'climate' we have to consider. Without any external change at all, if a major new mutation arises, and is favoured by natural selection, all the other genes in the genome will experience it as a change in the internal 'genetic climate'. No less than a shift in the weather, it is a change to which they have to adjust. Natural selection has to come along afterwards, adjusting to compensate for a major change in genetic 'climate', exactly as it would if a change had occurred in the external climate. The initial shift from a four-legged to a two-legged gait could even have been 'internally' generated rather than engendered by a shift in the external environment. Either way, it would have initiated a complicated cascade of consequences, each one of which necessitated compensatory adjustments of 'trim'.

'Unintelligent design' would have been a good title for this chapter. It might well, indeed, be a worthy banner for a whole book on the imperfections of life as a cogent indicator of the lack of deliberate design, and more than one author has independently seized upon it. Of these, because I love the robust irreverence of Australian English ('So where did Intelligent Design spring from, like a boil on a bum?') I homed in on the delightful book of Robyn Williams, doyen of Sydney science broadcasters. After complaining of the agony his own back gives him every morning, in terms that wouldn't come amiss from a whingeing Pom (don't get me wrong, I sympathize profoundly), Williams goes on, 'nearly all backs could make an instant claim on the warranty, if there were one. If [God] *were* responsible for back design, you'll have to concede that it wasn't one of His best moments and must have been a deadline rush job at the end of the Six Days.' The problem, of course, is that our ancestors walked for hundreds of millions of years with the backbone held more-or-less horizontally, and it doesn't take kindly to the sudden readjustment imposed by the last few million. And the point, once again, is that a real designer of an upright-walking primate would have gone back to the drawing board and done the job properly, instead of starting with a quadruped and tinkering.

Williams next mentions the pouch of that iconic Australian animal the koala, which – not a great idea in an animal that spends its time clinging to tree trunks – opens downwards, instead of upwards as in a kangaroo. Once again, the reason is a legacy of history. Koalas are descended from a wombat-like ancestor. Wombats are champion diggers,

> flinging great paws full of soil backwards like an excavator digging out a tunnel. Had this ancestor's pouch pointed forwards, its babies would have had eyes and teeth permanently filled with grit. So backwards it was and, when one day the creature moved up a tree, perhaps to exploit a fresh food source, the 'design' came with it, too complicated to change.

As with the recurrent laryngeal nerve, it might theoretically be possible to change the embryology of the koala to turn its pouch the other way up. But – I'm guessing – the embryological upheaval attendant on such a major change would render the intermediates even worse off than koalas coping with the existing state of affairs.

Another consequence of our own shift from quadruped to biped concerns the sinuses, which give such grief to many of us (including me at the moment of writing) because their drainage hole is in the very last place a sensible designer would have chosen. Williams quotes an Australian colleague, Professor Derek Denton:* 'The big maxillary sinuses or cavities are behind the cheeks on either side of the face. They have their drainage hole in their top, which is not much of an idea in terms of using gravity to assist drainage of fluid.' In a quadruped, the 'top' is not the top at all but the front, and the position of the drainage hole makes much more sense: the legacy of history, yet again, is written all over us.

Williams goes on to quote another Australian colleague, who shares the national gift for chucking a bonzer phrase, on the Ichneumonid wasps, whose designer, if there were one, 'must have been a sadistic bastard'. Darwin, although he visited Australia as a young man, expressed the same sentiment in staider, less antipodean terms: 'I cannot persuade myself that a beneficent and omnipotent God would have designedly created the Ichneumonidae with the express intention of their feeding within the living bodies of caterpillars.' The legendary cruelty of ichneumon wasps (also the related digger wasps and tarantula wasps) is a leitmotif which will recur in the final two chapters of the book.

I find it hard to articulate what I am about to say, but it is something that I have been thinking for a while, and it came to a head during that memorable day of the dissection of the giraffe. When we look at animals from the outside, we are overwhelmingly impressed by the elegant illusion of design. A browsing giraffe, a soaring albatross, a

---

* Not to be confused with another Australian, Michael Denton, beloved of creationists who conveniently overlook the fact that, in his second book, *Nature's Destiny*, he recanted his earlier anti-evolutionary stance, while remaining theistic.

diving swift, a swooping falcon, a leafy sea dragon invisible among the seaweed, a sprinting cheetah at full stretch after a swerving, pronking gazelle – the illusion of design makes so much intuitive sense that it becomes a positive effort to put critical thinking into gear and overcome the seductions of naïve intuition. That's when we look at animals from the outside. When we look inside, the impression is opposite. Admittedly, an *impression* of elegant design is conveyed by simplified diagrams in textbooks, neatly laid out and colour-coded like an engineer's blueprint. But the reality that hits you when you see an animal opened up on a dissecting table is very different. I think it would be an instructive exercise to ask an engineer to draw an improved version of, say, the arteries leaving the heart. I imagine the result would be something like the exhaust manifold of a car, with a neat line of pipes coming off in orderly array, instead of the haphazard mess that we actually see when we open a real chest.

My purpose in spending a day with the anatomists dissecting a giraffe was to study the recurrent laryngeal nerve as an example of evolutionary imperfection. But I soon realized that, where imperfection is concerned, the recurrent laryngeal is just the tip of the iceberg. The fact that it takes such a long detour drives the point home with peculiar force. That is the aspect that would finally provoke a Helmholtz to send it back. But the overwhelming impression you get from surveying any part of the innards of a large animal is that it is a mess! Not only would a designer never have made a mistake like that nervous detour; a decent designer would never have perpetrated *anything* of the shambles that is the criss-crossing maze of arteries, veins, nerves, intestines, wads of fat and muscle, mesenteries and more. To quote the American biologist Colin Pittendrigh, the whole thing is nothing but a 'patchwork of makeshifts pieced together, as it were, from what was available when opportunity knocked, and accepted in the hindsight, not the foresight, of natural selection'.

CHAPTER 12

ARMS RACES AND 'EVOLUTIONARY THEODICY'

E YES and nerves, sperm tubes, sinuses and backs are poorly designed from the point of view of individual welfare, but the imperfections make perfect sense in the light of evolution. The same applies to the larger economy of nature. An intelligent creator might be expected to have designed not just the bodies of individual animals and plants but also whole species, entire ecosystems. Nature might be expected to be a planned economy, carefully designed to eliminate extravagance and waste. It isn't, and this chapter will show it.

THE SOLAR ECONOMY

The natural economy is solar-powered. Photons from the sun rain down upon the entire daytime surface of the planet. Many photons do nothing more useful than heat up a rock or a sandy beach. A few find their way into an eye – yours, or mine, or the compound eye of a shrimp or the parabolic reflector eye of a scallop. Some may happen to fall on a solar panel – either a man-made one like those that, in a fit of green zeal, I have just installed on my roof to heat the bathwater, or a green leaf, which is nature's solar panel. Plants use solar energy to drive 'uphill' chemical syntheses, manufacturing organic fuels, primarily sugars. 'Uphill' means that the synthesis of sugar needs energy to drive it; by the same token, the sugar can later be 'burned' in a 'downhill' reaction that releases (a fraction of) the energy again to do useful work, for example muscular work, or the work of building a great tree trunk. The 'downhill' and 'uphill' analogy is with water flowing downhill from a high tank and driving water wheels to do useful work; or being energetically pumped uphill into the high tank, so that it can later be used to drive water wheels when it flows downhill again. At every stage of the energy economy, whether uphill or downhill, some energy is lost – no energy transaction is ever perfectly efficient. That is why patent offices don't need even to look at designs for perpetual motion machines: they are implacably and forever impossible. You can't use the downhill energy from a water wheel to pump the same amount of water uphill again so that it can drive the water wheel. There must always be some energy fed in from outside to compensate for the wastage – and that is where the sun comes in. I'll return to this important theme in Chapter 13.

Much of the land surface of the Earth is covered by green leaves, which constitute a many-layered catchment for photons. If a photon is not caught by one leaf, it has a good chance of being caught by the one below. In a dense forest, not many photons make it to the ground uncaught, which is exactly why mature forests are such dark places in which to walk. Most of the photons that constitute our planet's minute share of the sun's rays hit water, and the surface layers of the sea swarm with single-celled green plants to catch them. Whether at sea or on land, the chemical process that traps photons and uses them to drive 'uphill' energy-consuming chemical reactions, manufacturing convenient energy-storage molecules such as sugars and starch, is called photosynthesis. It was invented, more than a billion years ago, by bacteria; and green bacteria still underlie most photosynthesis. I can say this because the chloroplasts – tiny green photosynthetic engines that actually do the business of photosynthesis in all leaves – are themselves the direct descendants of green bacteria. Indeed, since they still autonomously reproduce themselves after the manner of bacteria, within plant cells, we can justly say that they still are bacteria, albeit heavily dependent on the leaves that house them and to which they give their colour. It appears that originally free-living green bacteria were hijacked into plant cells, where they eventually evolved into what we now call chloroplasts.

And it is a neatly symmetrical fact that, just as the uphill chemistry of life is mostly taken care of by green bacteria thriving inside plant cells, so too the downhill chemistry of metabolism – the slow burning of sugars and other fuels to release energy in cells of both animals and plants – is the special expertise of another class of bacteria, once free-living but now reproducing themselves in larger cells, where they are known as mitochondria. Mitochondria and chloroplasts, descended from different kinds of bacteria, each built up their complementary chemical wizardries billions of years before the existence of any living organism visible to the naked eye. Both were later shanghaied for their chemical skills, and today they multiply inside the liquid interiors of the much larger and more complicated cells of creatures big enough for us to see and touch – plant cells in the case of chloroplasts, plant and animal cells in the case of mitochondria.

The solar energy captured by chloroplasts in plants lies at the base of complicated food chains, in which the energy passes from plants through herbivores, which may be insects, through carnivores, which may be insects or insectivores as well as wolves and leopards, through scavengers such as vultures and dung beetles, and eventually to agents of decay such as fungi and bacteria. At every stage of these food chains, some of the energy is wasted as heat as it passes through, while some of it is used to drive biological processes such as muscle contraction. No new energy is added after the initial input from the sun. With a few interesting but minor exceptions such as the denizens of deep ocean 'smokers' whose energy comes from volcanic sources, all the energy that drives life comes ultimately from sunlight, trapped by plants.

Look at a single tall tree standing proud in the middle of an open area. Why is it so tall? Not to be closer to the sun! That long trunk could be shortened until the crown of the tree was splayed out over the ground, with no loss in photons and huge savings in cost. So why go to all that expense of pushing the crown of the tree up towards the sky? The answer eludes us until we realize that the natural habitat of such a tree is a forest. Trees are tall to overtop rival trees – of the same and other species. Don't be misled when you see a tree in an open field or garden that has leafy branches all the way down to the ground. It has that well-rounded shape so beloved of sergeant instructors because it *is* in an open field or garden.* You are seeing it out of its natural habitat, which is a dense forest. The natural shape of a forest tree is tall and bare-trunked, with most of the branches and leaves near the top – in the canopy which bears the brunt of the photon rain. And now, here's an odd thought. If only all the trees in the forest could come to some agreement – like a trades union restrictive practice – to grow no higher than, say, 10 feet, every one would benefit. The entire community – the entire ecosystem – could gain from the savings in wood, and energy, which are consumed in building up those towering and costly trunks.

The difficulty of cultivating such agreements of mutual restraint is well known, even in human affairs where we can potentially deploy the gift of foresight. A familiar example is a suggested agreement to sit, rather than stand, when watching a spectacle such as a horse race. If everybody sat, tall people would still get a better view than short people, just as they would if everybody stood, but with the advantage that sitting is more comfortable for everybody. The problems start when one short person sitting behind a tall person stands, to get a better view. Immediately, the person sitting behind him stands, in order to see anything at all. A wave of standing sweeps around the field, until everybody is standing. In the end, everybody is worse off than they would be if they had all stayed sitting.

In a typical mature forest, the canopy can be thought of as an aerial meadow, just like a rolling grassland prairie, but raised on stilts. The canopy is gathering solar energy at much the same rate as a grassland prairie would. But a substantial proportion of the energy is 'wasted' by being fed straight into the stilts, which do nothing more useful than loft the 'meadow' high in the air, where it picks up exactly

---

* 'In the army, we has three kinds of trees: fir, poplar, and bushy top.'

the same harvest of photons as it would – at far lower cost – if it were laid flat on the ground.

And this brings us face to face with the difference between a designed economy and an evolutionary economy. In a designed economy there would be no trees, or certainly no very tall trees: no forests, no canopy. Trees are a waste. Trees are extravagant. Tree trunks are standing monuments to futile competition – futile if we think in terms of a planned economy. But the natural economy is not planned. Individual plants compete with other plants, of the same and other species, and the result is that they grow taller and taller, far taller than any planner would recommend. Not indefinitely taller, however. There comes a point when growing another foot taller, although it confers a competitive advantage, costs so much that the individual tree doing it actually ends up worse off than its rivals that forgo the extra foot. It is the balance of costs and benefits to the individual trees that finally determines the height to which trees are pressed to grow, not the benefits that a rational planner could calculate for the trees as a group. And of course the balance ends up at a different maximum in different forests. The Pacific Coast redwoods (see them before you die) have probably never been exceeded.

Imagine the fate of a hypothetical forest – let's call it the Forest of Friendship – in which, by some mysterious concordat, all the trees have somehow managed to achieve the desirable aim of lowering the entire canopy to 10 feet. The canopy looks just like any other forest canopy except that it is only 10 feet high instead of 100 feet. From the point of view of a planned economy, the Forest of Friendship is more efficient *as a forest* than the tall forests with which we are familiar, because resources are not put into producing massive trunks that have no purpose apart from competing with other trees.

But now, suppose one mutant tree were to spring up in the middle of the Forest of Friendship. This rogue tree grows marginally taller than the 'agreed' norm of 10 feet. Immediately, this mutant secures a competitive advantage. Admittedly, it has to pay the cost of the extra length of trunk. But it is more than compensated, *as long as all other trees obey the self-denying ordinance*, because the extra photons gathered more than pay the extra cost of lengthening the trunk. Natural selection therefore favours the genetic tendency to break out of the self-denying ordinance and grow a bit taller, say to 11 feet. As the generations go by, more and more trees break the embargo on height. When, finally, all the trees in the forest are 11 feet tall, they are all worse off than they were before: all are paying the cost of growing the extra foot. But they are not getting any extra photons for their trouble. And now natural selection favours any mutant tendency to grow to, say 12 feet. And so the trees go on getting taller and taller. Will this futile climb towards the sun ever come to an end? Why not trees a mile high, why not Jack's beanstalk? The limit is set at the height where the marginal cost of growing another foot outweighs the gain in photons from growing that extra foot.

We are talking individual costs and benefits throughout this argument. The forest would look very different if its economy had been designed for the benefit of the forest *as a whole*. In fact, what we actually see is a forest in which each tree species evolved through natural selection favouring *individual* trees that out-competed rival individual trees, whether of their own or another species. Everything about trees is compatible with the view that they were not designed – unless, of course, they were designed to supply us with timber, or to delight our eyes and flatter our cameras in the New England Fall. And history is not short of those who would believe just that, so let's turn to a parallel case, where the benefits to humanity are harder to allege: the arms race between hunters and hunted.

RUNNING TO STAY IN THE SAME PLACE

The five fastest runners among mammal species are the cheetah, the pronghorn (often called 'antelope' in America although it is not closely related to the 'true' antelopes of Africa), the gnu (or wildebeest, a true antelope although it doesn't look much like the others), the lion, and the Thomson's gazelle (another true antelope, which really does look like a standard antelope, a small one). Note that these top-ranked runners are a mixture of hunted and hunters, and my point is that this is no accident.

Cheetahs are said to be capable of accelerating from 0 to 60 mph in three seconds, which is right up there with a Ferrari, a Porsche or a Tesla. Lions, too, have formidable acceleration, even better than gazelles, who have more stamina and the ability to jink. Cats generally are built for sprinting, and springing on prey taken unawares; dogs, such as the Cape hunting dog or the wolf, for endurance, for wearing down their prey. Gazelles and other antelopes have to cope with both types of predator, and they perhaps have to compromise. Their acceleration is not quite so good as a big cat's, but their endurance is better. By jinking, a Tommy can sometimes throw a cheetah off its stride, thereby postponing matters until the cheetah has gone beyond its maximum acceleration phase into the exhausted phase, where its poor stamina starts to count. Successful cheetah hunts usually end soon after they start, the cheetah relying on surprise and acceleration. Unsuccessful cheetah hunts also end early, with the cheetah giving up to save energy when its initial sprint fails. All cheetah hunts, in other words, are brief!

Never mind the details of top speeds and accelerations, stamina and jinking, surprise and sustained pursuit. The salient fact is that the fastest animals include both those that hunt and those that are hunted. Natural selection drives predator species to become ever better at catching prey, and it simultaneously drives prey species to become ever better at escaping them. Predators and prey are engaged in an evolutionary arms race, run in evolutionary time. The result has been a steady escalation in the quantity of economic resources that animals, on both sides, spend on the arms race, at the expense of other departments of their bodily economy. Hunters and hunted alike get steadily better equipped to outrun (surprise, outwit, etc.) the other side. But improved equipment to outrun doesn't obviously translate into improved success in outrunning – for the simple reason that the other side in the arms race is upgrading its equipment too: that is the hallmark of an arms race. You could say, as the Red Queen said to Alice, that they have to run as fast as they can just to stay in the same place.

Darwin was well aware of evolutionary arms races, although he didn't use the phrase. My colleague John Krebs and I published a paper on the subject in 1979, in which we attributed the phrase 'armament race' to the British biologist Hugh Cott. Perhaps significantly, Cott published his book, *Adaptive Coloration in Animals*, in 1940, in the depths of the Second World War:

> Before asserting that the deceptive appearance of a grasshopper or butterfly is unnecessarily detailed, we must first ascertain what are the powers of perception and discrimination of the insects' natural enemies. Not to do so is like asserting that the armour of a battle-cruiser is too heavy, or the range of her guns too great, without inquiring into the nature and effectiveness of the enemy's armament. The fact is that in the primeval struggle of the jungle, as in the refinements of civilized warfare,* we see in progress a great evolutionary armament race – whose results, for defence, are manifested in such devices as speed, alertness, armour, spinescence, burrowing habits, nocturnal habits, poisonous secretions, nauseous taste, and procryptic, aposematic, and mimetic coloration; and for offence, in such counter-attributes as speed, surprise, ambush, allurement, visual acuity, claws, teeth, stings, poison fangs, and anticryptic and alluring coloration. Just as greater speed in the pursued has developed in relation to increased speed in the pursuer; or defensive armour in relation to aggressive weapons; so the perfection of concealing devices has evolved in response to increased powers of perception.

---

* An oxymoron if ever there was one.

Note that the arms race is run in evolutionary time. It is not to be confused with the race between an individual cheetah, say, and a gazelle, which is run in real time. The race in evolutionary time is a race to build up equipment for races run in real time. And what that actually means is that genes for making the equipment to outsmart or outrun the other side build up in the gene pools on the two sides. Second – and this is a point that Darwin himself knew well – the equipment for running fast is used to outrun *rivals* of the same species, who are fleeing from the same predator. The well-known joke, which has an almost Aesopian ring to it, about the running shoes and the bear is apposite.* When a cheetah chases a herd of gazelles, it may be more important for an individual gazelle to outrun the slowest member of the herd than to outrun the cheetah.

Now that I have introduced the terminology of the arms race, you can see that trees in a forest, too, are engaged in one. Individual trees are racing towards the sun, against their immediate neighbours in the forest. This race is particularly keen when an old tree dies and leaves a vacant slot in the canopy. The echoing crash of an old tree falling is the starting gun for a race, in real time (although a slower real time than we animals are accustomed to), between saplings that have been waiting for just such a chance. And the winner is likely to be an individual tree that is well equipped, by genes that prospered through ancestral arms races in evolutionary time, to grow fast and high.

The arms race between species of forest trees is a symmetrical race. Both sides are trying to achieve the same thing: a place in the canopy. The arms race between predators and prey is an asymmetric arms race: an arms race between weapons of attack and weapons of defence. The same is true of the arms race between parasites and hosts. And there are even, though it may seem surprising, arms races

---

* Two hikers are pursued by a bear. One hiker runs away, the other stops to put on his running shoes. 'Are you mad? Even with running shoes, you can't outrun a grizzly.' 'No, but I can outrun you.'

between males and females within a species, and between parents and offspring.

One thing about arms races that might worry enthusiasts for intelligent design is the heavy dose of futility that loads them down. If we are going to postulate a designer of the cheetah, he has evidently put every ounce of his designing expertise into the task of perfecting a superlative killer. One look at that magnificent running machine leaves us in no doubt. The cheetah, if we are going to talk design at all, is superbly designed for killing gazelles. But the very same designer has equally evidently strained every nerve to design a gazelle that is superbly equipped to escape from those very same cheetahs. For heaven's sake, whose side is the designer on? When you look at the cheetah's taut muscles and flexing backbone, you must conclude that the designer wants the cheetah to win the race. But when you look at the sprinting, jinking, dodging gazelle, you reach exactly the opposite conclusion. Does the designer's left hand not know what his right hand is doing? Is he a sadist, who enjoys the spectator sport and is forever upping the ante on both sides to increase the thrill of the chase? Did He who made the lamb make thee?

Is it really part of the divine plan that the leopard shall lie down with the kid, and the lion eat straw like the ox? In that case, what price the formidable carnassial teeth, the murderous claws of the lion and the leopard? Whence the breathtaking speed and agile escapology of the antelope and the zebra? Needless to say, no such problems arise on the evolutionary interpretation of what is going on. Each side is struggling to outwit the other because, on both sides, those individuals who succeed will automatically pass on the genes that contributed to their success. Ideas of 'futility' and 'waste' spring to our minds because we are human, and capable of looking at the welfare of the whole ecosystem. Natural selection cares only for the survival and reproduction of individual genes.

It's like the trees in the forest. Just as each tree has an economy, in which goods that are put into trunks are not available for fruits or leaves, so cheetahs and gazelles each have their own internal economy.

Running fast is costly, not just in energy ultimately wrung from the sun but in the materials that go into the making of muscles, bones and sinews – the machinery of speed and acceleration. The food that a gazelle ingests in the form of plant material is finite. Whatever is spent on muscles and long legs for running has to be taken away from some other department of life, such as making babies, on which the animal might ideally 'prefer' to spend its resources. There is an extremely complicated balance of compromises to be micro-managed. We can't know all the details but we do know (It is an unbreakable law of economics) that it is possible to spend *too much* on one department of life, thereby taking resources away from some other department of life. An individual that puts more than the ideal amount into running may save its own skin. But in the Darwinian stakes it will be out-competed by a rival individual of the same species, who skimps a little on running speed and hence incurs a greater risk of being eaten, but who gets the balance right and ends up with more descendants to pass on the genes for getting the balance right.

It isn't just energy and costly materials that have to be correctly balanced. There's also risk: and risk, too, is no stranger to the calculations of economists. Legs that are long and thin are good at running fast. Inevitably, they are also good at breaking. All too regularly a racehorse will break a leg in the heat of a race, and usually is promptly executed. As we saw in Chapter 3, the reason they are so vulnerable is that they have been overbred to be fast, at the expense of everything else. Gazelles and cheetahs have also been selectively bred for speed – naturally, not artificially selected – and they too would be vulnerable to fractures if nature were to overbreed them for speed. But nature never overbreeds for anything. Nature gets the balance right. The world is full of genes for getting the balance right: that is why they are there! What it means in practice is that individuals with a genetic tendency to develop exceptionally long and spindly legs, which are admittedly superior for running, are less likely to pass on their genes, on average, than slightly slower individuals whose less spindly legs are less likely to break. This is just one hypothetical example of the many hundreds of trade-offs and compromises that all animals and plants juggle. They juggle with risks and they juggle with economic trade-offs. It is, of course, not the individual animals and plants that do the juggling and balancing. It is the relative numbers of alternative genes in gene pools that are juggled and balanced, by natural selection.

As you would expect, the optimum compromise in a trade-off is not fixed. In gazelles, the trade-off between running speed and other demands within the economy of the body will shift its optimum depending upon the prevalence of carnivores in the area. It's the same story as for the guppies of Chapter 5. If there are few predators around, the gazelle's optimum leg length will shorten: the most successful individuals will be the ones whose genes predispose them to shunt some energy and material away from legs and into, say, making babies, or laying down fat for the winter. These are also the individuals who are less likely to break their legs. Conversely, if the number of predators increases, the optimum balance will shift towards longer legs, greater danger of fractures, and less energy and material to spend on those aspects of the body's economy that are not concerned with running fast.

And just the same kinds of implicit calculation will balance up the optimum compromises in the predators. A cheetah who breaks her leg will undoubtedly die of starvation, and so will her cubs. But, depending on how difficult it is to find a meal, the risk of failing to catch enough food if she runs too slowly may outweigh the risk of breaking a leg through being equipped with the wherewithal to run too fast.

Predators and prey are locked in an arms race in which each side is unwittingly pressing the other to shift its optimum – in the economic and risk compromises of life – further and further in the same direction: either literally in the same direction, for example towards increased running speed; or in the same direction in the looser sense of being aimed at the predator/prey arms race rather than some other department of life such as milk production. Given that both sides have to balance the risks of, say, running too fast (breaking legs or skimping on the other parts of the bodily economy) against the risks of running too slowly (failing to catch prey, or failing to escape, respectively), each side is pushing the other in the same direction, in a sort of grim *folie à deux*.

Well, perhaps *folie* (madness) doesn't quite do justice to the seriousness of the matter, for the penalty of failure on either side is death – murder on the side of the prey, starvation on the side of the predator. But *à deux* captures handily the feeling that, if only hunter and hunted could sit down together and hammer out a sensible agreement, everybody would be better off. Just as with the trees in the Forest of Friendship, it is easy to see how such a compact would benefit them, if only it could be made to stick. The same sense of futility as we encountered in the forest pervades the predator/prey arms race. Over evolutionary time, predators get better at catching prey, which prompts prey animals to get better at evading capture. Both sides in parallel improve their *equipment* to survive, but neither necessarily survives any better – because the other side is improving its equipment too.

On the other hand, it is easy to see how a central planner, with the welfare of the whole community at heart, might umpire an agreement in the following terms, along the lines of the Forest of Friendship. Let both sides 'agree' to scale down their armoury: both sides shift resources to other departments of life, and all will do better as a result. Just the same, of course, can happen in a human arms race. We wouldn't need our fighters if you didn't have your bombers. You wouldn't need your missiles if we didn't have ours. We could both save billions if we halved our armaments spending and put the money into ploughshares. And now, having halved our arms budget and reached a stable stand-off, let's halve it again. The trick is to do it in synchrony with each other, so that each side remains exactly as well equipped to counter the other's steadily de-escalating arms budget. Such planned de-escalation has to be just that – planned. And, once again, planned is precisely what evolution is not. Just as with the trees in the forest, escalation is inevitable, right up until the moment when it no longer pays a typical individual to escalate any further. Evolution, unlike a designer, never stops to consider whether there might be a better way – a mutualistic way – for all concerned, rather than bilateral escalation for a selfish advantage: an advantage that is neutralized precisely because the escalation *is* mutual.

The temptation to think like a planner has long been rife among 'pop ecologists', and even academic ecologists sometimes come perilously close to it. The tempting notion of 'prudent predators', for example, was dreamed up not by some tree-hugging airhead but by a distinguished American ecologist.

The idea of prudent predators is this. Everybody knows that, from the point of view of humanity as a whole, we'd be better off if we all refrained from overfishing an important food species, such as the cod, to extinction. That is why governments and NGOs in stately conclave meet to draw up quotas and restrictions. That is why the precise mesh size of fishing nets is minutely specified by government decree, and that is why gunboats patrol the seas in pursuit of dissenting trawlermen. We humans, on our good days and when properly policed, are 'prudent predators'. Therefore – or so it seems to certain ecologists – shouldn't we expect wild predators, like wolves or lions, to be prudent predators too? The answer is no. No. No. No. And it is worthwhile understanding why, because it's an interesting point, one that the forest trees and this whole chapter should have prepared us for.

A planner – an ecosystem designer with the welfare of the whole community of wild animals at heart – could indeed calculate an optimum culling policy, which lions, for example, should ideally adopt. Don't take more than a certain quota from any one species of antelope. Spare pregnant females, and don't take young adults full of reproductive potential. Avoid eating members of rare species, which might be in danger of extinction and might come in useful in future, if conditions change. If only all the lions in the country would abide by the agreed norms and quotas, carefully calculated to be 'sustainable', wouldn't that be nice? And so sensible. If only!

Well, it would be sensible, and it is what a designer would

prescribe, at least if he had the welfare of the ecosystem as a whole at heart. But it isn't what natural selection would prescribe (mainly because natural selection, lacking foresight, cannot *prescribe* at all) and it isn't what happens! Here's why, and it is again the same story as for the trees in the forest. Imagine that, by some quirk of leonine diplomacy, a majority of lions in an area somehow managed to agree to limit their hunting to sustainable levels. But now, suppose that in this otherwise restrained and public-spirited population, a mutant gene arose that caused an individual lion to break away from the agreement and exploit the prey population to the uttermost, even at the risk of driving the prey species extinct. Would natural selection penalize the rebellious selfish gene? Alas, it would not. Offspring of the rebel lion, possessors of the rebel gene, would out-compete and out-reproduce their rivals in the lion population. Within a few generations, the rebel gene would spread through the population and nothing would be left of the original amicable compact. He* who gets the lion's share passes on the genes for doing so.

But, the planning enthusiast will protest, when all the lions are behaving selfishly and over-hunting the prey species to the point of extinction, *everybody* is worse off, even the individual lions that are the most successful hunters. Ultimately, if all the prey go extinct, the entire lion population will too. Surely, the planner insists, natural selection will step in to stop that happening? Once again alas, and once again no. The problem is that natural selection doesn't 'step in', natural selection doesn't look into the future,† and natural

---

* Or she. The particular case of lions is complicated by the fact that females do most of the hunting, but males tend to get 'the lion's share' in any case. Don't get hung up on 'lions' in my hypothetical example. Think of a generalized predator species, and imagine 'prudent' individuals who refrain from over-hunting, and 'imprudent' individuals who break away from the agreement.

† Loose talk about Darwinian adaptation frequently founders on the fallacious assumption (not made explicit, and the more pernicious in consequence) that evolution has foresight. Sydney Brenner, hero of the *Caenorhabditis* section of Chapter 8, has a sardonic wit to match his scientific brilliance. I once heard him lampoon the 'evolutionary foresight' fallacy by imagining a species in the Cambrian that retained in its gene pool an otherwise useless protein, because 'It might come in handy in the Cretaceous.'

selection doesn't choose between rival groups. If it did, there would be some chance that prudent predation could be favoured. Natural selection, as Darwin realized much more clearly than many of his successors, chooses between rival individuals within a population. Even if the entire population is diving to extinction, driven down by individual competition, natural selection will still favour the most competitive individuals, right up to the moment when the last one dies. Natural selection can drive a population to extinction, while constantly favouring, to the bitter end, those competitive genes that are destined to be the last to go extinct. The hypothetical planner that I have imagined is a certain kind of economist, a welfare economist calculating an optimum strategy for a whole population, or an entire ecosystem. If we must make economic analogies, we should think instead of Adam Smith's 'invisible hand'.

EVOLUTIONARY THEODICY?

But now I want to leave economics altogether. We shall stay with the idea of a planner, a designer, but our planner will be a moral philosopher rather than an economist. A beneficent designer might – you'd idealistically think – seek to minimize suffering. This is not incompatible with economic welfare, but the system created will differ in detail. And, once again, it unfortunately doesn't happen in nature. Why should it? Terrible but true, the suffering among wild animals is so appalling that sensitive souls would best not contemplate it. Darwin knew whereof he spoke when he said, in a letter to his friend Hooker, 'What a book a devil's chaplain might write on the clumsy, wasteful, blundering low and horridly cruel works of nature.' The memorable phrase 'devil's chaplain' gave me my title for one of my previous books, and in another I put it like this:

> [N]ature is neither kind nor unkind. She is neither against suffering, nor for it. Nature is not interested in suffering one

way or the other unless it affects the survival of DNA. It is easy to imagine a gene that, say, tranquillises gazelles when they are about to suffer a killing bite. Would such a gene be favoured by natural selection? Not unless the act of tranquillising a gazelle improved that gene's chances of being propagated into future generations. It is hard to see why this should be so and we may therefore guess that gazelles suffer horrible pain and fear when they are pursued to the death – as most of them eventually are. The total amount of suffering per year in the natural world is beyond all decent contemplation. During the minute that it takes me to compose this sentence, thousands of animals are being eaten alive, others are running for their lives, whimpering with fear, others are being slowly devoured from within by rasping parasites, thousands of all kinds are dying of starvation, thirst and disease. It must be so. If there is ever a time of plenty, this very fact will automatically lead to an increase in population until the natural state of starvation and misery is restored.

Parasites probably cause even more suffering than predators, and understanding their evolutionary rationale adds to, rather than mitigates, the sense of futility we experience when we contemplate it. I fulminate against it every time I get a cold (I have one now, as it happens). Maybe it is only a minor inconvenience, but it is so *pointless*! At least if you are eaten by an anaconda you can feel that you have contributed to the well-being of one of the lords of life. When you are eaten by a tiger, perhaps your last thought could be, What immortal hand or eye could frame thy fearful symmetry? (In what distant deeps or skies, burnt the fire of thine eyes?) But a virus! A virus has pointless futility written into its very DNA – actually, RNA in the case of the common cold virus, but the principle is the same. A virus exists for the sole purpose of making more viruses. Well, the same is ultimately true of tigers and snakes, but there it doesn't *seem* so futile. The tiger and the snake may be DNA-replicating machines but they are beautiful, elegant, complicated, expensive DNA-replicating machines. I've given money to preserve the tiger, but who would think of giving money to preserve the common cold? It's the futility of it that gets to me, as I blow my nose yet again and gasp for breath.

Futility? What nonsense. Sentimental, human nonsense. Natural selection is *all* futile. It is all about the survival of self-replicating instructions for self-replication. If a variant of DNA survives through an anaconda swallowing me whole, or a variant of RNA survives by making me sneeze, then that is all we need by way of explanation. Viruses and tigers are both built by coded instructions whose ultimate message is, like a computer virus, 'Duplicate me.' In the case of the cold virus, the instruction is executed rather directly. A tiger's DNA is also a 'duplicate me' program, but it contains an almost fantastically large digression as an essential part of the efficient execution of its fundamental message. That digression is a tiger, complete with fangs, claws, running muscles, stalking and pouncing instincts. The tiger's DNA says, 'Duplicate me by the round-about route of building a tiger first.' At the same time, antelope DNA says, 'Duplicate me by the round-about route of building an antelope first, complete with long legs and fast muscles, complete with timorous instincts and finely honed sense organs tuned to the danger from tigers.' Suffering is a by-product of evolution by natural selection, an inevitable consequence that may worry us in our more sympathetic moments but cannot be expected to worry a tiger – even if a tiger can be said to worry about anything at all – and certainly cannot be expected to worry its genes.

Theologians worry about the problems of suffering and evil, to the extent that they have even invented a name, 'theodicy' (literally, 'justice of God'), for the enterprise of trying to reconcile it with the presumed beneficence of God. Evolutionary biologists see no problem, because evil and suffering don't count for anything, one way or the other, in the calculus of gene survival. Nevertheless, we do need to consider the problem of pain. Where, on the evolutionary view, does it come from?

Pain, like everything else about life, we presume, is a Darwinian device, which functions to improve the sufferer's survival. Brains are built with a rule of thumb such as, 'If you experience the sensation of pain, stop whatever you are doing and don't do it again.' It remains a matter for interesting discussion why it has to be so damned painful. Theoretically, you'd think, the equivalent of a little red flag could painlessly be raised somewhere in the brain, whenever the animal does something that damages it: picks up a red-hot cinder, perhaps. An imperative admonition, 'Don't do that again!' or a painless change in the wiring diagram of the brain such that, as a matter of fact, the animal *doesn't* do it again, would seem, on the face of it, enough. Why the searing agony, an agony that can last for days, and from which the memory may never shake itself free? Perhaps grappling with this question is evolutionary theory's own version of theodicy. Why so painful? What's wrong with the little red flag?

I don't have a decisive answer. One intriguing possibility is this. What if the brain is subject to opposing desires and impulses, and there is some kind of internal tussle between them? Subjectively, we know the feeling well. We may be in a conflict between, say, hunger and a desire to be slim. Or we may be in a conflict between anger and fear. Or between sexual desire and a shy fear of rejection, or a conscience that urges fidelity. We can literally feel the tug of war within us, as our conflicting desires battle it out. Now, back to pain and its possible superiority over a 'red flag'. Just as the desire to be slim can over-rule hunger, it is clearly possible to over-rule the desire to escape pain. Torture victims may succumb eventually, but they often go through a phase of enduring considerable pain rather than, say, betray their comrades or their country or their ideology. In so far as natural selection can be said to 'want' anything, natural selection doesn't want individuals to sacrifice themselves for the love of a country, or for the sake of an ideology or a party or a group or a species. Natural selection is 'against' individuals over-ruling the warning sensations of pain. Natural selection 'wants' us to survive, or more specifically, to reproduce, and be blowed to country, ideology or their non-human equivalents. As far as natural selection is concerned, little red flags will be favoured only if they are never over-ruled.

Now, despite philosophical difficulties, I think that instances where pain was over-ruled for non-Darwinian reasons – reasons of loyalty to country, ideology, etc. – would be more frequent if we had a 'red flag' in the brain rather than real, full-on, intolerable pain. Suppose genetic mutants arose who could not feel the excruciating agony of pain but relied upon a 'red flag' system to keep them away from bodily damage. It would be so easy for them to resist torture, they'd promptly be recruited as spies. Except that it would be so easy to recruit agents prepared to bear torture that torture would simply stop being used as a method of extortion. But, in a wild state, would such pain-free, red-flag mutants survive better than rival individuals whose brains do pain in earnest? Would they survive to pass on the genes for red-flag pain substitutes? Even setting aside the special circumstance of torture, and the special circumstances of loyalty to ideologies, I think we can see that the answer might be no. And we can imagine non-human equivalents.

As a matter of interest, there are aberrant individuals who cannot feel pain, and they usually come to a bad end. 'Congenital insensitivity to pain with anhidrosis' (CIPA) is a rare genetic abnormality in which the patient lacks pain receptor cells in the skin (and also – that's the 'anhidrosis' – doesn't sweat). Admittedly, CIPA patients don't have a built-in 'red flag' system to compensate for the breakdown of the pain system, but you'd think they could be taught to be cognitively aware of the need to avoid bodily damage – a learned red flag system. At all events, CIPA patients succumb to a variety of unpleasant consequences of their inability to feel pain, including burns, breakages, multiple scars, infections, untreated appendicitis and scratches to the eyeballs. More unexpectedly, they also suffer serious damage to their joints because, unlike the rest of us, they don't shift their posture when they have been sitting or lying in one position for a long time. Some patients set timers to remind themselves to change position frequently during the day.

Even if a 'red flag' system in the brain could be made effective, there seems to be no reason why natural selection would positively favour it over a real pain system just because it is less unpleasant. Unlike our hypothetically beneficent designer, natural selection is indifferent to the intensity of suffering – except in so far as it affects survival and reproduction. And, just as we should expect if the survival of the fittest, rather than design, underlies the world of nature, the world of nature seems to take no steps at all to reduce the sum total of suffering. Stephen Jay Gould reflected on such matters in a nice essay on 'Nonmoral nature'. I learned from it that Darwin's famous revulsion at the Ichneumonidae, which I quoted at the end of the previous chapter, was far from unique among Victorian thinkers.

Ichneumon wasps, with their habit of paralysing but not killing their victim, before laying an egg in it with the promise of a larva gnawing it hollow from within, and the cruelty of nature generally, were major preoccupations of Victorian theodicy. It's easy to see why. The female wasps lay their eggs in live insect prey, such as caterpillars, but not before carefully seeking out with their sting each nerve ganglion in turn, in such a way that the prey is paralysed, but still stays alive. It must be kept alive to provide fresh meat for the growing wasp larva feeding inside. And the larva, for its part, takes care to eat the internal organs in a judicious order. It begins by taking out the fat bodies and digestive organs, leaving the vital heart and nervous system till last – they are necessary, you see, to keep the caterpillar alive. As Darwin so poignantly wondered, what kind of beneficent designer would have dreamed *that* up? I don't know whether caterpillars can feel pain. I devoutly hope not. But what I do know is that natural selection would in any case take no steps to dull their pain, if the job could be accomplished more economically by simply paralysing their movements.

Gould quotes the Reverend William Buckland, a leading nine-teenth-century geologist, who found consolation in the optimistic spin that he managed to confer on the suffering caused by carnivores:

The appointment of death by the agency of carnivora, as the ordinary termination of animal existence, appears therefore in its main results to be a dispensation of benevolence; it deducts much from the aggregate amount of the pain of universal death; it abridges, and almost annihilates, throughout the brute creation, the misery of disease, and accidental injuries, and lingering decay; and imposes such salutary restraint upon excessive increase of numbers, that the supply of food maintains perpetually a due ratio to the demand. The result is, that the surface of the land and depths of the waters are ever crowded with myriads of animated beings, the pleasures of whose life are coextensive with its duration; and which throughout the little day of existence that is allotted to them, fulfill with joy the functions for which they were created.

Well, isn't that nice for them!

CHAPTER 13

THERE IS GRANDEUR IN THIS VIEW OF LIFE

Unlike his evolutionist grandfather Erasmus, whose scientific verse was (somewhat surprisingly, I have to say) admired by Wordsworth and Coleridge, Charles Darwin was not known as a poet, but he produced a lyrical crescendo in the last paragraph of *On the Origin of Species*.

Thus, from the war of nature, from famine and death,* the most exalted object which we are capable of conceiving, namely, the production of the higher animals, directly follows. There is grandeur in this view of life, with its several powers, having been originally breathed into a few forms or into one; and that, whilst this planet has gone cycling on according to the fixed law of gravity, from so simple a beginning endless forms most beautiful and most wonderful have been, and are being, evolved.

---

* Darwin told us that he derived his original inspiration for natural selection from Thomas Malthus, and perhaps this particular phrase of Darwin was prompted by the following apocalyptic paragraph, called to my attention by my friend Matt Ridley: 'Famine seems to be the last, the most dreadful resource of nature. The power of population is so superior to the power in the earth to produce subsistence for man, that premature death must in some shape or other visit the human race. The vices of mankind are active and able ministers of depopulation. They are the precursors in the great army of destruction, and often finish the dreadful work themselves. But should they fail in this war of extermination, sickly seasons, epidemics, pestilence, and plague, advance in terrific array, and sweep off their thousands and ten-thousands. Should success be still incomplete, gigantic inevitable famine stalks in the rear, and with one mighty blow, levels the population with the food of the world.'

There's a lot packed into this famous peroration, and I want to sign off by taking it line by line.

'FROM THE WAR OF NATURE, FROM FAMINE AND DEATH'

Clear-headed as ever, Darwin recognized the moral paradox at the heart of his great theory. He didn't mince words – but he offered the mitigating reflection that nature has no evil intentions. Things simply follow from 'laws acting all around us', to quote an earlier sentence from the same paragraph. He had said something similar at the end of Chapter 7 of *The Origin*:

> it may not be a logical deduction, but to my imagination it is far more satisfactory to look at such instincts as the young cuckoo ejecting its foster-brothers, – ants making slaves, – the larvae of ichneumonidae feeding within the live bodies of caterpillars, – not as specially endowed or created instincts, but as small consequences of one general law, leading to the advancement of all organic beings, namely, multiply, vary, let the strongest live and the weakest die.

I've already mentioned Darwin's revulsion – widely shared by his contemporaries – in the face of the female ichneumon wasp's habit of stinging its victim to paralyse but not kill it, thereby keeping the meat fresh for its larva as it eats the live prey from within. Darwin, you'll remember, couldn't persuade himself that a beneficent creator would conceive such a habit. But with natural selection in the driving seat, all becomes clear, understandable and sensible. Natural selection cares naught for any comfort. Why should it? For something to happen in nature, the only requirement is that the same happening in ancestral times assisted the survival of the genes promoting it. Gene survival is a sufficient explanation for the cruelty of wasps and the callous indifference of all nature: sufficient – and satisfying to the intellect if not to human compassion.

Yes, there is grandeur in this view of life, and even a kind of grandeur in nature's serene indifference to the suffering that inexorably follows in the wake of its guiding principle, survival of the fittest. Theologians may here wince at this echo of a familiar ploy in theodicy, in which suffering is seen as an inevitable correlate of free will. Biologists, for their part, will find 'inexorably' by no means too strong when they reflect – perhaps along the lines of my 'red flag' meditation of the previous chapter – on the biological function of the capacity to suffer. If animals aren't suffering, somebody isn't working hard enough at the business of gene survival.

Scientists are human, and they are as entitled as anyone to revile cruelty and abhor suffering. But good scientists like Darwin recognize that truths about the real world, however distasteful, have to be faced. Moreover, if we are going to admit subjective considerations, there is a fascination in the bleak logic that pervades all of life, including wasps homing in on the nerve ganglia down the length of their prey, cuckoos ejecting their foster brothers ('Thow mortherer of the heysugge on y braunche'), slave-making ants, and the single-minded – or rather zero-minded – indifference to suffering shown by all parasites and predators. Darwin was bending over backwards to console when he concluded his chapter on the struggle for survival with these words:

> All that we can do, is to keep steadily in mind that each organic being is striving to increase at a geometrical ratio; that each at some period of its life, during some season of the year, during each generation or at intervals, has to struggle for life, and to suffer great destruction. When we reflect on this struggle, we may console ourselves with the full belief, that the war of nature is not incessant, that no fear is felt,* that death is generally prompt, and that the vigorous, the healthy, and the happy survive and multiply.

* I wish I could believe that.

Shooting the messenger is one of humanity's sillier foibles, and it underlies a good slice of the opposition to evolution that I mentioned in the Introduction. 'Teach children that they are animals, and they'll behave like animals.' Even if it were true that evolution, or the teaching of evolution, encouraged immorality, that would not imply that the theory of evolution was false. It is quite astonishing how many people cannot grasp this simple point of logic. The fallacy is so common it even has a name, the *argumentum ad consequentiam* – X is true (or false) because of how much I like (or dislike) its consequences.

'THE MOST EXALTED OBJECT WHICH WE ARE CAPABLE OF CONCEIVING'

Is 'the production of the higher animals' really 'the most exalted object which we are capable of conceiving'? *Most* exalted? Really? Are there not more exalted objects? Art? Spirituality? *Romeo and Juliet*? General Relativity? The Choral Symphony? The Sistine Chapel? Love?

You have to remember that, for all his personal modesty, Darwin nursed high ambitions. On his world-view, everything about the human mind, all our emotions and spiritual pretensions, all arts and mathematics, philosophy and music, all feats of intellect and of spirit, are themselves productions of the same process that delivered the higher animals. It is not just that without evolved brains spirituality and music would be impossible. More pointedly, brains were naturally selected to increase in capacity and power for utilitarian reasons, until those higher faculties of intellect and spirit emerged as a by-product, and blossomed in the cultural environment provided by group living and language. The Darwinian world-view does not denigrate the higher human faculties, does not 'reduce' them to a plane of indignity. It doesn't even claim to explain them at the sort of level that will seem particularly satisfying, in the way that, say, the Darwinian explanation of a snake-mimicking caterpillar is satisfying. It does, however, claim to have wiped out the impenetrable – not even worth trying to penetrate – mystery that must have dogged all pre-Darwinian efforts to understand life.

But Darwin doesn't need any defence from me, and I'll pass over the question of whether the production of the higher animals is the most exalted object we can conceive, or merely a very exalted object. What, however, of the predicate? Does the production of the higher animals 'directly follow' from the war of nature, from famine and death? Well, yes, it does. It directly follows if you understand Darwin's reasoning, but nobody understood it until the nineteenth century. And many still don't understand it, or perhaps are reluctant to do so. It is not hard to see why. When you think about it, our own existence, together with its post-Darwinian explicability, is a candidate for the most astonishing fact that any of us are called upon to contemplate, in our whole life, ever. I'll come to that shortly.

'HAVING BEEN ORIGINALLY BREATHED'

I have lost count of the irate letters I have received from readers of a previous book, taking me to task for, as the writers think, deliberately omitting the vital phrase, 'by the Creator' after 'breathed'? Am I not wantonly distorting Darwin's intention? These zealous correspondents forget that Darwin's great book went through six editions. In the first edition, the sentence is as I have written it here. Presumably bowing to pressure from the religious lobby, Darwin inserted 'by the Creator' in the second and all subsequent editions. Unless there is a very good reason to the contrary, when quoting *On the Origin of Species* I always quote the first edition. This is partly because my own copy of that historic print run of 1,250 is one of my most precious possessions, given me by my benefactor and friend Charles Simonyi. But it is also because the first edition is the most historically important. It is the one that thumped the Victorian solar plexus and drove out the wind of centuries. Moreover, later editions, especially the sixth, pandered to more than public opinion. In an attempt to respond to various learned but misguided critics of the first edition, Darwin backtracked and even reversed his position on a number of important points that he had actually got right in the first place. So, 'having been originally breathed' it is, with no mention of any Creator.

It seems that Darwin regretted this sop to religious opinion. In a letter of 1863 to his friend the botanist Joseph Hooker, he said, 'But I have long regretted that I truckled to public opinion, and used the Pentateuchal term of creation, by which I really meant "appeared" by some wholly unknown process.' The 'Pentateuchal term' Darwin is referring to here is the word 'creation'. The context, as Francis Darwin explains in his 1887 edition of his father's letters, was that Darwin was writing to thank Hooker for the loan of a review of a book by Carpenter, in which the anonymous reviewer had spoken of 'a creative force . . . which Darwin could only express in Pentateuchal terms as the primordial form "into which life was originally breathed"'. Nowadays, we should dispense even with the 'originally breathed'. What is it that is supposed to have been breathed into what?* Presumably the intended reference was to some kind of breath of life,* but what might that mean? The harder we look at the border between life and non-life, the more elusive does the distinction become. Life, the animate, was supposed to have some sort of vibrant, throbbing quality, some vital essence – made to sound yet more mysterious when dropped into French: *élan vital*.[†] Life, it seemed, was made of a special living substance, a witch's brew called 'protoplasm'. Conan Doyle's Professor Challenger, a fictional character even more preposterous than Sherlock Holmes, discovered that the Earth was living, a kind of giant sea urchin whose shell was the crust that we see, and whose core consisted of pure protoplasm. Right up to the middle of the

---

* Religious traditions have long identified life with breath. 'Spirit' comes from the Latin for 'breath'. Genesis has God first making Adam and then firing him up by breathing into his nostrils. The Hebrew word for 'soul' is *ruah* or *ruach* (cognate *ruh* in Arabic), which also means 'breath', 'wind', 'inspiration'.

† The term was coined in 1907 by the French philosopher Henri Bergson. I've always treasured Julian Huxley's sarcastic deduction that railway trains must be propelled by *élan locomotif*.

twentieth century, life was thought to be qualitatively beyond physics and chemistry. No longer. The difference between life and non-life is a matter not of substance but of *information*. Living things contain prodigious quantities of information. Most of the information is digitally coded in DNA, and there is also a substantial quantity coded in other ways, as we shall see presently.

In the case of DNA, we understand pretty well how the information content builds up over geological time. Darwin called it natural selection, and we can put it more precisely: the non-random survival of information that encodes embryological recipes for that survival. Self-evidently it is to be expected that recipes for their own survival will tend to survive. What is special about DNA is that it survives not in its material self but in the form of an indefinite series of copies. Because there are occasional errors in the copying, new variants may survive even better than their predecessors, so the database of information encoding recipes for survival will improve as time goes by. Such improvements will be manifest in the form of better bodies and other contrivances and devices for the preservation and propagation of the coded information. On the ground, the preservation and propagation of DNA information will normally mean the survival and reproduction of bodies containing it. It was at the level of bodies, their survival and reproduction, that Darwin himself worked. The coded information within them was implicit in his world-view, but not made explicit until the twentieth century.

The genetic database will become a storehouse of information about the environments of the past, environments in which ancestors survived and passed on the genes that helped them to do so. To the extent that present and future environments resemble those of the past (and mostly they do), this 'genetic book of the dead' will turn out to be a useful manual for survival in the present and future. The repository of that information will, at any one moment, reside in individual bodies, but in the longer term, where reproduction is sexual and DNA is shuffled from body to body, the database of survival instructions will be the gene pool of a species.

Each individual's genome, in any one generation, will be a sample from the species database. Different species will have different databases because of their different ancestral worlds. The database in the gene pool of camels will encode information about deserts and how to survive in them. The DNA in mole gene pools will contain instructions and hints for survival in dark, moist soil. The DNA in predator gene pools will increasingly contain information about prey animals, their evasive tricks and how to outsmart them. The DNA in prey gene pools will come to contain information about predators and how to dodge and outrun them. The DNA in all gene pools contains information about parasites and how to resist their pernicious invasions.

Information on how to handle the present so as to survive into the future is necessarily gleaned from the past. Non-random survival of DNA in ancestral bodies is the obvious way in which information from the past is recorded for future use, and this is the route by which the primary database of DNA is built up. But there are three further ways in which information about the past is archived in such a way that it can be used to improve future chances of survival. These are the immune system, the nervous system, and culture. Along with wings, lungs and all the other apparatus for survival, each of the three secondary information-gathering systems was ultimately prefigured by the primary one: natural selection of DNA. We could together call them the four 'memories'.

The first memory is the DNA repository of ancestral survival techniques, written on the moving scroll that is the gene pool of the species. Just as the inherited database of DNA records the recurrent details of ancestral environments and how to survive them, the immune system, the 'second memory', does the same thing for diseases and other insults to the body during the individual's own lifetime. This database of past diseases and how to survive them is unique to each individual and is written in the repertoire of proteins that we call antibodies – one population of antibodies for each pathogen (disease-causing organism), precisely tailored by past 'experience' with the proteins that characterize the pathogen. Like many children of my generation, I had measles and chickenpox. My body 'remembers' the 'experience', the memories being embodied in antibody proteins, along with the rest of my personal database of previously vanquished invaders. I have fortunately never had polio, but medical science has cleverly devised the technique of vaccination for planting false memories of diseases never suffered. I shall never contract polio, because my body 'thinks' it has done so in the past, and my immune system database is equipped with the appropriate antibodies, 'fooled' into making them by the injection of a harmless version of the virus. Fascinatingly, as the work of various Nobel Prize-winning medical scientists has shown, the immune system's database is itself built up by a quasi-Darwinian process of random variation and non-random selection. But in this case the non-random selection is selection not of bodies for their capacity to survive, but of proteins *within* the body for their capacity to envelop or otherwise neutralize invading proteins.

The third memory is the one we ordinarily think of when we use the word: the memory that resides in the nervous system. By mechanisms that we don't yet fully understand, our brains retain a store of past experiences to parallel the antibody 'memory' of past diseases and the DNA 'memory' (for so we can regard it) of ancestral deaths and successes. At its simplest, the third memory works by a trial-and-error process that can be seen as yet another analogy to natural selection. When searching for food, an animal may 'try' various actions. Though not strictly random, this trial stage is a reasonable analogy to genetic mutation. The analogy to natural selection is 'reinforcement', the system of rewards (positive reinforcement) and punishments (negative reinforcement). An action such as turning over dead leaves (trial) turns out to yield beetle larvae and woodlice hiding under the leaves (reward). The nervous system has a rule that says, 'Any trial action that is followed by reward should be repeated. Any trial action that is followed by nothing, or, worse, followed by punishment, for example pain, should not be repeated.'

But the brain's memory goes much further than this quasi-Darwinian process of non-random survival of rewarded actions, and elimination of punished actions, in the animal's repertoire. The brain's memory (no need for inverted commas here, because it is the primary meaning of the word) is, at least in the case of human brains, both vast and vivid. It contains detailed scenes, represented in an internal simulacrum of all five senses. It contains lists of faces, places, tunes, social customs, rules, words. You know it well from the inside, so there is no need for me to spend my words evoking it, except to note the remarkable fact that the lexicon of words at my disposal for writing, and the identical, or at least heavily overlapping, dictionary at your disposal for reading, all reside in the same vast neuronal database, along with the syntactic apparatus for arranging them into sentences and deciphering them.

Furthermore, the third memory, the one in the brain, has spawned a fourth. The database in my brain contains more than just a record of the happenings and sensations of my personal life – although that was the limit when brains originally evolved. Your brain includes collective memories inherited non-genetically from past generations, handed down by word of mouth, or in books or, nowadays, on the internet. The world in which you and I live is richer by far because of those who went before us and inscribed their impacts on the database of human culture: Newton and Marconi, Shakespeare and Steinbeck, Bach and the Beatles, Stephenson and the Wright brothers, Jenner and Salk, Curie and Einstein, von Neumann and Berners-Lee. And, of course, Darwin.

All four memories are part of, or manifestations of, the vast super-structure of apparatus for survival which was originally, and primarily, built up by the Darwinian process of non-random DNA survival.

'INTO A FEW FORMS OR INTO ONE'

Darwin was right to hedge his bets, but today we are pretty certain that all living creatures on this planet are descended from a single ancestor. The evidence, as we saw in Chapter 10, is that the genetic code is universal, all but identical across animals, plants, fungi, bacteria, archaea and viruses. The 64-word dictionary, by which three-letter DNA words are translated into twenty amino acids and one punctuation mark, which means 'start reading here' or 'stop reading here', is the same 64-word dictionary wherever you look in the living kingdoms (with one or two exceptions too minor to undermine the generalization). If, say, some weird, anomalous microbes called the harumscaryotes were discovered, which didn't use DNA at all, or didn't use proteins, or used proteins but strung them together from a different set of amino acids from the familiar twenty, or which used DNA but not a triplet code, or a triplet code but not the same 64-word dictionary – if any of these conditions were met, we might suggest that life had originated twice: once for the harumscaryotes and once for the rest of life. For all Darwin knew – indeed, for all anyone knew before the discovery of DNA – some existing creatures might have had the properties I have here attributed to the harumscaryotes, in which case his 'into a few forms' would have been justified.

Is it possible that two independent origins of life could both have hit upon the same 64-word code? Very unlikely. For that to be plausible, the existing code would have to have strong advantages over alternative codes, and there would have to be a gradual ramp of improvement towards it, a ramp for natural selection to climb up. Both these conditions are improbable. Francis Crick early suggested that the genetic code is a 'frozen accident', which, once in place, was difficult or impossible to change. The reasoning is interesting. Any mutation in the genetic code itself (as opposed to mutations in the genes that it encodes) would have an instantly catastrophic effect, not just in one place but throughout the whole organism. If any word in the 64-word dictionary changed its meaning, so that it came to specify a different amino acid, just about every protein in the body would instantaneously change, probably in many places along its length. Unlike an ordinary mutation, which might, say, slightly lengthen a leg, shorten a wing or darken an eye, a change in the genetic code would change everything at once, all over the body, and this would spell disaster. Various theorists have come up with ingenious suggestions for special ways in which the genetic code might evolve: ways in which, to quote one of their papers, the frozen accident might be 'thawed'. Interesting as these are, I think it is all but certain that every living creature whose genetic code has been looked at is descended from one common ancestor. No matter how elaborate and different the high-level programs that underlie the various life forms, all are, at bottom, written in the same machine language.

Of course we cannot rule out the possibility that other machine languages may have arisen in yet other creatures that are now extinct – the equivalent of my harumscaryotes. And the physicist Paul Davies has made the reasonable point that we haven't actually looked very hard to see if there are any harumscaryotes (he doesn't use the word, of course) that are not extinct but still lurking in some extreme redoubt of our planet. He admits that it is not very likely, but argues – somewhat along the lines of the man who searches for his keys under a street lamp rather than where he lost them – that it is a lot easier and cheaper to look thoroughly on our planet than to travel to other planets and look there. Meanwhile, I don't mind recording my private expectation that Professor Davies won't find anything, and that all surviving life forms on this planet use the same machine code and are all descended from a single ancestor.

'WHILST THIS PLANET HAS GONE CYCLING ON ACCORDING TO THE FIXED LAW OF GRAVITY'

Humans were aware of the cycles that govern our lives long before we understood them. The most obvious cycle is the day/night cycle. Objects floating in space, or orbiting other objects under the law of gravity, have a natural tendency to spin on their own axis. There are exceptions, but our planet is not one of them. Its period of rotation is now twenty-four hours (it used to spin faster) and we experience it, of course, as night follows day.

Because we live on a relatively massive body, we think of gravity primarily as a force that pulls everything towards the centre of that body, which we experience as 'down'. But gravity, as Newton was the first to understand, has a ubiquitous effect, which is to keep bodies throughout the universe in semi-permanent orbit around other bodies. We experience this as the yearly cycle of seasons, as our planet orbits the sun.* Because the axis on which our planet spins is tilted relative to the axis of rotation around the sun, we experience longer days and shorter nights during the half of the year when the hemisphere on which we happen to live is tilted sunwards, the period that climaxes in summer. And we experience shorter days and longer nights during the other half of the year, the period that, at its extreme, we call winter. During our hemisphere's winter, the sun's rays, when they strike us at all, do so at a shallower angle. The glancing angle spreads a winter sunbeam more thinly over a wider area than the same beam would cover in summer. On the receiving end of fewer photons per square inch, it feels colder. Fewer photons per green leaf means less photosynthesis. Shorter days and longer nights have the same effect. Winter and summer, day and night, our lives are governed by cycles, just as Darwin said – and Genesis before him: 'While the earth remaineth, seedtime and harvest, and cold and heat, and summer and winter, and day and night shall not cease.'

Gravity mediates other cycles that also matter to life, although they are less obvious. Unlike other planets that have many satellites, often relatively small, Earth happens to have a single large satellite, which we call the moon. It is large enough to exert a significant gravitational effect in its own right. We experience this principally in the cycle of tides: not just the relatively fast cycle as tides come in and out daily,

---

* It is with horrified fascination that I return, as if scratching an itch or pressing a toothache, to the poll, documented in the Appendix, suggesting that 19% of British people don't know what a year is, and think the Earth orbits the sun once per month. Even of those who understand what a year is, a larger percentage has no understanding of what causes seasons, presuming, with rampant Northern Hemisphere chauvinism, that we are closest to the sun in June and furthest away in December.

but the slower monthly cycle of spring tides and neap tides, which is caused by interactions between the sun's gravitational effect and that of the monthly orbiting moon. These tidal cycles are especially important for marine and coastal organisms, and people have rather implausibly wondered whether some kind of species memory of our marine ancestry survives in our monthly reproductive cycles. That may be far-fetched, but it is a matter for intriguing speculation how different life would be if we had no orbiting moon. It has even been suggested, again implausibly in my opinion, that life without the moon would be impossible.

What if our planet didn't spin on its axis? If it kept one face permanently towards the sun, as the moon does towards us, the half with permanent day would be a roasting hell, while the half with permanent night would be insufferably cold. Could life survive in the twilight hinterland between, or perhaps buried deep in the ground? I doubt if it would have originated in such unfriendly conditions, but if Earth gradually spun down to a halt there would be plenty of time to accommodate, and it is not implausible that at least some bacteria would succeed.

What if Earth spun, but on an axis that was not tilted? I doubt if that would rule life out. There would be no summer/winter cycle. Summer and winter conditions would be a function of latitude and altitude but not time. Winter would be the permanent season experienced by creatures living close to either of the two poles, or up high mountains. I don't see why that should rule life out, but life without seasons would be less interesting. There would be no incentive to migrate, or to breed at any particular time of the year rather than any other, or to shed leaves or to moult or hibernate.

If the planet were not in orbit around a star at all, life would be completely impossible. The only alternative to orbiting a star is hurtling through the void – dark, close to absolute zero temperature, alone and far from the source of energy that enables life to trickle upstream, temporarily and locally, against the thermodynamic torrent. Darwin's phrase 'cycling on according to the fixed law of gravity' is more than just a poetic device to express the relentless and unimaginably extended passage of time.

Being in orbit around a star is the only way a body can remain a relatively fixed distance away from a source of energy. In the vicinity of any star – and our sun is typical – there is a finite zone bathed in heat and light, where the evolution of life is possible. As you move away from a star into space, this habitable zone dwindles rapidly, following the famous inverse square law. That is, light and heat diminish not in direct proportion to the distance from the star, but in proportion to the square of the distance. It is easy to see why this must be so. Imagine concentric spheres of increasing radius centred on a star. The energy radiating outwards from the star falls on the inside of a sphere and is 'shared' evenly by every square inch of the internal area of the sphere. The surface area of a sphere is proportional to the square of the radius (ESK).* So if sphere A is twice as far from the star as sphere B, the same number of photons has to be 'shared' over an area four times as great. This is why Mercury and Venus, the innermost planets of our solar system, are scorching hot, while the outer ones, such as Neptune and Uranus, are cold and dark, although still not as cold and dark as deep space.

The Second Law of Thermodynamics states that, although energy can be neither created nor destroyed, it can – must, in a closed system – become more impotent to do useful work: that is what it means to say that 'entropy' increases. 'Work' includes things like pumping water uphill or – the chemical equivalent – extracting carbon from atmospheric carbon dioxide and using it in plant tissues. As already spelled out in Chapter 12, both those feats can be achieved only if energy is fed into the system, for example electrical energy to drive the water pump, or solar energy to drive the synthesis of sugar and starch in a green plant. Once the water has been pumped to the top of the hill, it will then tend to flow downhill, and some of the energy

---

* 'Every Schoolboy Knows' (and every schoolgirl can prove it by Euclidean geometry).

of its downward flow can be used to drive a water wheel, which can generate electricity, which can drive an electric motor to pump some of the water uphill again: but only some! Some of the energy is always lost – though never destroyed. Perpetual motion machines (you can't say it too dogmatically) are impossible.

In life's chemistry, the carbon extracted from the air by sun-driven 'uphill' chemical reactions in plants can be burned to release some of the energy. We can literally burn it in the form of coal, which you can think of as stored solar energy, for it was put there by the solar panels of long-dead plants in the Carboniferous age and other past times. Or the energy may be released in a more controlled way than actual combustion. Inside living cells, either of plants or of animals that eat plants, or of animals that eat animals that eat plants (etc.), sun-made carbon compounds are 'slow-burned'. Instead of literally bursting into flames, they give up their energy in a serviceable trickle, where it works in a controlled manner to drive 'uphill' chemical reactions. Inevitably, some of this energy is wasted as heat – if it were not, we'd have a perpetual motion machine, which is (you can't say it too often) impossible.

Almost all the energy in the universe is steadily being degraded from forms that are capable of doing work to forms that are incapable of doing work. There is a levelling off, a mixing up, until eventually the entire universe will settle into a uniform, (literally) uneventful 'heat death'. But while the universe as a whole is hurtling downhill towards its inevitable heat death, there is scope for small quantities of energy to drive little local systems in the opposite direction. Water from the sea is lifted into the air as clouds, which later deposit their water on mountaintops, from which it runs downhill in streams and rivers, which can drive water wheels or electric power stations. The energy to lift the water (and hence to drive the turbines in the power stations) comes from the sun. This is not a violation of the Second Law, for energy is constantly being fed in from the sun. The sun's energy is doing something similar in green leaves, driving chemical reactions locally 'uphill' to make sugar and starch and cellulose and plant tissues. Eventually the plants die, or they may be eaten by animals first. The trapped solar energy has the opportunity to trickle down through numerous cascades, and through a long and complex food chain culminating in bacterial or fungal decay of the plants, or of the animals that prolong the food chain. Or some of it may be sequestered underground, first as peat and then as coal. But the universal trend towards ultimate heat death is never reversed. In every link of the food chain, and through every trickle-down cascade within every cell, some of the energy is degraded to uselessness. Perpetual motion machines are ... all right, that's enough repetition, but I *won't* apologize for quoting, as I have done in at least one previous book, the marvellous saying of Sir Arthur Eddington on the subject:

> If someone points out to you that your pet theory of the universe is in disagreement with Maxwell's equations – then so much the worse for Maxwell's equations. If it is found to be contradicted by observation – well, these experimentalists do bungle things sometimes. But if your theory is found to be against the second law of thermodynamics I can give you no hope; there is nothing for it but to collapse in deepest humiliation.

When creationists say, as they frequently do, that the theory of evolution contradicts the Second Law of Thermodynamics, they are telling us no more than that they don't understand the Second Law (we already knew that they don't understand evolution). There is no contradiction, because of the sun!

The whole system, whether we are talking about life, or about water rising into the clouds and falling again, is finally dependent on the steady flow of energy from the sun. While never actually disobeying the laws of physics and chemistry – and certainly never disobeying the Second Law – energy from the sun powers life, to coax and stretch the laws of physics and chemistry to evolve prodigious feats of complexity, diversity, beauty, and an uncanny illusion of statistical improbability and deliberate design. So compelling is that illusion that it fooled our greatest minds for centuries, until Charles Darwin burst on to the scene. Natural selection is an improbability pump: a process that generates the statistically improbable. It systematically seizes the minority of random changes that have what it takes to survive, and accumulates them, step by tiny step over unimaginable timescales, until evolution eventually climbs mountains of improbability and diversity, peaks whose height and range seem to know no limit, the metaphorical mountain that I have called 'Mount Improbable'. The improbability pump of natural selection, driving living complexity up 'Mount Improbable', is a kind of statistical equivalent of the sun's energy raising water to the top of a conventional mountain.* Life evolves greater complexity only because natural selection drives it locally away from the statistically probable towards the improbable. And this is possible only because of the ceaseless supply of energy from the sun.

'FROM SO SIMPLE A BEGINNING'

We know a great deal about how evolution has worked ever since it got started, much more than Darwin knew. But we know little more than Darwin did about how it got started in the first place. This is a book about evidence, and we have no evidence bearing upon the momentous event that was the start of evolution on this planet. It could have been an event of supreme rarity. It only had to happen once, and as far as we know it did happen only once. It is even possible that it happened only once in the entire universe, although I doubt that. One thing we can say, on a basis of pure logic rather than evidence, is that Darwin was sensible to say 'from so simple a beginning'. The opposite

---

* It is no accident that Claude Shannon, when developing his metric of 'information', which is itself a measure of statistical improbability, lit upon exactly the same mathematical formula that Ludwig Boltzman had developed for entropy in the previous century.

of simple is statistically improbable. Statistically improbable things don't spontaneously spring into existence: that is what statistically improbable *means*. The beginning had to be simple, and evolution by natural selection is still the only process we know whereby simple beginnings can give rise to complex results.

Darwin didn't discuss how evolution began in *On the Origin of Species*. He thought the problem was beyond the science of his day. In the letter to Hooker that I quoted earlier, Darwin went on to say, 'It is mere rubbish, thinking at present of the origin of life; one might as well think of the origin of matter.' He didn't rule out the possibility that the problem would eventually be solved (indeed, the problem of the origin of matter largely has been solved) but only in the distant future. 'It will be some time before we see "slime, protoplasm, etc" generating a new animal.'

At this point in his edition of his father's letters, Francis Darwin inserted a footnote telling us,

> On the same subject my father wrote in 1871: 'It is often said that all the conditions for the first production of a living organism are now present, which could ever have been present. But if (and oh! what a big if!) we could conceive in some warm little pond, with all sorts of ammonia and phosphoric salts, light, heat, electricity, etc, present, that a proteine compound was chemically formed ready to undergo still more complex changes, at the present day such matter would be instantly devoured or absorbed, which would not have been the case before living creatures were formed.'

Charles Darwin was here doing two rather distinct things. On the one hand he was presenting his only speculation on how life might have originated (the famous 'warm little pond' passage). On the other hand, he was disabusing present-day science of the hope of ever seeing the event replicated before our eyes. Even if 'the conditions for the first production of a living organism' are still present, any such new production would be 'instantly devoured or absorbed' (presumably by bacteria, we would today have good reason to add), 'which would not have been the case before living creatures were formed'.

Darwin wrote this seven years after Louis Pasteur had said, in a lecture at the Sorbonne, 'Never will the doctrine of spontaneous generation recover from the mortal blow struck by this simple experiment.' The simple experiment was the one in which Pasteur showed, contrary to popular expectation at the time, that broth sealed off from access by micro-organisms would not spoil.

Demonstrations such as Pasteur's are sometimes cited by creationists as evidence in their favour. The false syllogism runs as follows: 'Spontaneous generation is never nowadays observed. Therefore the origin of life is impossible.' Darwin's 1871 remark was precisely designed as a riposte to that kind of illogicality. Evidently, the spontaneous generation of life is a very rare event, but it must have happened once, and this is true whether you think the original spontaneous generation was a natural or a supernatural event. The question of just how rare an event the origin of life was is an interesting one to which I shall return.

The first serious attempts to think about how life might have originated, those of Oparin in Russia and (independently) Haldane in England, both began by denying that the conditions for the first production of life are still with us. Oparin and Haldane suggested that the early atmosphere would have been very different from the present one. In particular, there would have been no free oxygen, and the atmosphere was thus – as chemists mysteriously call it – a 'reducing' atmosphere. We now know that all the free oxygen in the atmosphere is the product of life, specifically plants – obviously, not a part of the antecedent conditions in which life arose. Oxygen flooded into the atmosphere as a pollutant, even a poison, until natural selection shaped living things to thrive on the stuff and, indeed, suffocate without it. The 'reducing' atmosphere inspired the most famous experimental attack on the problem of the origin of life, Stanley Miller's flask full of simple ingredients, which bubbled and sparked for only a week before yielding amino acids and other harbingers of life.

Darwin's 'warm little pond', together with the witch's brew concocted by Miller that it inspired, are nowadays often rejected as a preamble to advancing some favoured alternative. The truth is that there is no overwhelming consensus. Several promising ideas have been suggested, but there is no decisive evidence pointing unmistakably to any one. In previous books I have attended to various interesting possibilities, including the inorganic clay crystals theory of Graham Cairns-Smith, and the more recently fashionable view that the conditions under which life first arose were akin to the Hadean habitat of today's 'thermophilous' bacteria and archaea, some of which thrive and reproduce in hot springs that are literally boiling. Today, a majority of biologists are moving towards the 'RNA World theory', and for a reason that I find quite persuasive.

We have no evidence about what the first step in making life was, but we do know the *kind* of step it must have been. It must have been whatever it took to get natural selection started. Before that first step, the sorts of improvement that only natural selection can achieve were impossible. And that means the key step was the arising, by some process as yet unknown, of a self-replicating entity. Self-replication spawns a population of entities, which compete with each other to be replicated. Since no copying process is perfect, the population will inevitably come to contain variety, and if variants exist in a population of replicators those that have what it takes to succeed will come to predominate. This is natural selection, and it could not start until the first self-replicating entity came into existence.

Darwin, in his 'warm little pond' paragraph, speculated that the key event in the origin of life might have been the spontaneous arising of a protein, but this turns out to be less promising than most of Darwin's ideas. This isn't to deny that proteins are vitally important for life. We saw in Chapter 8 that they have the very special property of coiling themselves up to form three-dimensional objects, whose exact shape is specified by the one-dimensional sequence of their constituents, the amino acids. We also saw that the same exact shape confers on them the ability to catalyse chemical reactions with great specificity, speeding particular reactions up perhaps a trillionfold. The specificity of enzymes makes biological chemistry possible, and proteins seem almost indefinitely flexible in the range of shapes that they can assume. That, then, is what proteins are good at. They are very, very good at it indeed, and Darwin was quite right to mention them. But there is something that proteins are outstandingly bad at, and this Darwin overlooked. They are completely hopeless at replication. They can't make copies of themselves. This means that the key step in the origin of life cannot have been the spontaneous arising of a protein. What, then, was it?

The best-replicating molecule that we know is DNA. In the advanced forms of life with which we are familiar, DNA and proteins are elegantly complementary. Protein molecules are brilliant enzymes but lousy replicators. DNA is exactly the reverse. It doesn't coil up into three-dimensional shapes, and therefore doesn't work as an enzyme. Instead of coiling up it retains its open, linear form, and this is what makes it ideal both as a replicator and as a specifier of amino-acid sequences. Protein molecules, precisely because they coil up into 'closed' shapes, do not 'expose' their sequence information in a way that might be copied or 'read'. The sequence information is buried inaccessibly inside the coiled-up protein. But in the long chain of DNA the sequence information is exposed and available to act as a template.

The 'Catch-22' of the origin of life is this. DNA can replicate, but it needs enzymes in order to catalyse the process. Proteins can catalyse DNA formation, but they need DNA to specify the correct sequence of amino acids. How could the molecules of the early Earth break out of this bind and allow natural selection to get started? Enter RNA.

RNA belongs to the same family of chain molecules as DNA, the polynucleotides. It is capable of carrying what amount to the same four code 'letters' as DNA, and it indeed does so in living cells, carrying genetic information from DNA to where it can be used. DNA acts as the template for RNA code sequences to build up. And then protein sequences build up using RNA, not DNA, as their template. Some viruses have no DNA at all. RNA is their genetic molecule, solely responsible for carrying genetic information from generation to generation.

Now for the key point of the 'RNA World theory' of the origin of life. In addition to stretching out in a form suitable for passing on sequence information, RNA is also capable of self-assembling, like our magnetic necklace of Chapter 8, into three-dimensional shapes, which have enzymatic activity. RNA enzymes do exist. They are not as efficient as protein enzymes, but they do work. The RNA World theory suggests that RNA was a good enough enzyme to hold the fort until proteins evolved to take over the enzyme role, and that RNA was also a good enough replicator to muddle along in that role until DNA evolved.

I find the RNA World theory plausible, and I think it quite likely that chemists will, within the next few decades, simulate in the laboratory a full reconstruction of the events that launched natural selection on its momentous way four billion years ago. Fascinating steps in the right direction have already been taken.

Before leaving the subject, however, I must repeat the warning I have given in earlier books. We don't actually need a plausible theory of the origin of life, and we might even be a little bit anxious if a too plausible theory were to be discovered! This glaring paradox arises from the famous 'Where is everybody?' question, which was posed by the physicist Enrico Fermi. Enigmatic as his question sounds, Fermi's companions, fellow physicists at the Los Alamos Laboratory, were attuned enough to know exactly what he meant. Why haven't we been visited by living creatures from elsewhere in the universe? If not visited in person, at least visited by radio signals (which is vastly more probable).

It is now possible to estimate that there are upwards of a billion planets in our galaxy, and about a billion galaxies. This means that, although it is possible that ours is the only planet in the galaxy that has life, in order for that to be true, the probability of life arising on a planet would have to be not much greater than one in a billion. The theory that we seek, of the origin of life on this planet, should therefore positively *not* be a plausible theory! If it were, life should be common in the galaxy. Maybe it is common, in which case a plausible theory is what we want. But we have no evidence that life exists outside this planet, and at very least we are entitled to be satisfied with an implausible theory. If we take the Fermi question seriously, and interpret the lack of visitations as evidence that life is exceedingly rare in the galaxy, we should move towards positively expecting that no plausible theory of the origin of life exists. I have developed the argument more fully in *The Blind Watchmaker*, and shall leave it there. My guess, for what it is worth (not much, because there are too many unknowns), is that life is very rare, but that the number of planets is so large (more are being discovered all the time) that we are probably not alone, and there may be millions of islands of life in the universe. Nevertheless, even millions of islands could still be so far apart that they have almost no chance of ever encountering one another, even by radio. Sadly, as far as practicalities are concerned, we might as well be alone.

'ENDLESS FORMS MOST BEAUTIFUL AND MOST WONDERFUL HAVE BEEN, AND ARE BEING, EVOLVED'

I'm not sure what Darwin meant by 'endless'. It could have been just a superlative, deployed to soup up 'most beautiful' and 'most wonderful'. I expect that was part of it. But I like to think that Darwin meant something more particular by 'endless'. As we look back on the history of life, we see a picture of never-ending, ever-rejuvenating novelty. Individuals die; species, families, orders and even classes go extinct. But the evolutionary process itself seems to pick itself up and resume its recurrent flowering, with undiminished freshness, with unabated youthfulness, as epoch gives way to epoch.

Dr Clare D'Alberto's interest in the diversity of life is more than skin-deep.

(a) When a tree-climbing animal needs a fifth limb it doesn't grow it afresh but presses into service what is already there. Spider monkey of the South American forests. (b) The colugo or 'flying lemur' of the south-east Asian forests is not a lemur at all, but occupies its own unique corner of the mammal tree. It doesn't really fly but glides from tree to tree. Unlike 'flying squirrels' (rodents) and 'flying phalangers' (marsupials), colugos incorporate the tail into the gliding membrane. (c) Egyptian fruit bat, whose translucent wings beautifully display the skeleton's homology with our hand.

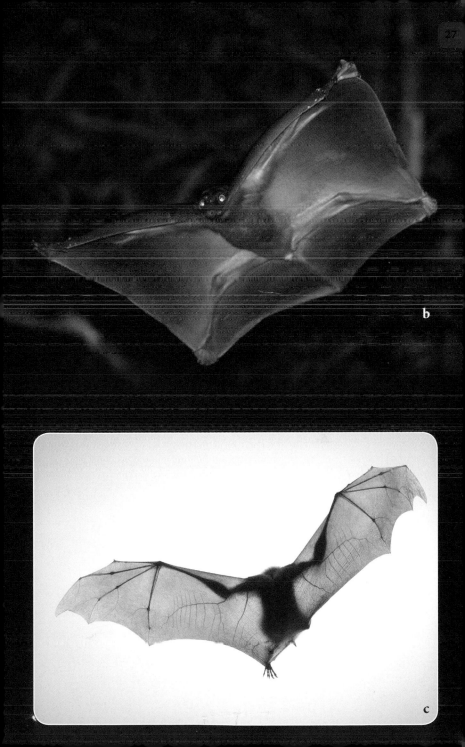

b

c

The stubby wings of these flightless birds
unequivocally betray their descent from flying
ancestors. Ostriches (**a**) still use their wings,
but only for balance and social purposes.
The Galapagos flightless cormorant (**b**) still
hangs its useless wings out to dry, like its
more familiar flying cousins. It is an expert
underwater fisher (**c**) but, unlike penguins, it
doesn't use its wings to swim but propels itself
along with vigorous thrusts of its large, webbed
feet. (**d**) 'Sadly,' according to Douglas Adams,
'not only has the kakapo forgotten how to fly,
but it has also forgotten that it has forgotten
how to fly. Apparently a seriously worried
kakapo will sometimes run up a tree and jump
out of it, whereupon it flies like a brick and
lands in a graceless heap on the ground.'

c

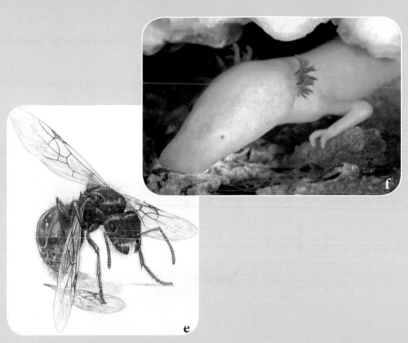

(f)

e

(e) Wings are a burden underground, which is probably why worker ants don't grow them. Poignant testimony to this is provided by queen ants, who use their wings once only, to fly out of the natal nest, find a mate, and then settle down to dig a hole for a new nest. As they begin their new life underground the first thing they do is lose their wings, in some cases by literally biting them off. (f) Cave-dwelling animals, like this salamander, are very often white. But, even though they don't use them in the dark caves, why do they 'bother' to reduce their eyes? See text p. 351. Mammalian dolphins (g) superficially resemble large, fast-swimming fish, like the 'dolphin fish' or dorado, because they make their living in a similar way.

g

Products of evolutionary arms races.
(**a**) 'Thow mortherer of the heysugge on y braunche.'
Baby cuckoo instinctively murders a foster-sibling
before it can hatch and compete for food. (**b**) This
kudu has lost its individual race with a lioness
and its life will soon end, but the arms race
between the gene pools of the two species is
run in evolutionary time. (**c**) A parasitoid
wasp laid its eggs in this caterpillar, and the
wasp larvae are here bursting out, full of
life, to contribute their genes to the next
generation. (**d**) In the forest economy, light
is a precious commodity. There is little
to be had below the canopy, because the
canopy itself is parcelled out between
individual trees, with almost no gaps.

It is no accident that we see green almost wherever we look . . . Without green plants to outnumber us at least ten to one there would be no energy to power us.

Let me briefly return to the computer models of artificial selection that I described in Chapter 2: the 'safari park' of computer biomorphs, including arthromorphs and the conchomorphs that showed how the great variety of mollusc shells might have evolved. In that chapter, I introduced these computer creatures as an illustration of how artificial selection works and how powerful it is, given enough generations. Now I want to use these computer models for a different purpose.

My overwhelming impression, while staring into the computer screen and breeding biomorphs, whether coloured or black, and when breeding arthromorphs, was that it never became boring. There was a sense of endlessly renewed strangeness. The program never seemed to get 'tired', and nor did the player. This was in contrast to the 'D'Arcy' program that I briefly described in Chapter 10, the one in which the 'genes' tugged mathematically at the coordinates of a virtual rubber sheet on which an animal had been drawn. When doing artificial selection with the D'Arcy program, the player seems, as time goes by, to move further and further away from a reference point where things made sense, out into a no-man's-land of mis-shapen inelegance, where sense seems to decrease the further we travel from the starting point. I have already hinted at the reason for this. In the biomorph, arthromorph and conchomorph programs, we have the computer equivalent of an embryological process – three different embryological processes, all in their different ways biologically plausible. The D'Arcy program, by contrast, doesn't simulate embryology at all. Instead, as I explained in Chapter 10, it manipulates the distortions by which one adult form may be transformed into another adult form. This lack of an embryology deprives it of the 'inventive fertility' that the biomorphs, arthromorphs and conchomorphs display. And the same inventive fertility is displayed by real-life embryologies, which is a minimal reason why evolution generates 'endless forms most beautiful and most wonderful'. But can we go beyond the minimal?

In 1989 I wrote a paper called 'The evolution of evolvability' in which I suggested that not only do animals get better at surviving, as the generations go by: lineages of animals get better at *evolving*. What does it mean to be 'good at evolving'? What kinds of animals are good at evolving? Insects on land and crustaceans in the sea seem to be champions at diversifying into thousands of species, parcelling up the niches, changing costumes through evolutionary time with frolicsome abandon. Fish, too, show amazing evolutionary fecundity, so do frogs, as well as the more familiar mammals and birds.

What I suggested in my 1989 paper is that evolvability is a property of embryologies. Genes mutate to change an animal's body, but they have to work through processes of embryonic growth. And some embryologies are better than others at throwing up fruitful ranges of genetic variation for natural selection to work upon, and might therefore be better at evolving. 'Might' seems too weak. Isn't it almost obvious that some embryologies *must* be better than others at evolving, in this sense? I think so. It is less obvious, but nevertheless I think a case can be made, that there might be a kind of higher-level natural selection in favour of 'evolvable embryologies'. As time goes by, embryologies improve their evolvability. If there is a 'higher-level selection' of this kind, it would be rather different from ordinary natural selection, which chooses individuals for their capacity to pass on genes successfully (or, equivalently, chooses genes for their capacity to build successful individuals). The higher-level selection that improves evolvability would be of the kind that the great American evolutionary biologist George C. Williams called 'clade selection'. A clade is a branch of the tree of life, like a species, a genus, an order or a class. We could say that clade selection has occurred when a clade, such as the insects, spreads, diversifies and populates the world more successfully than another clade such as the pogonophora (no, you probably haven't heard of these obscure, worm-like creatures, and there's a reason: they are an unsuccessful clade!). Clade selection doesn't imply that clades have to compete with each other. The insects don't compete, at least not directly, with the pogonophora for food or space or any other resource. But the world is full of insects, and almost devoid of pogonophora, and we are rightly tempted to attribute the success of the insects to some feature that they possess. I am conjecturing that it is something about their embryology that makes them evolvable. In the chapter of *Climbing Mount Improbable* entitled 'Kaleidoscopic Embryos' I offered various suggestions for specific features that make for evolvability, including constraints of *symmetry*, and including modular architectures such as a *segmented* body plan.

Perhaps partly because of its segmentally modular architecture, the arthropod clade* is good at evolving, at throwing up variation in multiple directions, at diversifying, at opportunistically filling niches as they become available. Other clades may be similarly successful because their embryologies are constrained to mirror-image development in various planes.† The clades that we see peopling the lands and the seas are the clades that are good at evolving. In clade selection, unsuccessful clades go extinct, or fail to diversify to meet varying challenges: they wither and perish. Successful clades blossom and flourish as leaves on the phylogenetic tree. Clade selection sounds seductively like Darwinian natural selection. The seduction should be resisted, or should at least ring alarm bells. Superficial resemblances can be actively misleading.

The fact of our own existence is almost too surprising to bear. So is the fact that we are surrounded by a rich ecosystem of animals that more or less closely resemble us, by plants that resemble us a little less and on which we ultimately depend for our nourishment, and by bacteria that resemble our remoter ancestors and to which we shall all return in decay when our time is past. Darwin was way ahead of his time in understanding the magnitude of the problem of our existence, as well as in tumbling to its solution. He was ahead

---

* Insects, crustaceans, spiders, centipedes, etc.

† For example, a mutation in the leg of a millipede will be mirrored on both sides, and probably repeated the length of the body as well. Although it is a single mutation, embryological processes constrain it to be repeated many times left and right. It may at first seem paradoxical that a constraint should increase the evolutionary versatility of a clade. The reason is spelled out in the same chapter of *Climbing Mount Improbable*, 'Kaleidoscopic Embryos'.

of his time, too, in appreciating the mutual dependencies of animals and plants and all other creatures, in relationships whose intricacy staggers the imagination. How is it that we find ourselves not merely existing but surrounded by such complexity, such elegance, such endless forms most beautiful and most wonderful?

The answer is this. It could not have been otherwise, given that we are capable of noticing our existence at all, and of asking questions about it. It is no accident, as cosmologists point out to us, that we see stars in our sky. There may be universes without stars in them, universes whose physical laws and constants leave the primordial hydrogen evenly spread and not concentrated into stars. But nobody is observing those universes, because entities capable of observing anything cannot evolve without stars. Not only does life need at least one star to provide energy. Stars are also the furnaces in which the majority of the chemical elements are forged, and you can't have life without a rich chemistry. We could go through the laws of physics, one by one, and say the same thing of all of them: it is no accident that we see . . .

The same is true of biology. It is no accident that we see green almost wherever we look. It is no accident that we find ourselves perched on one tiny twig in the midst of a blossoming and flourishing tree of life; no accident that we are surrounded by millions of other species, eating, growing, rotting, swimming, walking, flying, burrowing, stalking, chasing, fleeing, outpacing, outwitting. Without green plants to outnumber us at least ten to one there would be no energy to power us. Without the ever-escalating arms races between predators and prey, parasites and hosts, without Darwin's 'war of nature', without his 'famine and death' there would be no nervous systems capable of seeing anything at all, let alone of appreciating and understanding it. We are surrounded by endless forms, most beautiful and most wonderful, and it is no accident, but the direct consequence of evolution by non-random natural selection – the only game in town, the greatest show on Earth.

THE
HISTORY-
DENIERS

A т irregular but frequent intervals since 1982, Gallup, America's best-known polling organization, has been sampling the national opinion on this question.

Which of the following statements comes closest to your views on the origin and development of human beings?

1   Human beings have developed over millions of years from less advanced forms of life, but God guided this process. (36%)

2   Human beings have developed over millions of years from less advanced forms of life, but God had no part in this process. (14%)

3   God created human beings pretty much in their present form at one time within the last 10,000 years or so. (44%)

The percentages I have inserted are from 2008. The figures for 1982, 1993, 1997, 1999, 2001, 2004, 2006 and 2007 are pretty much the same.

I am in what I am not surprised to see is a minority of 14% ticking the box for proposition 2. It is unfortunate that the wording of proposition 2, 'but God had no part in this process', seems calculated to bias religious people gratuitously against it. The real killer is the lamentably strong support for proposition 3. Forty-four per cent of

Americans deny evolution totally, whether it is guided by God or not, and the implication is that they believe the entire world is no more than 10,000 years old. As I have pointed out before, given that the true age of the world is 4.6 billion years, this is equivalent to believing that the width of North America is less than 10 yards. In none of the nine years sampled did the support for proposition 3 drop below 40%. In two of the sampling years, it hit 47%. More than 40% of Americans deny that humans evolved from other animals, and think that we – and by implication all of life – were created by God within the last 10,000 years. This book is necessary.

The questions posed by Gallup focused on human beings, and that could, it might be said, have upped the emotional ante and made it harder to accept the scientific view. In 2008 the Pew Forum published a similar poll of Americans which didn't specifically mention humans. The results were fully compatible with Gallup. The propositions on offer were as follows, with the percentages assenting to them:

Life on Earth has ...
Existed in its present form since the beginning of time 42%
Evolved over time 48%
    Evolution through natural selection 26%
    Evolution guided by supreme being 18%
    Evolved but don't know how 4%
Don't know 10%

The Pew questions didn't mention dates, so we don't know how many of the 42% who positively reject evolution also think the world is less than 10,000 years old, as Gallup's 44% presumably do. It seems likely that Pew's 42%, too, would go along with thousands of years rather than the scientists' date of about 4.6 billion years. To believe that life on Earth has existed in its present form for 4.6 billion years without any change at all would seem at least as absurd as to believe that it has existed in its present form for a few thousand years, and it is certainly unbiblical.

What about Britain? How do we compare? In 2006 the BBC's (comparatively) up-market science documentary series *Horizon** commissioned an Ipsos MORI poll among British people. Unfortunately, the key question was not well formulated. People were asked to choose one of the following three 'theories or explanations about the origin and development of life on earth'. After each option, I've put the percentage choosing it.

(a) The 'evolution theory' says that human kind has developed over millions of years from less advanced forms of life. God had no part in this process. (48%)

(b) The 'creationism theory' says that God created human kind pretty much in his/her present form at one time within the last 10,000 years. (22%)

(c) The 'intelligent design theory' says that certain features of living things are best explained by the intervention of a supernatural being, e.g. God. (17%)

(d) Don't know. (12%)

Regrettably, these choices could have left some people without their preferred option. They leave no room for '(a) but God played a part in this process'. Given the inclusion of the phrase, 'God had no part in this process', it is not surprising that the figure for (a) is as low as 48%. Option (b)'s tally of 22%, however, is alarmingly high, especially given the ludicrous age limit of 10,000 years. And, if we add (b) and (c) together to give the percentage who favour some form of creationism, we get 39%. This is still not as high as the American figure of more than 40%, especially bearing in mind that the American figure refers

---

* Similar to the US *Nova*, which often releases programs originally aired on *Horizon,* or enters into co-production arrangements with *Horizon.*

to young Earth creationists, whereas the British 39% presumably includes old Earth creationists, under (c).

The MORI poll posed a second question to the British sample, about education. Given the same three theories, people were asked whether they should or should not be taught in science classes. Disquietingly, only 69% positively thought that evolution should be taught in science classes at all – whether or not alongside some form of creation or intelligent design theory.

A more ambitious survey, which included Britain but not America, was conducted by Eurobarometer in 2005. This poll sampled opinions and beliefs about scientific matters in thirty-two European countries (including Turkey, which is the only substantially Islamic country to aspire to membership of the European Union). Table 1 shows the percentages in various countries assenting to the proposition that 'Human beings, as we know them today, developed from earlier species of animals.' Note that this is a more modest statement than (a) in the MORI poll, since it does not exclude the possibility that God played some part in the evolutionary process. I have ranked the countries by percentage assenting to the proposition, that is, the percentage giving the correct answer as judged by modern science. Thus 85% of the Icelandic sample think, as scientists do, that humans have evolved from other species. A paltry 27% of the Turkish population do. Turkey is the only country in the table where there appears to be a majority who think evolution is actually false. Britain is ranked fifth, with 13% actively denying evolution. The United States was not sampled in the European survey, but the deplorable fact that it comes out only just ahead of Turkey in such matters has been given much publicity of late.

Stranger are the results set out in Table 2, which shows the equivalent percentages for the proposition that 'The earliest humans lived at the same time as the dinosaurs.' Once again, I have ranked the countries by the percentage giving the correct answer,

TABLE 1: Responses to the proposition that 'human beings, as we know them today, developed from earlier species of animals'

Country	Total	True (%)	False (%)	Don't know (%)
Iceland	500	85	7	8
Denmark	1,013	83	13	4
Sweden	1,023	82	13	5
France	1,021	80	12	8
Britain	1,307	79	13	8
Belgium	1,024	74	21	5
Norway	976	74	18	8
Spain	1,036	73	16	11
Germany	1,507	69	23	8
Italy	1,006	69	20	11
Luxembourg	518	68	23	10
Netherlands	1,005	68	23	9
Ireland	1,008	67	21	12
Hungary	1,000	67	21	12
Slovenia	1,060	67	25	8
Finland	1,006	66	27	7
Czech Republic	1,037	66	27	7
Portugal	1,009	64	21	15
Estonia	1,000	64	19	17
Malta	500	63	25	13
Switzerland	1,000	62	28	10
Slovakia	1,241	60	29	12
Poland	999	59	27	14
Croatia	1,000	58	28	15
Austria	1,034	57	28	15
Greece	1,000	55	32	14
Romania	1,005	55	25	20
Bulgaria	1,008	50	21	29
Latvia	1,034	49	27	24
Lithuania	1,003	49	30	21
Cyprus	504	46	36	18
Turkey	1,005	27	51	22

Source: Eurobarometer, 2005.

which in this case is 'false'.* Again Turkey comes out bottom, with a full 42% believing that the earliest humans co-existed with dinosaurs, and only 30% prepared to deny it, compared with 87% of Swedes. Britain, I am sorry to say, is in the bottom half, with 28% of us apparently getting our scientific and historical knowledge from the Flintstones rather than from any educational source.

As a biological educator, I find myself pathetically consoled by another result from the Eurobarometer survey revealing the large number (19% in Britain) who believe it takes one month for the Earth to go around the sun. The figure is more than 20% for Ireland, Austria, Spain and Denmark. What, I wonder, do they think a year is? Why do the seasons come and go with such regularity? Are they not even *curious* about the reasons for such a salient feature of their world? These remarkable figures shouldn't really be consoling, of course. My emphasis was on 'pathetically'. I meant that we seem to be dealing with a general ignorance of science – which is bad enough, but at least it is better than the positive prejudice *against* one particular science, namely evolutionary science, which seems to be present in Turkey (and, one can't help guessing, in much of the Islamic world). Also, undeniably, in the United States of America, as we saw in the Gallup and Pew polls.

In October 2008 a group of about sixty American high-school teachers met at the Center for Science Education of Emory University, in Atlanta. Some of the horror stories they had to tell deserve wide attention. One teacher reported that students 'burst into tears' when

---

* I suppose, if I am to be pedantic, I have to acknowledge that modern zoologists classify birds as surviving dinosaurs. Strictly speaking, therefore, the correct answer is 'true', and the Turkish majority are right. I think we may safely assume, however, that when people are asked a question like that, they take 'the dinosaurs' to exclude birds and include only the extinct 'terrible lizards' that gave us the word.

TABLE 2: Responses to the proposition that 'the earliest humans lived at the same time as the dinosaurs'

Country	Total	True (%)	False (%)	Don't know (%)
Sweden	1,023	9	87	4
Germany	1,507	11	80	9
Denmark	1,013	14	79	6
Switzerland	1,000	9	79	12
Norway	976	13	79	7
Czech Republic	1,037	15	78	7
Luxembourg	518	15	77	9
Netherlands	1,005	14	75	10
Finland	1,006	21	73	7
Iceland	500	12	72	16
Slovenia	1,060	20	71	9
Belgium	1,024	24	70	6
France	1,021	21	70	9
Austria	1,034	15	69	15
Hungary	1,000	18	69	13
Estonia	1,000	20	66	14
Slovakia	1,241	18	65	18
Britain	1,307	28	64	8
Croatia	1,000	23	60	17
Lithuania	1,003	23	58	19
Spain	1,036	29	56	15
Ireland	1,008	27	56	17
Italy	1,006	32	55	13
Portugal	1,009	27	53	21
Poland	999	33	53	14
Latvia	1,034	27	51	21
Greece	1,000	29	50	21
Malta	500	29	48	24
Bulgaria	1,008	17	45	39
Romania	1,005	21	42	37
Cyprus	504	32	40	28
Turkey	1,005	42	30	28

Source: Eurobarometer, 2005.

told they would be studying evolution. Another teacher described how students repeatedly screamed 'No!' when he began talking about evolution in class. Another reported that pupils demanded to know why they had to learn about evolution, given that it was 'only a theory'. Yet another teacher described how 'churches train students to come to school with specific questions to ask to sabotage my lessons'. The Creation Museum in Kentucky is a lavishly financed institution entirely devoted to history-denial on this advanced scale. Children can ride on a model dinosaur with a saddle – and it's not just a bit of fun: the message is explicit and unequivocal that dinosaurs lived recently and coexisted with humans. It is run by Answers in Genesis, which is a tax-exempt organization. The taxpayer, in this case the American taxpayer, is subsidizing scientific falsehood, miseducation on a grand scale.

Such experiences are common throughout the United States, but also, though I am loath to admit it, becoming so in Britain. In February 2006 the *Guardian* reported that 'Muslim medical students in London distributed leaflets that dismissed Darwin's theories as false. Evangelical Christian students are also increasingly vocal in challenging the notion of evolution.' The Muslim leaflets are produced by the Al-Nasr Trust, a registered charity with tax-free status.* So the British taxpayer, too, is subsidizing the systematic distribution of a major and serious scientific falsehood to British educational institutions.

In 2006 the *Independent* reported Professor Steve Jones, of University College London, as saying:

---

* Tax-free status is easily obtained by almost any religious organization. Organizations that are not religious have to jump through hoops to demonstrate that they benefit humanity. I recently established a charitable foundation dedicated to promoting 'Reason and Science'. During the protracted, extremely expensive, and ultimately successful negotiations to obtain charitable status, I received a letter from the British Charity Commission dated 28 September 2006 which contained the following: 'It is not clear how the advancement of science tends towards the mental and moral improvement of the public. Please provide us with evidence of this or explain how it is linked to the advancement of humanism and rationalism.' Religious organizations, by contrast, are assumed to benefit humanity without any obligation to demonstrate it and even, apparently, if they are actively engaged in promoting scientific falsehood.

It's a real social change. For years, I've sympathised with my American colleagues, who have to cleanse creationism from their students' minds in their first few biology lectures. It's not a problem we've faced in Britain until now. I get feedback from Muslim schoolkids who say they are obliged to believe in creationism, because it's part of their Islamic identity, but the people I find more surprising are the other British kids who see creationism as a viable alternative to evolution. That's alarming. It shows how infectious the idea is.

The polls, then, suggest that at least 40% of Americans are creationists – that's dyed-in-the-wool, out-and-out, anti-evolution creationists, not believers in 'evolution but God sort of helped it along' (there were plenty of them too). The equivalent figures for Britain, and much of Europe, are slightly less extreme, but not much more encouraging. There are still no grounds for complacency.

STOP PRESS
Extended footnote to p.325

Some (but not all) bats and some (but not all) whales find their way around by 'sonar' – using the echoes of special cries that they utter for the purpose. I always presumed that they'd converged on the same clever trick by completely different routes, but the truth is even more amazing! Mammals have a gene called Prestin, which is involved in hearing. Two independent groups of researchers did a Penny-style analysis using the protein product of Prestin, and found that echolocating whales and bats come out *closer to each other* than either do to their non-echolocating counterparts! To pursue my 'voting' analogy, Prestin bucks the trend of the rest of the genome by casting its 'vote' for an outlandish tree, in which echolocating bats and toothed whales appear to be closely 'related' to each other! Of course Prestin is massively outvoted by the rest of the genome, confirming that bats and whales are not really closely related. What happened is that, while bats and toothed whales were independently evolving sonar, natural selection in both lines favoured parallel mutations in the same Prestin gene. Revealingly, the convergence shows itself if you look at the amino acid sequences of the Prestins but not if you look at the DNA sequences that specify them. There's no contradiction here, because the genetic code allows amino acids to be specified in several different ways by DNA. This provides a marvellous way to separate convergent from ancestrally derived resemblances. Thus, different species of bats derive their sonar, not convergently but from an echolocating ancestor. And this real cousinship, unlike the apparent cousinship with the whales, receives the 'vote' of the Prestin DNA sequence (as well as that of the amino acid sequence). This remarkable story bids us ask whether there are other examples. Dare we hope that the pill woodlouse and the pill millipede (see p. 300) have achieved rolling up into a ball by independently modifying the same gene as each other?

NOTES

PREFACE

p. vii   **offered an unfamiliar vision:** *The Selfish Gene* (1976; 30th anniversary edn 2006) and *The Extended Phenotype* (rev. edn 1999).

p. vii   **My next three books:** *The Blind Watchmaker* (1986), *River Out of Eden* (1995) and *Climbing Mount Improbable* (1996).

p. vii   **My largest book:** *The Ancestor's Tale* (2004).

CHAPTER 1: ONLY A THEORY?

p. 5   **In 2004 we wrote a joint article in the** *Sunday Times*: 'Education: questionable foundations', *Sunday Times*, 20 June 2004.

p. 12   **'Occasionally, I get a letter from someone':** Sagan (1996).

p. 13   **'We may all have come into existence five minutes ago':** Bertrand Russell, *Religion and Science* (Oxford: Oxford University Press, 1997), 70.

p. 14   **A famous example was prepared by Professor Daniel J. Simons at the University of Illinois:** Simons and Chabris (1999).

p. 16   **In Texas alone, thirty-five condemned people have been exonerated since DNA evidence became admissible in court:** The Innocence Project, http://www.innocenceproject.org.

p. 16   **during his six years as Governor, George W. Bush signed a death warrant once a fortnight on average:** 152 in total; see 'Bush's lethal legacy: more executions', *Independent*, 15 Aug. 2007.

p. 17   **Darwin explained in his autobiography:** Darwin (1887a), 83.

p. 17   **under the influence, Matt Ridley suspects:** Matt Ridley, 'The natural order of things', *Spectator*, 7 Jan. 2009.

CHAPTER 2: DOGS, COWS AND CABBAGES

p. 30   **'My dear Wallace':** Marchant (1916), 169–70.

p. 31   **fn. There is a persistent, but false, rumour:** The matter of Darwin's alleged knowledge of Mendel's research is taken up in Sclater (2003).

p. 34   **mongrels or mutts (as President Obama delightfully described himself):** 'Puppies and economy fill winner's first day', *Guardian*, 8 Nov. 2008.

p. 35   **Other genetic routes produce miniature breeds that retain the proportions of the original:** Fred Lanting, 'Pituitary dwarfism in the German Shepherd dog', *Dog World*, Dec. 1984, reproduced at http://www.fredlanting.org/2008/07/pituitary-dwarfism-in-the-german-shepherd-dog-part-1/.

CHAPTER 3: THE PRIMROSE PATH TO MACRO-EVOLUTION

p. 50   **'I have carefully measured the proboscis of a specimen':** Wallace (1871).

p. 57   **'The resemblance of *Dorippe* to an angry Japanese warrior':** Julian Huxley, 'Evolution's copycats', *Life*, 30 June 1952; also in Huxley (1957) as 'Life's improbable likenesses'.

p. 58   **I even found a website where you can vote:** Samurai crab poll from http://www.pollsb.com/polls/view/13022/the-heike-crab-seems-to-have-a-samurai-face-on-its-back-what-s-the-explanation.

p. 58   **as one authoritative sceptic has pointed out:** Martin (1993).

p. 65   **'My dear Darwin':** Marchant (1916), 170.

p. 67   **agronomists at the Illinois Experimental Station began the experiment rather a long time ago:** Dudley and Lambert (1992).

p. 67   **seventeen generations of rats, artificially selected for resistance to tooth decay:** Ridley (2004), 48.

p. 74   **'eager to establish human contact':** Trut (1999), 163.

p. 77   **The so-called spider orchid:** Some websites about these include http://www.arhomeandgarden.org/plantoftheweek/articles/orchid_red_spider_8-29-08.htm, http://www.orchidflowerhq.com/Brassiacare.php, http://www.absoluteastronomy.com/topics/Brassia, http://en.wikipedia.org/wiki/Brassia.

p. 78   **it can be seen in the recording of the lecture called 'The Ultraviolet Garden':** Available on the DVD *Growing Up in the Universe* from richard-dawkins.net.

p. 79   **Each species mixes a characteristic cocktail of substances gathered from various sources:** Eltz et al. (2005).

p. 81   **I have discussed the cleaning habit elsewhere:** Dawkins (2006), 186–7.

CHAPTER 4: SILENCE AND SLOW TIME

p. 92   **fn. Alas, the popular legend:** The legend that the periodic table came to Mendeleev in a dream is discussed in G. W. Baylor, 'What do we really know about Mendeleev's dream of the periodic table? A note on dreams of scientific problem solving', *Dreaming* 11: 2 (2001), 89–92.

p. 97   **A fortunate aspect of the way igneous rocks solidify is that they do so suddenly:** A sophisticated refinement of the method, 'Isochron dating', is fully described by Chris Stassen on the excellent 'Talk.Origins' website: www.talkorigins.org/faqs/isochron-dating.html.

p. 100   **Here's a direct quotation from a prizewinning creationist website:** From http://homepage.ntlworld.com/malcolmbowden/creat.htm.

p. 105   **The strip was divided into three parts:** Shroud of Turin dating from Damon et al. (1989).

p. 107   **And I haven't even mentioned various other dating methods:** For a full list of methods, see http://www.usd.edu/esci/age/current_scientific_clocks.html#.

CHAPTER 5: BEFORE OUR VERY EYES

p. 112   **The graph above shows data from the Uganda Game Department:** Brooks and Buss (1962).

p. 113   **In that year experimenters transported five pairs of *Podarcis sicula* from Pod Kopiste:** Research on the lizards of Pod Mrcaru from Herrel et al. (2008) and Herrel et al. (2004).

p. 117   **All this has been achieved with the bacterium *Escherichia coli*:** Lenski *E. coli* research from Lenski and Travisano (1994). In addition, Lenski group publications are collected at http://myxo.css.msu.edu/cgi-bin/lenski/prefman.pl?group=aad.

p. 131   **the celebrated scientific blogwit PZ Myers:** http://scienceblogs.com/pharyngula/2008/06/lenski_gives_conservapdia_a_lc.php

p. 133   **The experiments that John Endler recounted:** Guppies research from Endler (1980, 1983, 1986).

p. 138   **Among those who have taken it up is David Reznick of the University of California at Riverside:** Reznick et al. (1997).

p. 140   **some zoologists dispute *Lingula*'s claim to be an almost wholly unchanged 'living fossil':** e.g. Christian C. Emig, 'Proof that *Lingula* (Brachiopoda) is not a living fossil, and emended diagnoses of the Family Lingulidae', *Carnets de Géologie*, letter 2003/01 (2003).

CHAPTER 6: MISSING LINK? WHAT DO YOU MEAN, 'MISSING'?

p. 143   **What do you mean, 'missing'?:** www.talkorigins.org/faqs/faq-transitional/part2c.html#arti, http://web.archive.org/web/19990203140657/gly.fsu.edu/tour/article_7.html.

p. 147   **'already in an advanced state of evolution':** Dawkins (1986), 229.

p. 151   **'If people came from monkeys via frogs and fish, then why does the fossil record not contain a "fronkey"?':** 'Darwin's evolutionary theory is a tottering nonsense, built on too many suppositions', *Sydney Morning Herald,* 7 May 2006.

pp. 151–2   **long list of comments that follow an article in the *Sunday Times*(London):**http://www.timesonline.co.uk/tol/news/uk/education/article 4118420.ece.

p. 153   **Here is one called *Eomaia*:** Ji et al. (2002).

p. 154   ***Atlas of Creation*:** Incredibly there are now no fewer than three volumes of this infamous waste of expensive, glossy paper.

p. 154   a fishing lure as a 'caddis fly': This can be seen clearly at http://
www.grahamowengallery.com/fishing/more-fly-tying.html.

p. 163   'We went straight to the Museum': Smith (1956), 41.

p. 168   *Tiktaalik!* A name never to be forgotten: http://www90.homepage.
villanova.edu/lowell.gustafson/anthropology/tiktaalik.html

p. 172   *Pezosiren*, the 'walking manatee' fossil: Domning (2001).

p. 172   exciting news came in: Natalia Rybczynski, Mary Dawson and Richard
Tedford, 'A semi-aquatic Arctic mammalian carnivore from the Miocene
epoch and origin of Pinnipedia', *Nature* 458 (2009), pp. 1021–4. You can see
a short film of Natalia Rybczynski enthusiastically discussing the new fossil
at http://nature.ca/pujila/ne_vid_e.cfm.

p. 174   *Odontochelys semitestacea*: Li et al. (2008).

p. 175   The 'News and Views' commentary on the *Odontochelys* paper: Reisz
and Head (2008).

p. 177   *Proganochelys*: Joyce and Gauthier (2004).

p. 179–80 In another book I described DNA as 'the Genetic Book of the Dead':
Dawkins (1998), ch. 10.

CHAPTER 7: MISSING PERSONS? MISSING NO LONGER

p. 185   '*Pithecanthropus* [Java Man] was not a man': Dubois (1935), also quoted
in http://www.talkorigins.org/pdf/fossil-hominids.pdf.

p. 185   The creationist organization Answers in Genesis has, however, added it
to their list of discredited arguments: http://www.answersingenesis.org/
home/area/faq/dont_use.asp.

p. 186   'Georgian Man': http://www.talkorigins.org/faqs/homs/d2700.html.

p. 187   We are not descended from chimpanzees: http://www.talkorigins.org/
faqs/homs/chimp.html.

p. 189   The type specimen is the first individual of ⌐ new species to be named:
There is a useful list of hominid type specimens at http://www.talkorigins.
org/faqs/homs/typespec.html.

p. 190   fn. Arguably one of Darwin's greatest successors of the twentieth
century: See Hamilton's collected papers interspersed with his own
idiosyncratic reminiscences in Hamilton (1996, 2001), the second volume
of which also contains my own funerary tribute to him.

p. 193   the following range of names: http://www.mos.org/evolution/fossils/
browse.php.

p. 198   'The morning-after pill is a pedophile's best friend': 'Morning-after pill
blocked by politics', *Atlanta Journal-Constitution*, 24 June 2004.

CHAPTER 8: YOU DID IT YOURSELF IN NINE MONTHS

p. 212   fn. 'All things dull and ugly': lyrics reproduced with thanks by permission
of Python (Monty) Pictures. Thanks to Terry Jones and Eric Idle.

p. 218   There are some stunning films available on YouTube: For example, http://
www.youtube.com/watch?v=XH-groCeKbE.

pp. 219–20 **Calling it 'Boids', Craig Reynolds wrote a program along these lines:** http://www.red3d.com/cwr/boids/.

p. 229 **it has been deciphered by a group of scientists associated with the brilliant mathematical biologist George Oster:** Odell et al. (1980).

p. 233 **An early classic experiment by the Nobel Prize-winning embryologist Roger Sperry:** Meyer (1998).

p. 243 **the complete family tree of all 558 cells of a newly hatched larva:** *C. elegans* cell family tree from http://www.wormatlas.org/userguides. html/lineage.htm. The entire wormatlas.org site is a treasure trove of information on these minute creatures. I also strongly recommend the three Nobel Prize speeches on *C. elegans* from Sydney Brenner, H. Robert Horvitz and John Sulston – Brenner (2003), Horvitz (2003), Sulston (2003) – which are also available to read or view at http://nobelprize.org/nobel_ prizes/medicine/laureates/2002/index.html.

CHAPTER 9: THE ARK OF THE CONTINENTS

p. 253 **There are tiny nematode worms:** http://www.baycrcropscience.co.uk/ pdfs/nematodesguide.pdf.

p. 257 **Dr Ellen Censky, who led the original study:** Censky et al. (1998).

p. 260 **'Seeing this gradation and diversity of structure':** Darwin (1845), 380.

p. 261 **'It is a hideous looking creature':** Darwin (1845), 385–6.

p. 262 **'Hence we have the truly wonderful fact':** Darwin (1845), 396.

p. 262 **'My attention was first thoroughly aroused':** Darwin (1845), 394–5.

p. 263 **'that the tortoises differed from the different islands':** Darwin (1845), 394.

p. 267 **'Almost every rocky outcrop and island has a unique Mbuna fauna':** Owen et al. (1989).

p. 271 **'With respect to the absence of whole orders on oceanic islands':** Darwin (1859), 393.

p. 272 **'the naturalist in travelling':** Darwin (1859), 349.

pp. 282–3 **How do they cope with it? Very weirdly indeed:** At least some of them are confused. Others may be dishonest. The young Earth account at http:// www.answersingenesis.org/articles/am/v2/n2/a-catastrophic- breakup is refuted in detail by an old Earth creationist at http://www. answersincreation.org/rebuttal/aig/Answers/2007/answers_v2_n2_tectonics.htm.

p. 283 **the most authoritative recent book on speciation:** Coyne and Orr (2004).

CHAPTER 10: THE TREE OF COUSINSHIP

p. 317 **It was by this method, using rabbits:** Sarich and Wilson (1967).

p. 322 **The earliest large-scale study along these lines was done by a group of geneticists in New Zealand:** Penny et al. (1982).

p. 325 **Footnote on bats and whales:** Liu et al. (2010). *Current Biology*, 20, R53–R54. Liu et al. 2010. *Current Biology*, 20, R55–R56.

p. 330 **It's well worth downloading the Hillis tree from his website:** www. zo.utexas.edu/faculty/antisense/DownloadfilesToL.html.

p. 335   Yan Wong and I discussed them fully in 'The Epilogue to the Velvet Worm's Tale': Dawkins (2004).

CHAPTER 11: HISTORY WRITTEN ALL OVER US

p. 340   'Mr Sutton, the intelligent keeper in the Zoological Gardens': Darwin (1872), 95, 96, 97.

p. 341   In an 1845 communication to the Royal Society: Sibson (1848).

p. 346–7   J. W. S. Pringle . . . was mainly responsible for working out how halteres work: Pringle (1948).

p. 353   'If an optician wanted to sell me an instrument which had all these defects': Helmholtz (1881), 194.

p. 355   'For the eye has every possible defect that can be found in an optical instrument': Helmholtz (1881), 201.

p. 362   'Despite possession of a well developed larynx and a gregarious nature, the Giraffe is able to utter only low moans or bleats': Harrison (1980).

p. 370   'I cannot persuade myself': Darwin (1887b).

p. 370   fn. Not to be confused with another Australian: M. Denton, *Nature's Destiny* (New York: Free Press, 2002).

p. 371   'patchwork of makeshifts': C. S. Pittendrigh, 'Adaptation, natural selection, and behavior', in A. Roe and G. G. Simpson, eds, *Behavior and Evolution* (New Haven: Yale University Press, 1958).

CHAPTER 12: ARMS RACES AND 'EVOLUTIONARY THEODICY'

p. 380   The five fastest runners among mammal species: List from http://www.petsdo.com/blog/top-twenty-20-fastest-land-animals-including-humans.

p. 382   My colleague John Krebs and I published a paper on the subject in 1979: Dawkins and Krebs (1979).

p. 382   'Before asserting that the deceptive appearance': Cott (1940), 158–9.

pp. 383–4   And there are even, though it may seem surprising, arms races between males and females within a species, and between parents and offspring: See Dawkins (2006), chs 8 and 9, 'Battle of the generations' and 'Battle of the sexes'.

p. 390   'What a book a devil's chaplain might write': Darwin (1903).

p. 390   '[N]ature is neither kind nor unkind': Dawkins (1995), ch. 4, 'God's utility function'.

p. 394   As a matter of interest, there are aberrant individuals who cannot feel pain: For examples, see http://news.bbc.co.uk/2/hi/health/4195437.stm, http://www.msnbc.msn.com/id/6379795/.

p. 395   Stephen Jay Gould reflected on such matters in a nice essay on 'Non-moral nature': Reproduced in Gould (1983).

CHAPTER 13: THERE IS GRANDEUR IN THIS VIEW OF LIFE

p. 399   'Thus, from the war of nature': Darwin (1859), 490.

p. 400   'it may not be a logical deduction': Darwin (1859), 243.

p. 401   'All that we can do': Darwin (1859), 78.

p. 404   'But I have long regretted': Darwin (1887c).

p. 410   ways in which, to quote one of their papers, the frozen accident might be 'thawed': Söll and RajBhandary (2006).

p. 410   And the physicist Paul Davies has made the reasonable point: Davies and Lineweaver (2005).

p. 412   but it is a matter for intriguing speculation how different life would be if we had no orbiting moon: Comins (1993).

p. 417   'On the same subject my father wrote in 1871': Darwin (1887c).

p. 423   In 1989 I wrote a paper called 'The evolution of evolvability': Dawkins (1989).

p. 426   It is no accident, as cosmologists point out to us, that we see stars in our sky: See e.g. Smolin (1997).

APPENDIX: THE HISTORY-DENIERS

p. 429   At irregular but frequent intervals since 1982: Gallup poll numbers taken from 'Evolution, creationism, intelligent design', http://www.gallup.com/poll/21814/Evolution-Creationism-Intelligent-Design.aspx.

p. 430   In 2008 the Pew Forum published a similar poll: Pew poll numbers taken from 'Public divided on origins of life', conducted 17 July 2005, http://pewforum.org/surveys/origins/.

p. 431   What about Britain? How do we compare? Ipsos MORI poll numbers taken from 'BBC survey on the origins of life', conducted 5–10 Jan. 2006, http://www.ipsos-mori.com/content/bbc-survey-on-the-origins-of-life.ashx.

p. 432   A more ambitious survey: Eurobarometer 224 survey numbers taken from 'Europeans, science and technology', conducted Jan.–Feb. 2005, http://ec.europa.eu/public_opinion/archives/ebs/ebs_224_report_en.pdf.

p. 432   the deplorable fact that it comes out only just ahead of Turkey in such matters has been given much publicity: Miller et al. (2006).

p. 434   Some of the horror stories they had to tell deserve wide attention: 'Emory workshop teaches teachers how to teach evolution', Atlanta Journal-Constitution, 24 Oct. 2008.

p. 436   'Muslim medical students in London distributed leaflets that dismissed Darwin's theories as false': 'Academics fight rise of creationism at universities', Guardian, 21 Feb. 2006.

p. 437   'It's a real social change': 'Creationism debate moves to Britain', Independent, 18 May 2006.

BIBLIOGRAPHY AND
FURTHER READING

Adams, D. and Carwardine, M. 1991. *Last Chance to See*. London: Pan.

Atkins, P. W. 1984. *The Second Law*. New York: Scientific American.

Atkins, P. W. 1995. *The Periodic Kingdom*. London: Weidenfeld & Nicolson.

Atkins, P. W. 2001. *The Elements of Physical Chemistry: With Applications in Biology*. New York: W. H. Freeman.

Atkins, P. W. and Jones, L. 1997. *Chemistry: Molecules, Matter and Change*, 3rd rev. edn. New York: W. H. Freeman.

Ayala, F. J. 2006. *Darwin and Intelligent Design*. Minneapolis: Fortress.

Barash, D. P. and Barash, N. R. 2005. *Madame Bovary's Ovaries: A Darwinian Look at Literature*. New York: Delacorte.

Barlow, G. W. 2002. *The Cichlid Fishes: Nature's Grand Experiment in Evolution*, 1st pb edn. Cambridge, Mass.: Basic Books.

Berry, R. J. and Hallam, A. 1986. *The Collins Encyclopedia of Animal Evolution*. London: Collins.

Bodmer, W. and McKie, R. 1994. *The Book of Man: The Quest to Discover Our Genetic Heritage*. London: Little, Brown.

Brenner, S. 2003. 'Nature's gift to science', in T. Frängsmyr, ed., *Les Prix Nobel, The Nobel Prizes 2002: Nobel Prizes, Presentations, Biographies and Lectures*, 274–82. Stockholm: The Nobel Foundation.

Brooks, A. C. and Buss, I. O. 1962. 'Trend in tusk size of the Uganda elephant', *Mammalia*, 26, 10–34.

Browne, J. 1996. *Charles Darwin*, vol. 1: *Voyaging*. London: Pimlico.

Browne, J. 2003. *Charles Darwin*, vol. 2: *The Power of Place*. London: Pimlico.

Cain, A. J. 1954. *Animal Species and their Evolution*. London: Hutchinson.

Cairns-Smith, A. G. 1985. *Seven Clues to the Origin of Life: A Scientific Detective Story*. Cambridge: Cambridge University Press.

Carroll, S. B. 2006. *The Making of the Fittest: DNA and the Ultimate Forensic Record of Evolution*. New York: W. W. Norton.

Censky, E. J., Hodge, K. and Dudley, J. 1998. 'Over-water dispersal of lizards due to hurricanes', *Nature*, 395, 556.

Charlesworth, B. and Charlesworth, D. 2003. *Evolution: A Very Short Introduction*. Oxford: Oxford University Press.

Clack, J. A. 2002. *Gaining Ground: The Origin and Evolution of Tetrapods*. Bloomington: Indiana University Press.

Comins, N. F. 1993. *What If the Moon Didn't Exist? Voyages to Earths that Might Have Been.* New York: HarperCollins.

Conway Morris, S. 2003. *Life's Solution: Inevitable Humans in a Lonely Universe.* Cambridge: Cambridge University Press.

Coppinger, R. and Coppinger, L. 2001. *Dogs: A Startling New Understanding of Canine Origin, Behaviour and Evolution.* New York: Scribner.

Cott, H. B. 1940. *Adaptive Coloration in Animals.* London: Methuen.

Coyne, J. A. 2009. *Why Evolution is True.* Oxford: Oxford University Press.

Coyne, J. A. and Orr, H. A. 2004. *Speciation.* Sunderland, MA: Sinauer.

Crick, F. H. C. 1981. *Life Itself: Its Origin and Nature.* London: Macdonald.

Cronin, H. 1991. *The Ant and the Peacock: Altruism and Sexual Selection from Darwin to Today.* Cambridge: Cambridge University Press.

Damon, P. E.; Donahue, D. J.; Gore, B. H.; Hatheway, A. L.; Jull, A. J. T.; Linick, T. W.; Sercel, P. J.; Toolin, L. J.; Bronk, R.; Hall, E. T.; Hedges, R. E. M.; Housley, R.; Law, I. A.; Perry, C.; Bonani, G.; Trumbore, S.; Woelfli, W.; Ambers, J. C.; Bowman, S. G. E.; Leese, M. N.; and Tite, M. S. 1989. 'Radiocarbon dating of the Shroud of Turin', *Nature*, 337, 611–15.

Darwin, C. 1845. *Journal of researches into the natural history and geology of the countries visited during the voyage of H.M.S. Beagle round the world, under the Command of Capt. Fitz Roy, R.N.,* 2nd edn. London: John Murray.

Darwin, C. 1859. *On the Origin of Species by Means of Natural Selection,* 1st edn. London: John Murray.

Darwin, C. 1868. *The Variation of Animals and Plants under Domestication,* 2 vols. London: John Murray.

Darwin, C. 1871. *The Descent of Man, and Selection in Relation to Sex,* 2 vols. London: John Murray.

Darwin, C. 1872. *The Expression of the Emotions in Man and Animals.* London: John Murray.

Darwin, C. 1882. *The Various Contrivances by which Orchids are Fertilised by Insects.* London: John Murray.

Darwin, C. 1887a. *The Life and Letters of Charles Darwin,* vol. 1. London: John Murray.

Darwin, C. 1887b. *The Life and Letters of Charles Darwin,* vol. 2. London: John Murray.

Darwin, C. 1887c. *The Life and Letters of Charles Darwin,* vol. 3. London: John Murray.

Darwin, C. 1903. *More Letters of Charles Darwin: A Record of his Work in a Series of Hitherto Unpublished Letters,* 2 vols. London: John Murray.

Darwin, C. and Wallace, A. R. 1859. 'On the tendency of species to form varieties; and on the perpetuation of varieties and species by natural means of selection', *Journal of the Proceedings of the Linnaean Society (Zoology),* 3, 45–62.

Davies, N. B. 2000. *Cuckoos, Cowbirds and Other Cheats.* London: T. & A. D. Poyser.

Davies, P. C. W. 1998. *The Fifth Miracle: The Search for the Origin of Life.* London: Allen Lane, The Penguin Press.

Davies, P. C. W. and Lineweaver, C. H. 2005. 'Finding a second sample of life on earth', *Astrobiology*, 5, 154–63.

Dawkins, R. 1986. *The Blind Watchmaker*. London: Longman.

Dawkins, R. 1989. 'The evolution of evolvability', in C. E. Langton, ed., *Artificial Life*, 201–20. Reading, Mass.: Addison-Wesley.

Dawkins, R. 1995. *River Out of Eden*. London: Weidenfeld & Nicolson.

Dawkins, R. 1996. *Climbing Mount Improbable*. London: Viking.

Dawkins, R. 1998. *Unweaving the Rainbow*. London: Penguin.

Dawkins, R. 1999. *The Extended Phenotype*, rev. edn. Oxford: Oxford University Press.

Dawkins, R. 2004. *The Ancestor's Tale: A Pilgrimage to the Dawn of Life*. London. Weidenfeld & Nicolson.

Dawkins, R. 2006. *The Selfish Gene*, 30th anniversary edn. Oxford: Oxford University Press. (First publ. 1976.)

Dawkins, R. and Krebs, J. R. 1979. 'Arms races between and within species', *Proceedings of the Royal Society of London*, Series B, 205, 489–511.

de Panafieu, J.-B. and Gries, P. 2007. *Evolution in Action: Natural History through Spectacular Skeletons*. London: Thames & Hudson.

Dennett, D. 1995. *Darwin's Dangerous Idea: Evolution and the Meanings of Life*. London: Allen Lane.

Desmond, A. and Moore, J. 1991. *Darwin: The Life of a Tormented Evolutionist*. London: Michael Joseph.

Diamond, J. 1991. *The Rise and Fall of the Third Chimpanzee: Evolution and Human Life*. London: Radius.

Domning, D. P. 2001. 'The earliest known fully quadrupedal sirenian', *Nature*, 413, 625–7.

Dubois, E. 1935. 'On the gibbon-like appearance of *Pithecanthropus erectus*', *Proceedings of the Section of Sciences of the Koninklijke Akademie van Wetenschappen*, 38, 578–85.

Dudley, J. W. and Lambert, R. J. 1992. 'Ninety generations of selection for oil and protein in maize', *Maydica*, 37, 81–7.

Eltz, T.; Roubik, D. W.; and Lunau, K. 2005. 'Experience-dependent choices ensure species-specific fragrance accumulation in male orchid bees', *Behavioral Ecology and Sociobiology*, 59, 149–56.

Endler, J. A. 1980. 'Natural selection on color patterns in *Poecilia reticulata*', *Evolution*, 34, 76–91.

Endler, J. A. 1983. 'Natural and sexual selection on color patterns in poeciliid fishes', *Environmental Biology of Fishes*, 9, 173–90.

Endler, J. A. 1986. *Natural Selection in the Wild*. Princeton: Princeton University Press.

Fisher, R. A. 1999. *The Genetical Theory of Natural Selection: A Complete Variorum Edition*. Oxford: Oxford University Press.

Fortey, R. 1997. *Life: An Unauthorised Biography. A Natural History of the First Four Thousand Million Years of Life on Earth*. London: HarperCollins.

Fortey, R. 2000. *Trilobite: Eyewitness to Evolution*. London: HarperCollins.

Futuyma, D. J. 1998. *Evolutionary Biology*, 3rd edn. Sunderland, Mass.: Sinauer.

Gillespie, N. C. 1979. *Charles Darwin and the Problem of Creation*. Chicago: University of Chicago Press.

Goldschmidt, T. 1996. *Darwin's Dreampond: Drama in Lake Victoria*. Cambridge, Mass.: MIT Press.

Gould, S. J. 1977. *Ontogeny and Phylogeny*. Cambridge, Mass.: Harvard University Press.

Gould, S. J. 1978. *Ever since Darwin: Reflections in Natural History*. London: Burnett Books / Andre Deutsch.

Gould, S. J. 1983. *Hen's Teeth and Horse's Toes*. New York: W. W. Norton.

Grafen, A. 1989. *Evolution and its Influence*. Oxford: Clarendon Press.

Gribbin, J. and Cherfas, J. 2001. *The First Chimpanzee: In Search of Human Origins*. London: Penguin.

Haeckel, E. 1974. *Art Forms in Nature*. New York: Dover.

Haldane, J. B. S. 1985. *On Being the Right Size and Other Essays*. Oxford: Oxford University Press.

Hallam, A. and Wignall, P. B. 1997. *Mass Extinctions and their Aftermath*. Oxford: Oxford University Press.

Hamilton, W. D. 1996. *Narrow Roads of Gene Land*, vol. 1: *Evolution of Social Behaviour*. Oxford: W. H. Freeman / Spektrum.

Hamilton, W. D. 2001. *Narrow Roads of Gene Land*, vol. 2: *Evolution of Sex*. Oxford: Oxford University Press.

Harrison, D. F. N. 1980. 'Biomechanics of the giraffe larynx and trachea', *Acta Oto-Laryngology and Otology*, 89, 258–64.

Harrison, D. F. N. 1981. 'Fibre size frequency in the recurrent laryngeal nerves of man and giraffe', *Acta Oto-Laryngology and Otology*, 91, 383–9.

Helmholtz, H. von. 1881. *Popular Lectures on Scientific Subjects*, 2nd edn, trans. E. Atkinson. London: Longmans.

Herrel, A.; Huyghe, K.; Vanhooydonck, B.; Backeljau, T.; Breugelmans, K.; Grbac, I.; Van Damme, R.; and Irschick, D. J. 2008. 'Rapid large-scale evolutionary divergence in morphology and performance associated with exploitation of a different dietary resource', *Proceedings of the National Academy of Sciences*, 105, 4792–5.

Herrel, A.; Vanhooydonck, B.; and Van Damme, R. 2004. 'Omnivory in lacertid lizards: adaptive evolution or constraint?' *Journal of Evolutionary Biology*, 17, 974–84.

Horvitz, H. R. 2003. 'Worms, life and death', in T. Frängsmyr, ed., *Les Prix Nobel, The Nobel Prizes 2002: Nobel Prizes, Presentations, Biographies and Lectures*, 320–51. Stockholm: The Nobel Foundation.

Huxley, J. 1942. *Evolution: The Modern Synthesis*. London: Allen & Unwin.

Huxley, J. 1957. *New Bottles for New Wine: Essays*. London: Chatto & Windus.

Ji, Q.; Luo, Z.-X.; Yuan, C.-X.; Wible, J. R.; Zhang, J.-P.; and Georgi, J. A. 2002. 'The earliest known eutherian mammal', *Nature*, 416, 816–22.

Johanson, D. and Edgar, B. 1996. *From Lucy to Language*. New York: Simon & Schuster.

Johanson, D. C. and Edey, M. A. 1981. *Lucy: The Beginnings of Humankind.* London: Granada.

Jones, S. 1993. *The Language of the Genes: Biology, History and the Evolutionary Future.* London: HarperCollins.

Jones, S. 1999. *Almost Like a Whale: The Origin of Species Updated.* London: Doubleday.

Joyce, W. G. and Gauthier, J. A. 2004. 'Palaeoecology of Triassic stem turtles sheds new light on turtle origins', *Proceedings of the Royal Society of London,* Series B, 271, 1–5.

Keynes, R. 2001. *Annie's Box: Charles Darwin, his Daughter and Human Evolution.* London: Fourth Estate.

Kimura, M. 1983. *The Neutral Theory of Molecular Evolution.* Cambridge: Cambridge University Press.

Kingdon, J. 1990. *Island Africa.* London: Collins.

Kingdon, J. 1993. *Self-Made Man and his Undoing.* London: Simon & Schuster.

Kingdon, J. 2003. *Lowly Origin: Where, When, and Why our Ancestors First Stood Up.* Princeton and Oxford: Princeton University Press.

Kitcher, P. 1983. *Abusing Science: The Case Against Creationism.* Milton Keynes: Open University Press.

Leakey, R. 1994. *The Origin of Humankind.* London: Weidenfeld & Nicolson.

Leakey, R. and Lewin, R. 1992. *Origins Reconsidered: In Search of What Makes Us Human.* London: Little, Brown.

Leakey, R. and Lewin, R. 1996. *The Sixth Extinction: Biodiversity and its Survival.* London: Weidenfeld & Nicolson.

Lenski, R. E. and Travisano, M. 1994. 'Dynamics of adaptation and diversification: a 10,000 generation experiment with bacterial populations', *Proceedings of the National Academy of Sciences,* 91, 6808–14.

Li, C.; Wu, X.-C.; Rieppel, O.; Wang, L.-T.; and Zhao, L.-J. 2008. 'An ancestral turtle from the Late Triassic of southwestern China', *Nature,* 456, 497–501.

Lorenz, K. 2002. *Man Meets Dog,* 2nd edn. London: Routledge.

Malthus, T. R. 2007. *An Essay on the Principle of Population.* New York: Dover. (First publ. 1798.)

Marchant, J. 1916. *Alfred Russel Wallace: Letters and Reminiscences,* vol. 1. London: Cassell.

Martin, J. W. 1993. 'The samurai crab', *Terra,* 31, 30–4.

Maynard Smith, J. 2008. *The Theory of Evolution,* 3rd edn. Cambridge: Cambridge University Press.

Mayr, E. 1963. *Animal Species and Evolution.* Cambridge, Mass.: Harvard University Press.

Mayr, E. 1982. *The Growth of Biological Thought: Diversity, Evolution, and Inheritance.* Cambridge, Mass.: Harvard University Press.

Medawar, P. B. 1982. *Pluto's Republic.* Oxford: Oxford University Press.

Mendel, G. 2008. *Experiments in Plant Hybridisation.* New York: Cosimo Classics.

Meyer, R. L. 1998. 'Roger Sperry and his chemoaffinity hypothesis', *Neuropsychologia*, 36, 957–80.

Miller, J. D.; Scott, E. C.; and Okamoto, S. 2006. 'Public acceptance of evolution', *Science*, 313, 765–6.

Miller, K. R. 1999. *Finding Darwin's God: A Scientist's Search for Common Ground between God and Evolution.* New York: Cliff Street Books.

Miller, K. R. 2008. *Only a Theory: Evolution and the Battle for America's Soul.* New York: Viking.

Monod, J. 1972. *Chance and Necessity: An Essay on the Natural Philosophy of Modern Biology.* London: Collins.

Morris, D. 2008. *Dogs: The Ultimate Dictionary of Over 1,000 Dog Breeds.* London: Trafalgar Square.

Morton, O. 2007. *Eating the Sun: How Plants Power the Planet.* London: Fourth Estate.

Nesse, R. M. and Williams, G. C. 1994. *The Science of Darwinian Medicine.* London: Orion.

Odell, G. M.; Oster, G.; Burnside, B.; and Alberch, P. 1980. 'A mechanical model for epithelial morphogenesis', *Journal of Mathematical Biology*, 9, 291–5.

Owen, D. F. 1980. *Camouflage and Mimicry.* Oxford: Oxford University Press.

Owen, R. 1841. 'Notes on the anatomy of the Nubian giraffe (Camelopardalis)', *Transactions of the Zoological Society of London*, 2, 217–48.

Owen, R. 1849. 'Notes on the birth of the giraffe at the Zoological Society's gardens, and description of the foetal membranes and some of the natural and morbid appearances observed in the dissection of the young animal', *Transactions of the Zoological Society of London*, 3, 21–8.

Owen, R. B.; Crossley, R.; Johnson, T. C.; Tweddle, D.; Kornfield, I.; Davison, S.; Eccles, D. H.; and Engstrom, D. E. 1989. 'Major low levels of Lake Malawi and their implications for speciation rates in cichlid fishes', *Proceedings of the Royal Society of London*, Series B, 240, 519–53.

Oxford English Dictionary, 2nd edn, 1989. Oxford: Oxford University Press.

Pagel, M. 2002. *Encyclopedia of Evolution*, 2 vols. Oxford: Oxford University Press.

Penny, D.; Foulds, L. R.; and Hendy, M. D. 1982. 'Testing the theory of evolution by comparing phylogenetic trees constructed from five different protein sequences', *Nature*, 297, 197–200.

Pringle, J. W. S. 1948. 'The gyroscopic mechanism of the halteres of Diptera', *Philosophical Transactions of the Royal Society of London*, Series B, Biological Sciences, 223, 347–84.

Prothero, D. R. 2007. *Evolution: What the Fossils Say and Why It Matters.* New York: Columbia University Press.

Quammen, D. 1996. *The Song of the Dodo: Island Biogeography in an Age of Extinctions.* London: Hutchinson.

Reisz, R. R. and Head, J. J. 2008. 'Palaeontology: turtle origins out to sea', *Nature*, 456, 450–1.

Reznick, D. N.; Shaw, F. H.; Rodd, H.; and Shaw, R. G. 1997. 'Evaluation of the rate of evolution in natural populations of guppies (*Poecilia reticulata*)', *Science*, 275, 1934–7.

Ridley, Mark 1994. *A Darwin Selection*, 2nd rev. edn. London: Fontana.

Ridley, Mark 2000. *Mendel's Demon: Gene Justice and the Complexity of Life.* London: Weidenfeld & Nicolson.

Ridley, Mark 2004. *Evolution*, 3rd edn. Oxford: Blackwell.

Ridley, Matt 1993. *The Red Queen: Sex and the Evolution of Human Nature.* London: Viking.

Ridley, Matt 1999. *Genome: The Autobiography of a Species in 23 Chapters.* London: Fourth Estate.

Ruse, M. 1982. *Darwinism Defended: A Guide to the Evolution Controversies.* Reading, Mass.: Addison-Wesley.

Sagan, C. 1981. *Cosmos.* London. Macdonald.

Sagan, C. 1996. *The Demon-Haunted World: Science as a Candle in the Dark.* London: Headline.

Sarich, V. M. and Wilson, A. C. 1967. 'Immunological time scale for hominid evolution', *Science*, 158, 1200–3.

Schopf, J. W. 1999. *Cradle of Life: The Discovery of Earth's Earliest Fossils.* Princeton: Princeton University Press.

Schuenke, M.; Schulte, E.; Schumacher, U.; and Rude, J. 2006. *Atlas of Anatomy.* Stuttgart: Thieme.

Sclater, A. 2003. 'The extent of Charles Darwin's knowledge of Mendel', *Georgia Journal of Science*, 61, 134–7.

Scott, E. C. 2004. *Evolution vs. Creationism: An Introduction.* Westport, Conn.: Greenwood.

Shermer, M. 2002. *In Darwin's Shadow: The Life and Science of Alfred Russel Wallace.* Oxford: Oxford University Press.

Shubin, N. 2008. *Your Inner Fish: A Journey into the 3.5 Billion-Year History of the Human Body.* London: Allen Lane.

Sibson, F. 1848. 'On the blow-hole of the porpoise', *Philosophical Transactions of the Royal Society of London*, 138, 117–23.

Simons, D. J. and Chabris, C. F. 1999. 'Gorillas in our midst: sustained inattentional blindness for dynamic events', *Perception*, 28, 1059–74.

Simpson, G. G. 1953. *The Major Features of Evolution.* New York: Columbia University Press.

Simpson, G. G. 1980. *Splendid Isolation: The Curious History of South American Mammals.* New Haven: Yale University Press.

Skelton, P. 1993. *Evolution: A Biological and Palaeontological Approach.* Wokingham: Addison-Wesley.

Smith, J. L. B. 1956. *Old Fourlegs: The Story of the Coelacanth.* London: Longmans.

Smolin, L. 1997. *The Life of the Cosmos.* London: Weidenfeld & Nicolson.

Söll, D. and RajBhandary, U. L. 2006. 'The genetic code – thawing the "frozen accident"', *Journal of Biosciences*, 31, 459–63.

Southwood, R. 2003. *The Story of Life.* Oxford. Oxford University Press.

Stringer, C. and McKie, R. 1996. *African Exodus: The Origins of Modern Humanity.* London: Jonathan Cape.

Sulston, J. E. 2003. 'C. elegans: the cell lineage and beyond', in T. Frängsmyr, ed., Les Prix Nobel, The Nobel Prizes 2002: Nobel Prizes, Presentations, Biographies and Lectures, 363–81. Stockholm: The Nobel Foundation.

Sykes, B. 2001. The Seven Daughters of Eve: The Science that Reveals our Genetic Ancestry. London: Bantam.

Thompson, D. A. W. 1942. On Growth and Form. Cambridge: Cambridge University Press.

Thompson, S. P. and Gardner, M. 1998. Calculus Made Easy: Being a Very-Simplest Introduction to Those Beautiful Methods of Reckoning Which Are Generally Called by the Terrifying Names of the Differential Calculus and the Integral Calculus. Basingstoke: Palgrave Macmillan.

Thomson, K. S. 1991. Living Fossil: The Story of the Coelacanth. London: Hutchinson Radius.

Trivers, R. 2002. Natural Selection and Social Theory. Oxford: Oxford University Press.

Trut, L. N. 1999. 'Early canid domestication: the farm-fox experiment', American Scientist, 87, 160–9.

Tudge, C. 2000. The Variety of Life: A Survey and a Celebration of All the Creatures that Have Ever Lived. Oxford: Oxford University Press.

Wallace, A. R. 1871. Contributions to the Theory of Natural Selection: A Series of Essays. London: Macmillan.

Weiner, J. 1994. The Beak of the Finch: A Story of Evolution in our Time. London: Jonathan Cape.

Wickler, W. 1968. Mimicry in Plants and Animals. London: Weidenfeld & Nicolson.

Williams, G. C. 1966. Adaptation and Natural Selection: A Critique of Some Current Evolutionary Thought. Princeton: Princeton University Press.

Williams, G. C. 1992. Natural Selection: Domains, Levels, and Challenges. Oxford: Oxford University Press.

Williams, G. C. 1996. Plan and Purpose in Nature. London: Weidenfeld & Nicolson.

Williams, R. 2006. Unintelligent Design: Why God Isn't as Smart as She Thinks She Is. Sydney: Allen & Unwin.

Wilson, E. O. 1984. Biophilia. Cambridge, Mass.: Harvard University Press.

Wilson, E. O. 1992. The Diversity of Life. Cambridge, Mass.: Harvard University Press.

Wolpert, L. 1991. The Triumph of the Embryo. Oxford: Oxford University Press.

Wolpert, L.; Beddington, R.; Brockes, J.; Jessell, T.; Lawrence, P.; and Meyerowitz, E. 1998. Principles of Development. London and Oxford: Current Biology / Oxford University Press.

Young, M. and Edis, T. 2004. Why Intelligent Design Fails: A Scientific Critique of the New Creationism. New Brunswick, NJ: Rutgers University Press.

Zimmer, C. 1998. At the Water's Edge: Macroevolution and the Transformation of Life. New York: Free Press.

Zimmer, C. 2002. Evolution: The Triumph of an Idea. London: Heinemann.

PICTURE ACKNOWLEDGEMENTS

Special thanks go to the following who gave valuable advice and guidance on the accuracy
and suitability of the illustrations, in the text and in the colour sections: Larry Benjamin,
Catherine Bosivert, Philippa Brewer, Ralf Britz, Sandra Chapman, Jennifer Clack, Margaret Clegg,
Daryl P. Domning, Anthony Herrel, Zerina Johanson, Barrie Juniper, Paul Kenrick, Zhe-Xi Luo, Colin
McCarthy, David Martill, P. Z. Myers, Colin Palmer, Roberto Portela-Miguez, Mai Qaraman, Lorna
Steel, Chris Stringer, John Sulston and Peter Wellnhofer.

COLOUR SECTIONS

page 1: *The Earthly Paradise* by Jan Brueghel the Elder, 1607–8, Louvre, Paris: Lauros/Giraudon/The
Bridgeman Art Library.

pages 2–3: (a) Wild cabbage (*Brassica oleracea*), sea cliffs, Dorset: © Martin Fowler/Alamy; (b)
vegetable spiral: Tom Poland; (c) Bernard Lavery, holder of 14 world records, with one of his giant
cabbages in Spalding, Lincs., 1993: Chris Steele-Perkins/Magnum Photos; (d) sunflowers, Great
Sand Dunes National Monument, Colorado: © Chris Howes/Wild Places Photography/Alamy; (e)
sunflower field, Hokkaido: Mitsushi Okada/Getty Images; (f) Astucieux du Moulin de Rance, a British
Belgian Blue bull, presented by B. E. Newton: Yann Arthus-Bertrand/CORBIS; (g) Kathy Knott, the
winner in a posing routine at the 1996 British Bodybuilding Championships: © Barry Lewis/Corbis;
(h) Chihuahua and Great Dane: © moodboard/alamy.

pages 4–5: (background) summer meadow, Norfolk: © G&M Garden Images/Alamy; (a) comet orchid
(*Angraecum sesquipedale*), Perinet National Park, Madagascar: Pete Oxford/Nature Picture Library and
Xanthopan morgani praedicta: © the Natural History Museum/Alamy; (b) bucket orchid (*Coryanthes
speciosa*): © Custom Life Science Images/Alamy; (c) bee emerging from a bucket orchid: photolibrary/
Oxford Scientific Films; (d) Andean Emerald hummingbird (*Amazillia franciae*), Mindo, Ecuador:
Rolf Nussbaumer/Nature Picture Library; (e) South African sunbird, Cape Town, South Africa: © Nic
Bothma/epa/Corbis; (f) Hummingbird Hawk-moth (*Macroglossum stellatarum*), Switzerland: Rolf
Nussbaumer/Nature Picture Library; (g) hammer orchid and wasp, Western Australia: Babs and Bert
Wells/Oxford Scientific Films/photolibrary; (h) *Ophrys holosericea* orchid attracting male buff-tailed
bumble bee: blickwinkel/Alamy; (i, j) evening primrose (*Oenothera biennis*) in normal and ultraviolet
light: both Bjorn Rorslett/Science Photo Library; (k) spider orchid (*Brassia rex*), Papua New Guinea:
© Doug Steeley /Alamy

pages 6–7: (a) Pair of pheasants (*Phasianus colchius*): Richard Packwood/Oxford Scientific Films/
photolibrary; (b) guppies: Maximillian Winzieri/Alamy; (c) Malaysian orchid mantis (*Hymenopus
coronatus*), Malaysia: Thomas Minden/Minden Pictures/National Geographic Stock; (d) leaf mantis
nymph, Amazon rainforest, Ecuador: © Michael & Patricia Fogen/Corbis; (e) satanic leaf-tailed
gecko: © Jim Zuckerman/Corbis; (f) caterpillar mimicking snake, rainforest, Costa Rica: Stephen J.
Krasemann/Science Photo Library.

page 8: Gorilla experiment: Simons, D. J., & Chabris, C. F. (1999). Gorillas in our midst: Sustained
inattentional blindness for dynamic events. *Perception*, 28, 1059–1074. Crocoduck tie: courtesy of Josh
Timonen. Caddis fly: photo courtesy of Graham Owen.

page 9: *Darwinius masillae*: © Atlantic Productions Ltd/photo Sam Peach.

pages 10–11: (a) Devonian scene by Karen Carr: © Field Museum; (b) *Tiktaalik* fossil: © Ted Daeschler/
Academy of Natural Sciences/VIREO; (c) *Tiktaalik* model and photo: copyright Tyler Keillor;
(d) manatee and calves, ZooParc, Saint-Aignan, 2003: AFP/ Getty Images; (e) dugong at Sydney
Aquarium, 2008: AFP/Getty Images; (f) *Odontochelys*: Marlene Donnelly/courtesy of The Field Museum.

pages 12–13: (a, b) the enzyme hexokinase closes round a glucose molecule: courtesy Thomas A.
Steitz. (c) Cutaway artwork of an animal cell: Russell Kightley/Science Photo Library.

pages 14–15: (a) Fertilized human egg cell and (b) two-cell human embryo at 30 hours: both Edelmann/Science Photo Library; (c) eight-cell human embryo at 3 days and (d) sixteen-cell human embryo at 4 days: both Dr Yorgos Nikas/Science Photo Library; (e) embryo at 10 days inside the womb, just implanted in the uterine lining; (f) at 22 days, the embryo has a curved backbone and the neural tube is open at both ends; (g) at 24 days, the heart extends almost up to the head and the placenta links it to the uterus, and (h) at 25 days: all photo Lennart Nilsson © Lennart Nilsson; embryo (i) at 5-6 weeks; (j) at 7 weeks: both Edelmann/Science Photo Library; (k) foetus at 17 weeks, (l) at 22 weeks: both Oxford Scientific Films/photolibrary; (m) newborn baby: Getty Images/Steve Satushek.

page 16: Starling sequence: dylan.winter@virgin.net.

page 17: San Andreas Fault in the Carrizzo Plain, Central California: © Kevin Schafer/Alamy.

pages 18–19: (a) Diagram showing the age of the oceanic lithosphere, data source: R. D. Muller, M. Sdrolias, C. Gaina and W. R. Roest, 'Age spreading rates and spreading symmetry of the world's ocean crust', Geochem. Geophys. Geosyst. 9.Q04006. doi:10.1029/2007/GC001743. Image created by Elliot Lim, CIRES & NOAA/NGDC, Marine Geology and Geophysics Division. Data & images available from http://www.ngdc.noaa.gov/mgg/; (b) artwork showing the process of sea floor spreading: Gary Hincks/Science Photo Library; (c) artwork showing convection currents: © Tom Coulson/Dorling Kindersley.

pages 20–1: (a) Caldera of a volcano, Fernandina Island, Galapagos: Patrick Morris/Nature Picture Library; (b) Galapagos Islands from space: Jacques Descloitres, MODIS Land Rapid Response Team, NASA/GSFC; (c), (d), (f), (g) Diving pelican, Seymour Island; swimming marine iguana Fernandina Island; Galapagos tortoise, Santa Cruz; and pelican, penguin and Sally Lightfoot crabs, Santiago Island: all © Josie Cameron Ashcroft; (e) Espanola saddleback tortoise (*Geochelone elephantopus hoodensis*), Santa Cruz Island, Galapagos: Mark Jones/Oxford Scientific/photolibrary.

pages 22–3: (a) Eastern Grey kangaroo (*Macropus giganteus*), Murramarang National Park, New South Wales: Jean Paul Ferrero/Ardea; (b) open eucalyptus woodland, near Norseman, Western Australia: Brian Rogers/Natural Visions; (c) koala and joey: photo courtesy Wendy Blanshard/Lone Pine Koala Sanctuary; (d) duck-billed platypus (*Ornithorhynchus anatinus*), swimming underwater: © Dave Watts/Alamy; (e) ring-tailed lemur (*Lemur catta*), Berenty Reserve, Southern Madagascar: Hermann Brehm/Nature Picture Library; (f) baobab tree, (*Adansonia grandidieri*), Western Madagascar: Nick Garbutt/Nature Picture Library; (g) Verreaux's sifaka lemur (*Propithecus verreauxi*), Berenty Reserve, Southern Madagascar: (*left*) Kevin Schafer/Alamy; (*middle*) © Kevin Schafer/Corbis; (*right*) Heather Angel/Natural Visions.

page 24: Blue-footed booby (*Sula nebouxii*): (*main picture*) © Michael DeFreitas South America/Alamy; (*top to bottom*) © Westend 61/Alamy; © Fred Lord/Alamy; F1Online/photolibrary; (*bottom two*) Nick Garbutt/Photoshot.

page 25: Clare D'Alberto: © David Paul / dpimages 2009.

pages 26–7: (a) spider monkey, Belize, Central America: Cubolimages srl/Alamy; (b) male flying lemur, Borneo: Tim Laman/National Geographic Stock; (c) Egyptian fruit bat: © Tim Flach.

pages 28–9: (a) Ostrich (*Struthhio camelus*), running: © Juniors Bildarchiv/Alamy; (b) flightless cormorant (*Nannopterum harrisi*), Punta Espinosa, Fernandina, Galapagos: © Peter Nicholson/Alamy; (c) flightless cormorant (*Nannopterum harrisi*), diving, Fernandina, Galapagos: Pete Oxford/Nature Picture Library; (d) kakapo (*Strigops harboptilus*), New Zealand; (e) harvester ant removes her wings before giving birth, artwork by John Dawson: National Geographic/Getty Images; (f) cave salamander (*Proteus anguinus*): Francesco Tomasinelli/ Natural Visions; (g) short-beaked common dolphin (*Delphinus delphis*), Gulf of California, Mexico: François Gohiev/Ardea

pages 30–1: (a) European cuckoo ejecting host shrike (*Lanius senator*) egg from nest, Spain: © Nature Picture Library/Alamy; (b) lioness (*Panthera leo*), hunting young kudu, Etosha National Park, Namibia: © Martin Harvey/Alamy: (c) Large White (*Pieris brassicae*) caterpillar with larvae of parasitoid wasp (*Cotesia glomerata*) leaving to pupate: © WILDLIFE GmbH/Alamy; (d) canopy of Kapur trees, Selangor, Malaysia: © Hans Strand

page 32: (a) Amazon estuary, aerial view: © Stock Connection Distribution/Alamy; (b) wild garlic (*Allium ursinum*), Cornwall: © Tom Joslyn/Alamy; (c) hills and pastureland, Morgan Territory, California: © Brad Perks Lightscapes/Alamy; (d) moss (*Hookeria luscens*), leaf cells, a polarised light micrograph, showing two whole cells containing chloroplasts: Dr Keith Wheeler/Science Photo Library.

ILLUSTRATIONS IN TEXT

Figures in the text on the following pages were redrawn by HL Studios: 90, 149 (both), 153, 166, 167, 168 (both), 173 (both), 176, 177, 186, 187, 189, 191, 192 (both), 193, 196, 222, 225, 227, 228, 230, 231, 232, 287, 289, 290, 294 (both), 295 (both), 300 (all), 302, 304, 307, 343, 346, 347 (both), 350 (both), 354, 357 and 365.

Individual credits:

page 7: 'I still say it's only a theory', cartoon by David Sipress from the *New Yorker*, 23 May 2005: © The New Yorker Collection 2005 David Sipress from cartoonbank.com. All Rights Reserved.

pages 40 and 42: Computer-generated images courtesy the author.

page 55: Hamburgh fowl, Spanish fowl and Polish fowl, from Charles Darwin, *The Variation of Animals and Plants under Domestication*, 1868.

page 57: Kabuki mask of a samurai warrior, detail of a 19th-century woodblock by Utagawa Toyokuni III, photo courtesy Los Angeles Natural History Museum. *Heikea japonica*, a male collected in Ariake Bay, off Kyushyu, Japan, 1968, width 20.4mm, photo Dick Meier, courtesy Los Angeles Natural History Museum.

page 67: Two lines of maize selected for high and low oil content, from J. W. Dudley and R. G. Lambert, 'Ninety generations of selection for oil and protein in maize', *Maydica* 37 (1992) 81–7.

page 68: Two lines of rats, from H. R. Hunt, C. A. Hoppert and S. Rosen, 'Genetic factors in experimental rat caries', in R. F. Sognnaes, ed., *Advances in Experimental Caries Research* (Washington DC: American Association for the Advancement of Science, 1955), 66–81.

page 75: Dmitry Belyaev with laboratory foxes, Novosibirsk, Russia, March 1984, photo RIA Novosti; inset photo from D. K. Belyaev, 'Destabilizing selection as a factor in domestication', *Journal of Heredity* 70 (1979), 301–8.

page 112: Graph from A. C. Brooks and I. O. Buss, 'Trend in tusk size of the Uganda elephant', *Mammalia* 26: 1 (1962), 10–34.

page 115: Diagram from A. Herrel, B. Vanhooydonck and R. van Damme, 'Omnivory in lacertid lizards: adaptive evolution or restraint', Journal of Evolutionary Biology 17 (2004), 974–84.

page 116: Photograph of caecal valve, from A. Herrel, B. Vanhooydonck and R. van Damme, 'Omnivory in lacertid lizards: adaptive evolution or restraint', *Journal of Evolutionary Biology* 17 (2004), 974–84; photo courtesy Anthony Herrel.

pages 123 (both), 125 and 127: Lenski experiment, diagrams from R. E. Lenski and M. Travisano, 'Dynamics of adaptation and diversification: a 10,000-generation experiment with bacterial populations', *Proceedings of the National Academy of Sciences* 91 (1994), 6808–14.

page 140: *Lingula*: 'Recent specimen of the brachiopod *Lingula* with long pedicle emerging from the 5 cm long valves of the phosphatic shell', © Natural History Museum, London. *Lingulella*, engraving © Natural History Museum, University of Oslo.

page 153: *Eomaia scansoria*, Chinese Academy of Geological Sciences (CAGS), redrawn from Qiang Ji, Zhe-Xi Luo, Chong-Xi Yuan, John R. Wible, Jian-Ping Zhang and Justin A. Georgi, 'The earliest known eutherian mammal', *Nature* 416 (25 April 2002), 816–22.

page 166: *Eusthenopteron*, after S.M. Andrews and T.S. Westoll, 'The postcranial skeleton of *Eusthenopteron foordi* Whiteaves', *Transactions of the Royal Society of Edinburgh* 68 (1970), 207–329.

page 167: *Ichthyostega*, after Per Erik Ahlberg, Jennifer Clack and Henning Blom, 'The axial skeleton of the Devonian tetrapod *Ichthyostega*', *Nature* 437 (1 Sept. 2005), 137–40, fig. 1. *Acanthostega*, after J. A. Clack, 'The emergence of early tetrapods', *Palaeogeography, Palaeoclimatoogy, Palaeoecology* 232 (2006), 167–89.

page 168: *Panderichthys*, reconstruction after Jennifer A. Clack.

page 171: Diagram from D. R. Prothero, *Evolution: What the Fossils Say and Why it Matters*, copyright © 2007 Columbia University Press. Reprinted with permission from the publisher.

page 173 (below): Reconstructed composite skeleton of *Pezosiren portelli*. Lateral view, length roughly 2.1 m. Shaded elements are represented by fossils; unshaded elements . . . are not. The length of the tail, and the form and posture of the feet are partly conjectural. After D. P. Domning, 'The earliest known fully quadrupedal sirenian', *Nature* 413 (11 Oct. 2001), 626–7, fig. 1.

page 177: Diagram modified from W. G. Joyce and J. A. Gauthier, 'Palaeoecology of Triassic stem turtles sheds new light on turtle origins', *Proceedings of the Royal Society of London* 271 (2004), 1–5.

page 204: *Sahelanthropus tchadensis*, reconstruction by © Bone Clones.

page 205: Skull of a foetal chimpanzee, reconstruction by © Bone Clones.

page 206: Baby and adult chimpanzee, photos courtesy Stephen Carr, from Adolf Naef, 'Über die Urformen der Anthropomorphen und die Stammesgeschichte des Menschenschädels', *Die Naturwissenschaften* 14: 21 (1926), 472–7. Original photos by Herbert Lang taken during the American Natural History Museum Congo Expedition, 1909–15.

page 222: Three kinds of virus, after Neil. A. Campbell, Jane B. Reece and Lawrence G. Mitchell, *Biology*, 5th edn, fig. 18.2, p. 321. Copyright © 1999 by Benjamin/Cummings, an imprint of Addison Wesley Longman, Inc. Reprinted by permission of Pearson Education, Inc.

page 227: Neurulation diagram, courtesy PZ Myers.

pages 242–3: Cellular family tree of *Caenorhabditis elegans*, http://www.wormatlas.org.

page 259: Map of the Galapagos archipelago, from Charles Darwin, *Journal of Researches*, 1st illus. edn, 1890.

page 266: Forest trees on St Helena, by courtesy of Jonathan Kingdon.

page 275: 'South America Secedes', cartoon by John Holden from Robert S. Diets, 'More about continental drift', *Sea Frontiers*, magazine of the International Oceanographic Foundation, March–April 1967.

page 289: Pterodactyl skeleton, after P. Wellnhofer, *Pterosaurs* (London: Salamander Books, 1991).

page 292: Polydactylic horse, from O. C. Marsh, 'Recent polydactyle horses', *American Journal of Science*, April 1892.

page 295: Okapi skeleton, after a drawing by Jonathan Kingdon.

page 301: Thylacine skull, S. R. Sleightholme and N. P. Ayliffe, International Thylacine Specimen Database, Zoological Society of London (2005).

page 304: Bdelloid rotifer, after Marcus Hartog, 'Rotifera, gastrotricha, and kinorhyncha', *The Cambridge Natural History*, vol. II (1896)

page 309: 'Various species of crabs and crayfishes', from Ernst Haeckel, *Kunstformen der Natur* (1899–1904)

pages 311–12: diagrams from D'Arcy Wentworth Thompson, *On Growth and Form* (1917)

page 327: 'Hodgkin's Law', courtesy Jonathan Hodgkin.

page 329: Phylogenetic tree, from David Hillis, Derrick Zwickl and Robin Gutell, University of Texas at Austin, http://www.zo.utexas.edu/faculty/antisense/DownloadfilesToL.html.

page 347: *Anhanguera*: after John Sibbick.

page 349: Female *Thaumatoxena andreinii silvestri*, from R. H. L. Disney and D. H. Kistner, 'Revision of the termitophilous Thaumatoxeninae (Diptera: Phoridae)', *Journal of Natural History* (1992) 26: 953–91.

page 361: Diagram from R. J. Berry and A. Hallam, *The Collins Encyclopedia of Animal Evolution* (1986)

page 363: Giraffe dissection, photo Joy S. Reidenberg PhD.

page 364: Diagram after George C. Williams.

INDEX

Page nos in *italic* refer to illustrations in the text. Page nos in **bold** refer to illustrations in the colour sections.